STEM CELLS
A Short Course

Rob Burgess

WILEY Blackwell

Published by John Wiley & Sons, Inc., Hoboken, New Jersey.
Published simultaneously in Canada.

For general information on our other products and services or for technical support, please contact our Customer Care Department within the United States at (800) 762-2974, outside the United States at (317) 572-3993 or fax (317) 572-4002.

Wiley also publishes its books in a variety of electronic formats. Some content that appears in print may not be available in electronic formats. For more information about Wiley products, visit our web site at www.wiley.com.

Library of Congress Cataloging-in-Publication Data is available.

ISBN: 978-1-118-43919-7

Printed in the United States of America.

10 9 8 7 6 5 4 3 2 1

To my wife, Jane, daughter, Zoie, son, Bobby, mother, Lola, and father, Bob

TABLE OF CONTENTS

3 EMBRYONIC STEM, FETAL, AND AMNIOTIC STEM CELLS — 87

PREFACE TO THE PROFESSOR

My justification for writing this book comes from a thorough review of the scientific literature currently published in the area of stem cell research. While an enormous body of knowledge exists in this field, there has yet to be a comprehensive college-level textbook that thoroughly and properly addresses the material in a manner that allows for the student to grasp, retain, and get excited about the discipline. Many works highlight unique or ground-breaking findings or delve into the broader concept of stem cell identity, function, and usefulness to man, yet they do not address the field with the all-inclusive effort needed to give the student a strong understanding of the need for stem cell research. Second, most literary works in the area of stem cell research gloss over or fail to drive home key specific findings, definitions, concepts, and details that are required to give the student a thorough understanding of stem cells and their impact on therapeutics and medicine. Third, when key stem cell research and discoveries are described in teaching materials, the researchers involved are often not mentioned or given the proper credit for their findings. Thus, I have compiled a breakdown of stem cell research that highlights the history behind the discipline, emergent groundbreaking findings such as the discovery of embryonic stem cells and induced pluripotency, and given credit where credit is due. Within the body of each chapter, I have included Focus Boxes on key researchers and Case Studies highlighting seminal findings in stem cell research. For each Focus Box or Case Study, I have tried to provide an example of a true leader in the field or groundbreaking study. To ensure that the student absorbs the material at the end of each chapter, I have included a detailed summary of the text, key terms, review questions, and all cited references. The key term definitions and answers to select review questions can be found at the website www.stemcelltextbook.com. It is highly encouraged that the professors utilize the summaries, key terms, and review questions as study materials and ask students to periodically review the corresponding website for continuously updated information related to this text in particular and to the field of stem cell research in general.

The book is organized into eight chapters that gradually step from the history of stem cell research to the basic concepts behind cell culture, key stem cell-related findings, and applications of stem cells in drug discovery and therapy. Key topics I have addressed include:

1. The history, early studies, and the rapid advancement of the field of stem cell research
2. The basic requirements for a cell to be considered "stem" in nature
3. Types of stem cells and their capacities to differentiate into various mature lineages

4. Breakthroughs in early stem cell-based therapeutics endeavors
5. The differences between embryonic and adult stem cells at the basic and translational research levels
6. Induced pluripotency
7. Implications of stem cells on diagnostics and drug screening efforts
8. Stem cell-based cloning and its ethical considerations

Chapter 1 begins with an overview of the first concrete contemplation of the existence of stem cells in the early 1900s, and how that hypothesis rapidly evolved into accepted theory. It transitions by slowly introducing the concept of "stemness" via seminal discoveries in hematopoiesis, mouse, and human embryonic development, and even iPS technology. Optimization of cell culture conditions for various stem cell types and advances in cloning that utilizes stem cells as source material are also discussed. There is also a special section on reproductive verses. therapeutic cloning outlining the discovery of Dolly the sheep. Several examples of the use of stem cells as possible therapeutic regimens, such as for spinal cord injury repair, are also described. The chapter closes with a briefing on the limitations of stem cells in therapeutics due to immunorejection and efforts to save endangered species through iPS cell manipulation.

Chapter 2 transitions from the history of stem cell research and seminal findings discussed in Chapter 1 to how these studies were carried out. With an emphasis on historical discoveries and advancements, it reviews and describes key derived cell lines such a Hela cells and outlines the early refinement of general cell culture techniques. An important example described in this context is the discovery of hybridoma technology. Basic cell culture technology is expanded upon to describe the optimal conditions needed to either maintain stem cell pluripotency or drive differentiation towards one or more particular cellular fates. A basic overview of vertebrate embryonic development also provides a foundation for the next section, which is an outline and description of the basic various potency properties of stem cells. The most widely studied types of stem cells are also discussed, with particular attention paid to embryonic stem cells—given the recent advancements in, and controversy surrounding, this field of research. Signaling and transcriptional control pathways regulating stem cell fate are also described for each major stem cell type as they apply to cell culture applications. The chapter concludes with a brief introduction to stem cells as therapeutic or drug discovery medical platforms.

Chapter 3 expands upon the introduction of embryonic, fetal, and amniotic stem cells in Chapter 2 by providing a historical perspective on their discovery and properties. Embryonic development is emphasized in more detail, providing a solid foundation for the description of embryonic, amniotic, and adult stem cells as outlined in the remaining sections of the chapter. Again, signaling and regulatory pathways of stem cell fate are emphasized to give the student a firm understanding of what drives cellular behavior and identity. Numerous examples of different embryonic, amniotic, and adult stem cells are described, with each example emphasizing a potential therapeutic use.

I have focused all of Chapter 4 on the properties and study of adult stem cells. This is because, as the field has matured, it has become apparent that adult stem cells can provide as much or more value for therapeutic and diagnostic medical applications as embryonic stem cells, but without the need for controversial embryo destruction in the generation of embryonic stem cells. The discovery and basic properties of adult stem cells are described, and this is followed by a delineation of adult stem cell examples. As the discovery and characterization of hematopoietic stem cells have been seminal in advancing HSC as well as other stem cell applications in research and medicine, a considerable section on this cell type is included. This allows for the introduction of the "HSC niche,", which has common implications for other stem cell types in the developing embryo and adult. Adult stem cell types, such as those of muscle and endothelial origin, are chosen, as they hold the most promise as cell-based therapeutic agents. In each case, morphology and marker expression are described to give the student a visual perspective on cell identity at various stages of differentiation. In addition, different concepts in differentiation capabilities and cascades, such as osteogenesis, are described to lay the groundwork for the student to grasp the concept of cellular identity and fate transition. In this respect, key transcription factors involved in driving differentiation (or maintaining multipotency) are described for each cell type.

Chapter 5 addresses perhaps two of the most intriguing findings in stem cell research to date: the reality of somatic cell nuclear transfer (cloning or "SCNT") and induced pluripotency. Each phenomenon is related as they draw on the same capacity of a cell to become "reprogrammed" under the right environmental or internal conditions. The basic properties of SCNT and induced pluripotency are discussed as well as the types of cells most amenable to either technology. Throughout this chapter an emphasis is made regarding the value of SCNT or induced pluripotency as these technologies apply to autologous stem cell-based therapy and relieving the need for human embryo destruction. Various methods for implementing cloning, nuclear reprogramming, and the induction of pluripotency are discussed, with the latter being expanded upon via an outline of the key transcription factors necessary to drive dedifferentiation. Again, for cell fusion as a reprogramming technology, the generation of antibody-producing hybridomas is described. This is followed by a description of cloning, further describing the creation of Dolly the sheep, and review of induced pluripotency technologies including its origin in the study of *Drosophila melanogaster* by Walter Gehring. Various methods for the production of iPS cells and their biological properties are described, and the chapter finishes with a suggestion of why the utilization of iPS technology and cells might be superior to that involving other cell types.

The existence of cancer stem cells (CSCs) has long been controversial, and some researchers still have doubts, yet the concept that cancer emanates from a single progenitor has gained momentum over the last several years, and many of the properties inherent in stem cells could indeed drive uncontrollable cell division. As such, CSCs warrant thorough consideration, and Chapter 6 is dedicated solely to this topic. In a manner similar to that of other chapters, the discovery and origin of cancer stem cells are reviewed, including a description of the critical pathways involved in driving the stemness of a CSC and the transformation of a norma cell towards the CSC phenotype. Examples of cancer stem cells are also described such as those present in acute myeloid leukemia and colon cancer. The basic properties of cancer stem cells are covered in detail along with a comparison to normal stem cells. The chapter concludes with a mention of various developing therapeutic strategies which may be employed to treat cancer by directly attacking the cancer stem cell core.

Various different types of cell lines have long been utilized as drug screening platforms. Stem cells are no exception, and Chapter 7 denotes the value of implementing stem cell-based screening assays to define the therapeutic utility of various drugs or drug candidates. The chapter begins with a description of murine embryonic stem cells and their utility in defining gene function via the generation of genetically enineered mice;, including an example of the *paraxis* knockout mouse. This is followed by a description of the more direct use of stem cells in assays to screen for drug candidates. Various examples are described illustrating the utility of stem cell to provide virtually unlimited supplies of terminally differentiated lineages for use in drug screening. Particular attention in this respect is paid to neuronal lineages given their value in defining both drug efficacy and toxicity as related to the treatment of neurological disorders. Drug effects on cell division and differentiation are closely monitored to determine tumor-promoting or differentiation-inducing properties. Examples of embryonic, adult, and induced pluripotent stem cell sources for terminal lineages are described as well. Screening strategies using stem cells are also discussed that allow for the identification of key disease pathways or key small molecules that may efficiently promote the differentiation of a stem cell into a mature cell type. Given its importance, the chapter concludes with yet more examples on stem cells as both prenatal and postnatal toxicity screening tools.

Chapter 8 is perhaps the chapter of most interest for readers interested in stem cells as therapeutic agents. Again, I have provided a historical perspective, beginning with the history of tissue engineering followed by a foray into the use of stem cells for disease specific treatment. This chapter illustrates numerous examples of real-world patient trials, either completed or currently underway, which may have a significant impact on stem cell-based medicine in the years to come. In this respect, an overview of global regenerative medicine is given followed by human physiology or disease-specific stem cell-based therapeutics initiatives. Again, a major emphasis is placed on neurodegenerative diseases and neurological disorders given the potential of stem cells to treat or at least manage these anomalies. In addition, veterinary applications of stem cells are described and outlined, citing equine as the species most positively affected by stem cell transplantation efforts. Chapter 8 concludes with an overview of stem cells as an emerging industry, focused on the explosion in the number of companies applying business models centered on stem cell drug discovery or therapeutics initiatives.

In *Stem Cells: A Short Course*, I have attempted to provide a comprehensive text covering all aspects of stem cell biology, research, and medical application. To this end, I have addressed most, if not all, major areas of stem cell research focus, historical perspectives and futurstic concepts. I have given the student a wealth of information related to stem cell to absorb, and I have organized it in a manner that allows for the grasp of intricate details related to stem cell biology as well as an understanding of the bigger picture pertaining to the impact stem cells are having and will have on medicine. I always welcome feedback and comments. If you have suggestions for improvements on this text in future editions please do not hesitate to contact me.

Rob Burgess
www.stemcelltextbook.com
www.wiley.com/go/burgess/stemcells

PREFACE TO THE STUDENT

As of this book's completion there has been a significant lack of suitable educational materials to prepare future scientists and medical researchers in the discipline of stem cell biology. To date, no comprehensive text exists which effectively compiles a combination of the basic principles of stem cell biology with a delineation of the most cutting-edge research, scientific, and medical findings in this area. As such, students contemplating a career in the stem cell arena are left to cobble together research and reviews in the field. I have thus attempted to develop a comprehensive text that, if studied as outlined, will provide you with the knowledge base needed to either consider a career in this exciting area or further advance your basic understanding of stem cell biology and how the science may be translated into useful medical applications. It is my belief that if you take the time to read each chapter, review the chapter summaries, glossaries, and review questions, you will be thoroughly prepared to consider a career in this exciting field.

The text of the book is composed such that it draws on fact, beginning with the history of stem cell research and the basic biology of what it means to possess the property of "stemness." I have based the entire book on factual research and peer-reviewed publications in leading medical journals. No facet of the field of stem cell research is glossed over, including that of embryonic stem cells. In addition, I have not passed judgment regarding any aspect of the subject of stem cells. Rather, I have simply objectively conveyed the information in as comprehensive a manner as possible, and my goal is to allow you, the student, to both absorb material and to form your own opinions on the science and its impact on medicine. As the field is rapidly maturing, there are several instances when the central dogma of cell biology is countered and tested, and I do not gloss over or downplay findings that may not fit in with the general beliefs of researchers in the field. For example, Ramon y Cajal's "no new neurons" hypothesis was clearly disproven, and I have deneated this fact.

Regarding its organization, I have chosen to write this book in review format with key examples of research findings and data that drive home the points and principles of stem cell biology. The book begins with historical perspectives and an introduction to stemness, followed by chapter after chapter of detailed description pertaining to differing stem cell types, their differentiation capacities, and potential translational applications in either diagnostic or therapeutic endeavors. Each chapter is written in a self-contained format including an end-of-chapter summary, key terms, review questions, and cited references. In addition, I have included Focus Boxes introducing pioneering researchers in the field and

Case Studies summarizing some of the highest impact research articles published to date in the area of stem cells. Each chapter outlines a unique aspect of stem cell science, and this flow of content will allow you to have all the pertinent information you need for that particular subject matter should you choose to study only certain chapters. However, I recommend against this and urge you to take the time to study the entire text, as it will give you a firm understanding and grasp of the history, biology, and future of this exciting field.

My recommendations on how to successfully utilize this book as a learning tool are as follows: first, read the required material prior to lecture; start with a scanning of the chapter summary followed by a thorough read; familiarize yourself with each of the key terms at the end of the chapters and make an attempt to answer the review questions before checking them aginst the answers I have provided at the link www.stemcelltextbook.com; finally, scan the material again before lecture to refresh your memory of the main concepts. If you follow these steps, you will be prepared and knowledgeable in the subject matter, making the entire course a much more enjoyable and rewarding experience.

As always I am open to your praise, comments, and criticisms. There is no better reviewer of a college textbook than the student it is supposed to prepare and influence. Please contact me if you have suggestions on how to make this text more valuable to future students. It is my hope that you are inspired by this book.

Rob Burgess
www.stemcelltextbook.com
www.wiley.com/go/burgess/stemcells

ACKNOWLEDGMENTS

I would like to thank all the scientists, professors, and researchers who generously provided material for inclusion into this book. Your hard work and dedication to the field of stem cell research is inspiring and will have an enormous impact upon science and society.

I would also like to sincerely thank Emily Ann Neie for her critical proofreading of the book and efforts at obtaining reprint permissions, Connie Zhao for her excellent hand-drawn artwork, Mahendra Rao, M.D., Ph.D. for his critical review of the book, Stephanie Dollan of Wiley Blackwell for her undying dedication to pre-production content and formatting and John Wiley & Sons for making the publication of this book possible.

Rob Burgess

LIST OF FIGURES

LIST OF TABLES

LIST OF CASE STUDIES

LIST OF FOCUS BOXES

Chapter 1

A HISTORY OF STEM CELL RESEARCH

Chapter 1 outlines and describes the maturation of stem cell research, from early contemplations on the power of cell fate to cutting-edge clinical trials involving human embryonic stem cell (hESCs). A multitude of different stem cell types are described as they make their chronological appearance on the research front and key researchers as well as their findings and discoveries are highlighted throughout the chapter.

> "The possibility of obtaining a strain of cells in tissue culture which may become determined to differentiate in a variety of alternative ways is very attractive."
>
> Martin Evans, PhD (1972)—2007 Nobel Prize Winner in Physiology or Medicine

EARLY STUDIES

The existence of **stem cells**, which are defined as biological cells capable of self-renewal and the capacity to differentiate into a variety of cell types and are present within most if not all multicell organisms, has been contemplated for greater than 100 years. In fact, the concept of "stemness" can be traced back as far as ~300 BC when Aristotle disagreed with the generationally accepted hypothesis of spontaneous generation (Figure 1.1).

Russian-born medical doctor Alexander A. Maximow first coined the term "stem cell" in 1908, while addressing a hematologic society congress in Berlin (see Focus Box 1.1). Maximow was a scientist and histologist who spent several years around the turn of the 20th century contemplating the existence of a unique cell type that would allow for generation of many differentiated, mature phenotypes. Maximow's main focus was on blood cell type identity and what drives the generation of the terminally differentiated cells in the hematopoietic system. It was as a professor at the Imperial Military Academy in Saint Petersburg, Russia from 1903 to 1922, where he refined his theories on the existence of a common hematopoietic precursor cell. He is generally credited with the formulation of the **theory of hematopoiesis**, which states that all blood cellular components are derived from a common precursor stem cell. Maximow finished his career as a professor of anatomy at the University of Chicago.

Stem Cells: A Short Course, First Edition. Rob Burgess.
© 2016 John Wiley & Sons, Inc. Published 2016 by John Wiley & Sons, Inc.

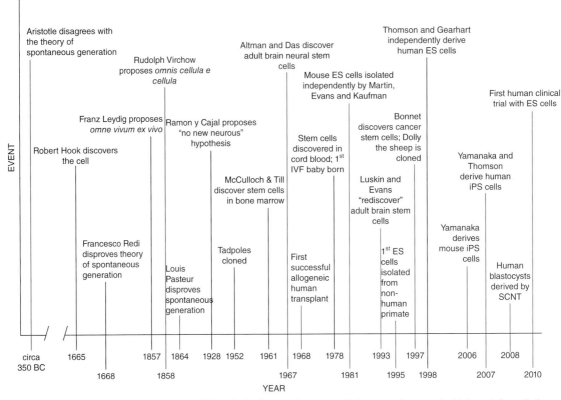

<u>Figure 1.1</u> Timeline of historical advances in stem cell theory and research. (Adapted from Rob Burgess, *Stem Cells Handbook* (Humana Press), 2nd Edition, Chapter 1.)

After the initial contemplation of the existence of hematopoietic stem cells (HSCs) in 1908, the field was relatively silent for more than 50 years. It was not until the early sixties when true scientific advancements in the area of stem cell research began to take place. Specifically, in 1963 researchers Ernest McCulloch and James Till (Altman and Das, 1967) of the University of Toronto demonstrated the existence of stem cells in the bone marrow (Figure 1.2). This was accomplished by injecting bone marrow cells into irradiated immune-deficient mice, which resulted in the growth of visible lumps termed **spleen colonies.** It was postulated that these colonies were the result of bone marrow-derived stem cells, and their clonal origin was confirmed. Published in the journal *Nature* that year, this finding is considered to be one of the most seminal discoveries in the field of stem cell research, laying the groundwork for virtually every major breakthrough in the discipline since.

Focus Box 1.1: Alexander A. Maximow and the Theory of Stem Cells

Alexander A. Maximow (1874–1928) was a Russian-born medical doctor and histologist and the first person to contemplate the existence of stem cells. His "theory of hematopoiesis" and histological textbook, which has been suggested to be the world's most respected textbook in histology, laid the groundwork for many of the stem cell discoveries impacting medicine today. (Photo courtesy Wikimedia Commons; reprinted with permission.)

Figure 1.2 The late Ernest McCulloch and James Till after accepting the 2005 Lasker Award for their studies on bone marrow-derived stem cells. Ernest McCulloch is at left. (Photograph courtesy Environmental Protection Agency; reprinted with permission.)

The lymphatic system was not the only area of hot pursuit for the identification and characterization of stem cells. In 1967, a key demonstration of **neurogenesis**, defined as the generation of neurons and glial cells, occurring in the adult brain was accomplished by Drs. Joseph Altman and Gopal Das of Massachusetts Institute of Technology (Prindull et al., 1978) (Figure 1.3). In these studies, an autoradiographic technique was employed to measure both mitotic activity and tag cells for tracking at later time points. To accomplish this, tritiated (^{3}H) thymidine was injected intraperitonially into 6-day-old guinea pigs and then monitored for incorporation into the cells of the cerebellar external germinal and cortical subependymal layers of the brain. Tritiated thymidine will incorporate into the DNA of mitotically active cells, thus marking cell division. In addition, it will remain in these cells long term as a tag for subsequent cell marking and characterization. Dr. Altman's group used this technique to reveal active mitosis in the brains of adult guinea pigs followed by confirmation that the tagged cells differentiate into identifiable small-caliber mature interneurons he termed "**microneurons.**" The findings of Altman and Das went against the **no new neurons** central dogma of leading neuroscientist Santiago Ramon y Cajal, and thus were largely dismissed by the scientific community (Altman and Das, 1967). It was only in the 1990s when adult neurogenesis was "rediscovered" that Altman's theories on adult brain neurogenesis were accepted by the scientific community. Dr. Altman and

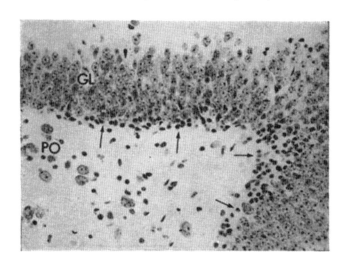

Figure 1.3 Discovery of active neurogenesis in the adult brain. The arrows denote ^{3}H-thymidine uptake in glial cells in rodent brain regions associated with trauma. Neurons and neuroblasts also demonstrated some staining, confirming mitosis and corresponding neurogenesis. (Photo courtesy *Nature* (Altman and Das, 1967); reprinted with permission.)

Figure 1.4 Dr. Robert Alan Goodwith, President Richard Nixon, and colleagues at the White House Conquest of Cancer Program in 1973. Dr. Goodwith is circled; President Nixon is second from the left. Also pictured is Dr. Robert L. Clark of the University of Texas M.D. Anderson Cancer Center. (Photo courtesy Nixon archives; reprinted with permission.)

his wife and colleague, researcher Shirley Altman-Bayer, still actively promote their research theories today and have a forthcoming book titled *MENTAL EVOLUTION: Origins of the Human Body, Brain, Behavior, Consciousness, and Culture.*

In 1968, a major therapeutic breakthrough based on the potential of stem cells present in bone marrow was realized when the first successful human bone marrow transplant was accomplished by the late American physician Dr. Robert Alan Goode while he was Professor in Pediatrics, Microbiology, and Pathology at the University of Minnesota Medical School (Figure 1.4). The transplant was performed between siblings for the treatment of **severe combined immunodeficiency syndrome (SCID)**, a genetic disorder in which both B and T cells of the immune system are severely compromised due to a defect in one of several possible genes. It was widely speculated at the time (and later confirmed) that bone marrow-derived stem cells from the healthy sibling aided in reconstructing the immune system of the recipient SCID patient. Dr. Good won the Albert Lasker Medical Research Award in 1970 and is generally accepted by the medical and research communities as the founder of modern immunology.

HEMATOPOIETIC STEM CELL DISCOVERY

Therapeutic advancements in bone marrow transplants set the groundwork for a related major discovery in 1978 when Dr. Gregor Prindull, Professor of Pediatric Hematology/ Oncology in the Department of Paediatrics at the University of Gottingen in Gottingen, West Germany (now retired), and colleagues B. Prindull and N. Meulen discovered the presence of HSCs in human umbilical cord blood. In this study, the researchers extracted cord blood from newborn infants 8–10 days old. Cells from the blood samples were processed by sedimentation and clearing of nonlymphoid cells and subsequently cultured in a methylcellulose cell culture system. By the 10th day of cell culture the researchers identified a small subpopulation of cells (1 in 1,678 on average) in a sample of nonadherent

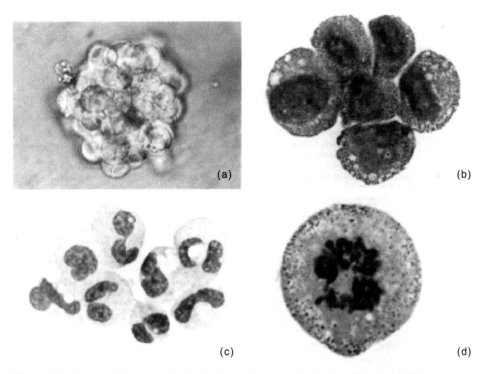

Figure 1.5 Hematopoietic stem cells isolated from human umbilical cord blood. (a) Colony cultured on methylcellulose. (b) Myelocytes and metamyelocytes. (c) Neutrophils. (d) Dividing myelocyte. (Photo courtesy Dr. G. Prindull and *Acta Paediatrica Scandinavica* (Prindull and Prindull, 1978); reprinted with permission.)

mononuclear cells that represented myelocytic **colony forming units (CFUs),** cells that have the ability to divide and form a clonal colony in tissue culture. These cells formed adherent colonies in a methylcellulose matrix (Prindull and Prindull, 1978). **Myelocytes** are of granulocytic origin and normally only present in bone marrow, yet given the high degree of proliferation during fetal development, as is evidenced by this study, they accumulate in the cord blood of newborn infants (Figure 1.5).

Focus Box 1.2: Gail R. Martin and the discovery of mouse embryonic stem cells

As a professor in the Department of Anatomy at the University of California, San Francisco, researcher Gail R. Martin is widely credited with the co-discovery of mouse embryonic stem cells, ushering in a new era of scientific research in embryonic development and the study of gene function. She is a member of the American Academy of Arts and Sciences, the National Academy of Sciences, and currently runs the Program in Biological Sciences at UCSF.

Source: Reproduced with permission from G. R. Martin.

MOUSE EMBRYONIC STEM CELL DISCOVERY

In 1981, the term "embryonic stem cell" was coined by University of California, San Francisco researcher Gail R. Martin, when she derived cells from the inner cell mass (ICM) (defined below) of 3.5-day-old mouse embryos and confirmed that these cells could give rise to a variety of mature, differentiated cell types. These cells also highly resembled embryonal teratocarcinoma cells which are known to be multipotent in nature (Bonnet and Dick, 1997) (see Focus Box 1.2 and Case Study 1.1). This seminal finding was simultaneously and independently accomplished by University of Cambridge professor Sir Martin Evans and researcher Matthew Kaufman. Evans later went on to receive the 2007 Nobel Prize in Physiology or Medicine for his contribution to rodent-based gene targeting technologies along with the University of Utah's Mario Capecchi and the University of Wisconsin's Oliver Smithies. The discovery of embryonic stem cells in mice and the development of corresponding gene targeting technologies are discussed in detail in the Chapter 7, section Embryonic Stem Cells and Animal Models of Gene Function.

Case Study 1.1: Isolation of a pluripotent cell line from early mouse embryos cultured in medium conditioned by teratocarcinoma stem cells

Gail R. Martin

In this study, Professor Gail R. Martin and colleagues at UCSF successfully isolated and cultured a clonal population of mitotically active pluripotent cells from the ICM of mouse 3.5 days post coitum (dpc) embryos. This was accomplished utilizing a special **"conditioned medium"** removed and concentrated from the culture of PSA-1 embryonal carcinoma (EC) cells. Given the inherent capacity of EC cells to differentiate into numerous cell types the medium was speculated to contain a growth factor or growth factors secreted by these cells capable of driving cell division and/or inhibiting differentiation. The embryo-derived stem cells exhibited a striking resemblance to EC cells and were demonstrated to have the capacity to differentiate into a wide variety of cell types in tissue culture (Martin, 1981 and Figure 1.6). As a final proof of pluripotency, Martin and colleagues showed that these cells could form teratocarcinomas when injected into mice (Martin, 1981).

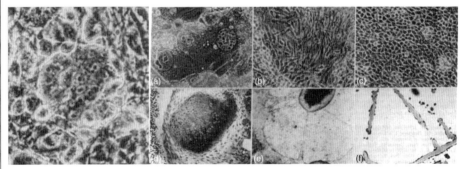

Figure 1.6 The discovery of mouse embryonic stem cells. (Left) The first published photo documentation of a mouse embryonic stem cell colony. (Right) Embryoid bodies demonstrating a variety of different cell types including (a) giant cells, (b) neuron-like cells, (c) endodermal cells, (d) cartilage, and (e) cells forming tubules. Source: Martin, 1981. Reproduced with permission from G. R. Martin.)

SUCCESSFUL NEURAL STEM CELL CULTURE

Despite Joseph Altman's speculation and demonstration regarding the existence of neural stem cells in the adult brain in 1967, it was only in 1992 when Brent A. Reynolds and Samuel Weiss in the Department of Pathology at the University of Calgary School of Medicine first successfully isolated neural progenitor and stem cells from the **subventricular zone** (a neurogenic region) of adult mouse brain tissue that their existence was accepted by the scientific community. In that same year researchers in the Department of Genetics at Harvard Medical School, led by Constance Cepko, isolated a multipotent cell line from adult mouse brain tissue and transformed it with **v-myc** —the viral homolog of c-myc—which is capable of cellular transformation, to create a stable indefinitely dividing neural stem cell population. Characterization of clonal cell populations revealed a common viral integration site, suggesting that individual lines actually originated from a common infected progenitor cell. Two cell lines exhibited extensive **process**-bearing morphology (a process with respect to neural or neuronal cell culture refers to either axon or dendrite-like protrusions). Three cell lines were demonstrated to have the capacity to differentiate into mature neurons and glia, and further subcloning of each line revealed the same morphological and molecular characteristics across these lines. For *in vivo* studies, the cells were marked for identification with a β-galactosidase (LacZ) genetic tag that allows for tracing of individual cell fate. β-**galactosidase** is an enzyme that catalyzes the hydrolysis of β-galactosides into monosaccharides. It also cleaves the organic compound X-gal to produce a characteristic blue dye for use in histology. When the researchers transplanted v-myc-transformed, LacZ-tagged neural cells into the cerebellum of newborn mice, the cells integrated in a non-tumorigenic fashion and properly differentiated into either neurons or glia depending upon the location of cerebellar integration (Snyder et al., 1992) (Figure 1.7).

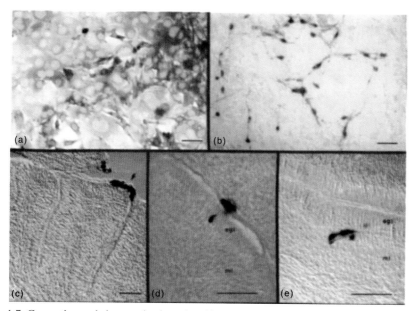

Figure 1.7 Generation and characterization of multipotent neural stem cells. (a) Non-cultured control and (b) 8-day coculture of transformed neural stem cells (stained in blue) with dissociated primary mouse cerebellum demonstrating process formation. (c–e) Sections of the cerebellar region of a mouse brain transplanted with LacZ tagged v-myc transformed neural stem cells. (c) Six hours post transplant; (d and e) 72 hours post transplant demonstrating proper migration into the molecular layer. (Photos courtesy Constance Cepko and *Cell* (Snyder et al., 1992); reprinted with permission.)

THE DISCOVERY OF CANCER STEM CELLS

Cancer can be defined as abnormal growth of cells that tend to proliferate in an uncontrolled way, and, in some cases, to **metastasize** (spread to other parts of the body). Perhaps the most intriguing key property of cancer cells is their ability to proliferate almost indefinitely. Cells exhibiting a cancerous phenotype have, in most cases, acquired multiple mutations resulting in stem cell-like properties such as active cell division, despite origins as mature, differentiated cell types. As such, it has been suggested that a subpopulation of cells, called cancer stem cells (CSCs—defined in more detail in Chapter 6), is not a prerequisite for tumorigenesis. Yet, given the stringent requirements of cancer cells to retain the genetics required for continuous mitotic activity, it has been widely speculated that CSCs exist within certain types of cancer for which relapse and metastasis are common. Initially in 1994 and then in a seminal research study in 1997, researchers in the Department of Genetics at the Hospital for Sick Children in Toronto, Canada led by John E. Dick identified a subpopulation of cells in a human acute myeloid leukemia (AML) sample that originated from a primitive HSC. In this study, a human cell type previously demonstrated as capable of initiating AML when introduced into NOD/SCID mice was characterized as having all the hallmarks of a primitive rather than committed progenitor cell. Termed a SCID-Leukemic Initiating Cell (SL-IC), the researchers confirmed the line's ability to indefinitely proliferate, self-renew, and differentiate into normal as well as cancerous leukemic cells of the immune system (Figure 1.8). Flow cytometric analyses revealed the SL-IC population to be exclusively CD34$^+$/CD38$^-$, which is a molecular hallmark of pluripotent, undifferentiated HSCs. Self-renewal properties were assessed through serial transplantation studies in which SL-IC cells were transplanted into primary and subsequent secondary recipients with no observable change in leukemic morphology or cell surface phenotype (Bonnet and Dick, 1997).

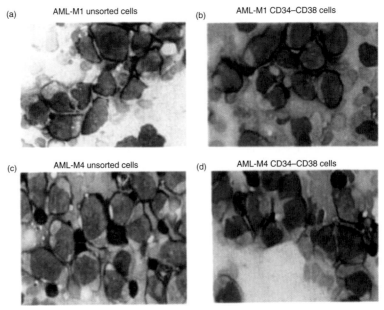

Figure 1.8 Differentiation capacity of SL-IC cancer stem cells. (a and c) Unsorted and (b and d) sorted CD34+/CD38- SL-ICs demonstrating colonization of the bone marrow of a recipient NOD/SCID mouse as assayed by the presence of the marker CD45 which is a transmembrane glycoprotein present on the cell surface of all cells of hematopoietic origin. (Photos courtesy John Dick and *Nature Medicine* (Bonnet and Dick, 1997); reprinted with permission.)

Focus Box 1.3: James Thomson and the discovery of hESCs

Dr. James A. Thomson of the University of Wisconsin is an American developmental biologist credited with deriving the first clonal hESC line in 1998. He later expanded upon this work in 2007 when he derived induced pluripotency stem (iPS) cells. He is currently Director of Regenerative Biology at the Morgridge Institute for Research in Madison, Wisconsin and a professor in the Department of Cell and Regenerative Biology at the University of Wisconsin School of Medicine and Public Health. He is a member of the National Academy of Sciences and was named one of *Time* magazine's 100 most influential people in 2008. (Photo courtesy UW; reprinted with permission.)

HUMAN EMBRYONIC STEM CELL DISCOVERY

The co-discovery of mouse embryonic stem cells by Gail Martin's group at UCSF and Martin Evan's team at the University of Cambridge immediately raised the speculation that a similar cell type might exist as a component of the ICM of developing human embryos. The existence of a hESC would have huge implications for medical research. For example, specific terminally differentiated cell types could be generated from hES cells to be used in drug screening assays or directly as cell-based therapeutics. The ability to derive embryonic stem cells from discarded human embryos was no easy feat, however, and it took a full 17 years of cell culture optimization before this was accomplished. Dr. James A. Thomson (see Focus Box 1.3) was the first to accomplish the isolation and characterization of a hESC line, which was published in the November 6, 1998 issue of *Science* magazine. In 1999, this discovery was featured again in *Science*'s "Breakthrough of the Year" article. In this study, Dr. Thomson's group obtained **cleavage-stage** (2-, 4-, 8-, and 16-cell stages containing blastomeres) human embryos produced by *in vitro* fertilization for clinical purposes. Embryos were cultured until fully formed blastocysts had developed and ICMs were further isolated for culture and characterization. Specific clonal isolates expressed high levels of telomerase activity (see Figure 1.9) and were designated as H1, H7, H9, H13, and H14 line with H7 and H9 of the female XX **karyotype** (the number and appearance of chromosomes in the nucleus of a cell) and the other lines male (XY) in origin. The **H9 cell line** is perhaps the most popular and widely studied of the original hES cell isolates, given its stable karyotype over extended passages and long periods of cell culture (Figure 1.9) (Thomson et al., 1998). Interestingly, the embryonic stem cell lines isolated in this study were not clonal (from a single cell), but were expanded from heterogeneous, uniform, undifferentiated colonies. Thomson's group went on to perform a number of analyses to determine the stability of each line and its capacity for differentiation. **Telomerase** is an enzymatic ribonucleoprotein which functions to add telomeric repeats to chromosomal ends and plays a critical role in extending the lifespan of a cell. Its expression and presence within a cell is directly correlated with cellular immortality. Each cell line isolate was shown to express high levels of telomerase activity. In addition, numerous cell surface markers previously identified in other embryonic stem cell lines such as those of mouse origin as defining pluripotency were expressed at high levels in each H line. These include SSEA3, SSEA4, Tra-1-60, and Tra-1-81. These markers will be discussed in detail in Chapter 3. Finally, in an experiment similar to that conducted by Gail Martin's group for the

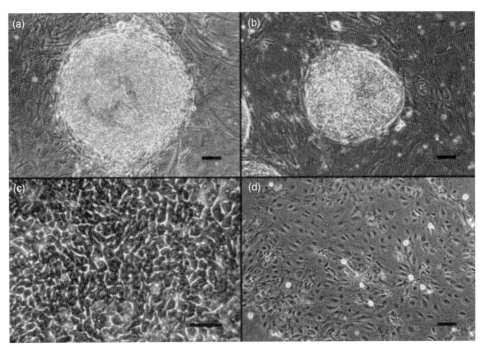

Figure 1.9 Derivation of the 1st clonal human embryonic stem cell line. (a) First inner cell mass colony cultured on a mouse feeder layer. (b) H9 clonal undifferentiated human ES cell colony. (c) High magnification of individual human ES cells. (d) Differentiated human ES cells cultured in the absence of a mouse feeder layer. (Photos courtesy Dr. James A. Thomson and *Science* (Thomson et al., 1998); reprinted with permission.)

characterization of mouse ES cells, cells from each isolate were confirmed to form tera- tomas in SCID-beige mice, a hallmark of pluripotency.

So how did Thomson's group successfully isolate true embryonic stem cells from human blastocysts when others had failed? First, it is speculated that the heterogeneity of each cell line population contribute to its pluripotent state, with hES cells secreting factors that keep neighboring hES cells undifferentiated. Second, the cell culture conditions implemented to isolate hES cells were notably different than that for mouse or other embryonic stem cells, such as those of primate origin. A conditioned medium prepared from feeder cells was not used for hES cell derivation and it can be speculated that differentiation inducing factors present in conditioned media were not present in the simpler formulation used by Thom- son's group. Third, it is evident from the successful isolation of the H9, H13, and H14 lines that mechanical dissociation of cellular clumps, rather than the use of enzymes such as trypsin, may result in less stress on the individual cells and thus promote the undifferenti- ated state. Subsequent use of a less harsh enzyme, crab collagenase, also appears to be a factor in maintaining cell pluripotency.

It should be noted that in 1998 John Gearhart and his research team at Johns Hopkins published the first report on the derivation of pluripotent stem cells from germ cells of the human embryo (Shamblott et al., 1998). In addition, serious ethical and moral issues have surrounded the derivation and study of hESCs. This topic will be discussed in detail in Chapter 3.

STEM CELLS AND CLONING

Cloning is defined as the process of creating genetically identical individuals from a single donor. In order to grasp the impact cloning has had on stem cell research it is important to understand the basic mechanistic procedure behind cloning. **Somatic cells** are defined as any biological cell forming the body of an organism other than a germ cell, gamete, gametocyte, or undifferentiated stem cell. **Somatic cell nuclear transfer (SCNT)** is defined as a technique for creating a clonal embryo by combining an ovum (egg) with a donor nucleus. In SCNT a donor nucleus is removed from a cell for which cloning is desired and inserted microsurgically into an enucleated (nucleus has been removed) egg. The "clonal" resulting cell may be either propagated in tissue culture or allowed to develop into an embryo and transplanted into a surrogate mother, a process known as **reproductive cloning** (Figure 1.10). It is important to note that, while reproductive cloning has not been carried out on humans, it has been successfully accomplished in other species such as mice, sheep, monkeys, and dogs, to name a few. **Therapeutic cloning** can be defined as reproduction of a genetically identical cell for purposes of cell therapy such as cellular replacement. Numerous research studies are currently underway to provide clonal populations of embryonic stem cells for either therapeutic or drug discovery initiatives.

Researchers at the Rosland Institute in Scotland, led by Sir Ian Wilmut, an English embryologist who is currently Director of the Medical Research Council Centre for Regenerative Medicine at the University of Edinburgh, successfully cloned the first mammal from an adult somatic cell in 1996. **Dolly the sheep** was cloned by combining the nucleus of a mammary cell from a Finn Dorset sheep with an enucleated egg from a Scottish Blackface ewe (Figure 1.11).

So if cloning technologies are based on the manipulation of somatic cell nuclei, why is cloning relevant from a stem cell perspective? Why would there be a desire to gen-

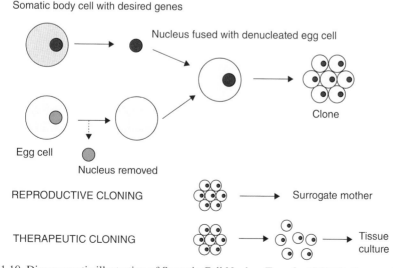

Figure 1.10 Diagrammatic illustration of Somatic Cell Nuclear Transfer (SCNT). See text for a detailed description. (Diagram courtesy Wikipedia.org; reprinted with permission.)

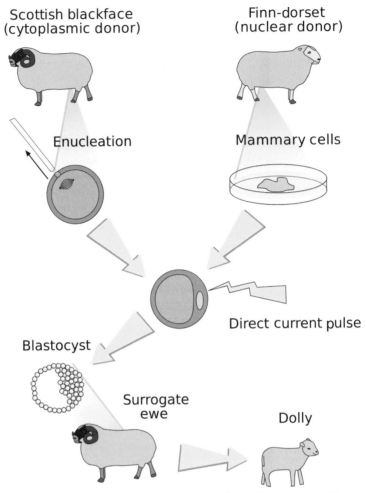

Figure 1.11 Diagram of the procedure undertaken for cloning Dolly the sheep. (Diagram courtesy Wikimedia Commons; reprinted with permission.)

erate a clonal population of stem cells from a stem cell? In a manner similar to that for somatic cells, cloning utilizing embryonic stem cells could allow for the generation of a virtually limitless supply of stem cells that are genetically identical, thus providing an extremely valuable therapeutic or perhaps even drug discovery platform as the population of cells would be genetically and phenotypically of an identical origin. Yet, having originated from adult cells, the population of cells would still have a composition of genetic material that has undergone the rigors of environmental influence and the aging process that so significantly damages that of adult somatic cells. As an example, there is much speculation that Dolly's premature death (she died in 2003 after only 7 years of life) may have been due to genetic abnormalities inherent in the aged mammary cell used as a nuclear donor. Scientific evidence indeed backs this theory as analysis of Dolly's chromosomal makeup revealed abnormally shortened **telomeres**, a

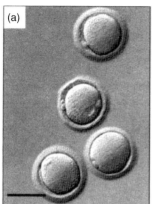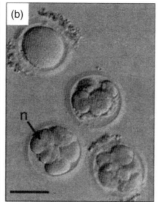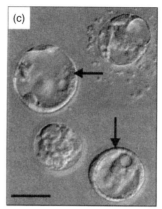

Figure 1.12 Advanced Cell Technology's parthenogenetically activated human embryos.
(a) **Isolated** unfertilized eggs. (b) 4–6 cell embryos 48 hours after activation of parthenogenesis.
(c) Day 6 revealing blastocoele cavities indicated by arrows. (Photos courtesy Jose B. Cibelli and
Scientific American (Cibelli et al., 2002); reprinted with permission.)

hallmark of aging, although the authors of this study deny that aging played a role in
Dolly's demise.

In 2001, the Worcester, Massachusetts-based company Advanced Cell Technology
(ACT) published a study focused on the generation of genetically identical hESCs. In
this study, researchers led by Jose B. Cibelli, who is now a Professor of Animal Biotechnology in the Departments of Animal Science and Physiology at Michigan State
University, created autologous embryos by two methods. First, they exploited a process known as parthenogenesis to produce the world's first cloned human embryos
from eggs. **Parthenogenesis** is defined as a form of asexual reproduction in females
where no fertilization from a male is required in order to reproduce. It is prevalent in
the plant kingdom and numerous natural examples exist in animals. In ACT's study,
parthenogenesis was artificially induced in 22 unfertilized eggs by various chemicals,
which resulted in a change in ion concentrations thus influencing parthenogenic development of single cell unfertilized eggs. Specific chemicals utilized included a calcium
ionophore and either puromycin or 6-dimethylaminopurine (DMAP) which have been
previously demonstrated to trigger not only **pronucleus** (the nucleus of an egg cell
during the process of fertilization) formation but early stage cleavage as well. A small
number of the eggs divided and developed into embryos containing a **blastocoele**,
which is defined as a cleavage cavity or segmentation cavity present in a developing
embryo (Figure 1.12) (Cibelli et al., 2002). Unfortunately these embryos did not
develop beyond the blastocoele stage, thus no ICMs were observed preventing the isolation of embryonic stem cells.

The second method pursued in this study for the generation of autologously derived
embryonic stem cells involved SCNT as discussed above. Nuclei from both human fibroblasts and **cumulus cells**, which are specialized granulosa cells that surround and nourish
a developing egg, were utilized in separate experiments to reconstitute human embryos via
SCNT. Some of these embryos developed to the 6-cell stage but did not reach the blastocyst
level required for isolation of embryonic stem cells (Figure 1.13).

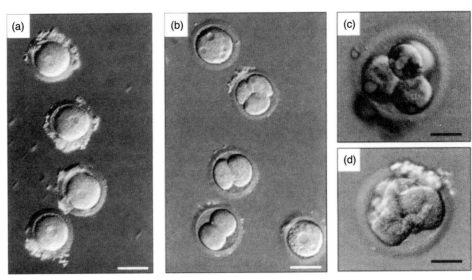

Figure 1.13 Somatic cell nuclear transfer cumulus cell-derived human embryos. (a) **12 hours**, (b) 36 hours (2 cell stage), (c) 72 hours (4-cell stage), and (d) 72 hours (6-cell stage) after nuclear transfer. (c) and (d) indicate nuclei stained with the fluorescent label bisbenzimide. (Photos courtesy Jose B. Cibelli and *Scientific American* (Cibelli et al., 2002); reprinted with permission.)

CORD BLOOD EMBRYONIC-LIKE STEM CELLS—AN ALTERNATIVE TO ES AND ADULT STEM CELLS

Given the controversial nature of utilizing hESCs for therapeutic purposes and their as yet unproven nature with respect to possible side effects such as tumorigenicity, many researchers have realized that commonly accepted use of hES cells in the clinic will not occur anytime soon. In addition, the large-scale production of a homogeneous population of hES cells or desired terminally differentiated lineages without the use of cell feeder layers represents a significant current technological hurdle. This issue, coupled with the near-universal immunological compatibility most recipients need, could hinder near-future clinical use. As such, numerous scientists have turned their attentions to the isolation and characterization of other ES cell-like cells that have similar multipotency capabilities but have not yet been isolated from developing embryos. For example, in 2005 researchers in the Stem Cell Therapy Programme at Kingston University in Surrey, United Kingdom led by Colin P. McGuckin isolated a population of stem cells—termed cord blood-derived embryonic-like stem cells (CBEs and defined in Chapter 4)—from human umbilical cord blood that bore striking resemblance to hES cells and possessed the capacity to differentiate into a variety of lineages. To accomplish CBE isolation, the researchers employed an immunomagnetic process to remove unwanted granulocytes, erythrocytes, and hematopoietic/myeloid/lymphoid progenitors from umbilical cord blood isolated by elective Cesarean section. This allowed for a concentration of the CBE population, given that typically only a limited number of all cell types exist in umbilical cord blood. Individual colonies were cultured for a period of 6 weeks followed by sub-cloning and an additional culture period for a minimum of 13 weeks. In addition, the cells were cultured in a liquid suspension environment and exhibited similar patterns of exponential growth as hESCs under these

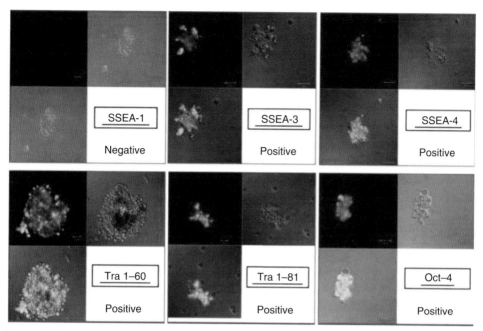

Figure 1.14 Marker characterization of cord blood embryonic-like stem cells. Cells were positive for the classical ES markers SSEA-3, SSEA-4, Tra 1-60, Tra 1-81 and Oct-4 yet, as is characteristic of embryonic stem cells, the CBEs did not express SSEA-1. (Photos courtesy Colin P. McGuckin and *Cell Proliferation* (McGuckin et al., 2005); reprinted with permission.)

conditions. Marker analysis was performed by immunofluorescence with antibodies specific for antigens well characterized in hESCs, revealing a similar profile to most pluripotent hES cell lines (Figure 1.14) (McGuckin et al., 2005).

McGuckin and colleagues also assessed the differentiation capacity of CBEs in a three-dimensional rotating cell culture microbioreactor. Single cells were introduced into the system and differentiation was initiated through the addition of hepatic cues for a period of 1 month. Three-dimensional clusters were immuno-phenotyped, revealing the presence of hepatic-specific antigens including cytokeratin-18, α-fetoprotein, and albumin, suggesting effective directed termination differentiation. The studies by McGuckin and colleagues laid the foundation for CBEs as a possible autologous stem cell-based therapeutic platform for a number of anomalies.

BREAKTHROUGH IN SPINAL CORD INJURY REPAIR

The promise of applying stem cells for real-world therapeutic intervention clearly is the primary focus of the widespread research and development efforts in this field. The adult central nervous system (CNS) has a very limited capacity to regenerate itself and, as a consequence, injuries to the spinal cord often result in partial or complete irreversible paralysis. Secondary degeneration of the CNS post injury also aggravates this scenario. Neural stem and progenitor cells have long been speculated as a potential source of therapeutic intervention by promoting tissue survival, growth, and the replacement of cells lost as a result of trauma or disease. In 2005, the first significant *in vivo* evidence of direct

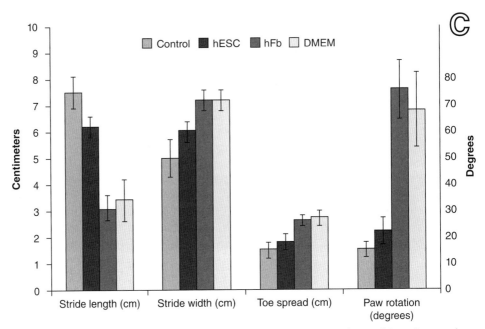

Figure 1.15 Human ES cell-derived OPCs improve locomotor recovery in rats. Note: decreased rear paw stride length and increased rear paw stride width, rear paw toe spread, and rear paw rotation are typical deficits in injured rats. These were largely corrected with hES-derived OPC transplants. (Graph courtesy Hans S. Keirstead and the *Journal of Neuroscience* (Keirstead et al., 2005); reprinted with permission.)

neural stem cell-based therapeutic applications came to light in a research study headed by Hans S. Keirstead, an Assistant Professor in the Department of Anatomy and Neurobiology, Reeve-Irvine Research Center at the University of California Medical School in Irvine. Dr. Keirstead's primary research focus is on the degeneration and regeneration of the spinal cord and efforts to repair spinal cord injuries. His team utilized induced rat models of human spinal cord injury, specializing in the study of **demyelination**, which is loss of the myelin sheath insulating the nerves. One particular study demonstrated an enhancement in myelination and improvement of locomotor recovery in injured rats upon intraspinal cord transplantation of hESC-derived **oligodendrocyte progenitor cells (OPCs)**. OPCs are defined as immature myelin-producing stem cells. When transplanted, the cells drove partial reconstitution of the myelin sheath through remyelination, which is speculated to have partially restored neuronal cell function (Figure 1.15) (Keirstead et al., 2005). These findings illustrate that it is possible to at least partially restore locomotor function in spinal cord injured vertebrates using stem cell-based transplant technologies.

THE GENERATION OF IPS CELLS

The controversy surrounding the use of embryonic stem cells as well as the restricted multi-lineage differentiation capacity of adult stem cells has driven researchers to search for yet other alternative cell-based therapeutics platforms. It has long been known by cell and developmental biologists that in certain instances adult differentiated cells may be driven to lose their morphological and molecular identities and transform into other cell types. The process is referred to as **transdifferentiation**, which is defined as a non-stem cell transforming into a different cell type, or when a differentiated stem cell generates other cell types outside of its normal realm of multipotency. If it can be harnessed and controlled, the phenomenon of transdifferentiation provides an exciting alternative to the use of embryonic or adult stem cells to produce the mature cell phenotypes needed for cell replacement therapy. In addition, in an artificial setting, differentiated cells can be reprogrammed to an embryonic-like state, a process known as **dedifferentiation**, which can be accomplished by either nuclear transfer into oocytes or fusion with other ES cells. Until 2006, it was not possible to drive this process with individual factors. It was in this year that researchers in the Department of Stem Cell Biology at the Institute for Frontier Medical Sciences in Kyoto, Japan led by Shinya Yamanaka managed to induce adult mouse fibroblast cells to dedifferentiate into stem cells and become pluripotent (See Case Study 1.2). The same researchers followed this groundbreaking study in 2007 with a demonstration of pluripotency induction of human fibroblasts. That same year, James Thomson's group at the University of Wisconsin published a similar finding separately and independently (see Figure 1.18).

The groundbreaking studies by Yamanaka's group in 2006 on the induction of pluripotent properties in mouse adult fibroblasts (see Case Study 1.2) provided a solid framework for similar efforts on human cells. From a clinical perspective, the production of iPS stem cells from human adult cell types would be of enormous benefit for several reasons. First, as the cells could be derived from a patient's own cell sample—for example, a skin punch biopsy—it would allow for the generation of patient-specific stem cells, thus eliminating any possibilities of immunorejection. Second, due to their pluripotent properties, it would allow for the development of individual cell types that could be valuable for the treatment of specific diseases. In 2007, two groups separately and independently accomplished the creation of human iPS cells from adult human fibroblasts. In a manner similar to that for the induction of iPS properties in mouse fibroblasts in 2006 (see Case Study 1.2), the introduction of four key transcription factors into human fibroblasts proved critical to drive dedifferentiation and the induction of pluripotency. Yamanaka's group employed retroviral transduction to introduce genes encoding the four key transcription factors, Oct 3/4, Sox2, Klf4, and c-Myc to human dermal fibroblasts (HDFs). They demonstrated a striking resemblance of the iPS cells to hESCs with respect to a number of characteristics including morphology, proliferation properties, cell surface markers, telomere length/telomerase activity, and differentiation capacities (Figure 1.17) (Takashi and Yamanaka, 2006).

Interestingly, in a study headed by James Thomson and colleagues at the Genome Center of Wisconsin and the Wisconsin National Primate Research Center at the University of Wisconsin in Madison the four factors Oct3/4, Sox2, Nanog, and Lin28 were sufficient for the induction of pluripotency in human somatic cells. The difference in the identity of transcription factors utilized for iPS induction in comparison to the research of Yamanaka and colleagues is intriguing, and suggests that there is not one universal transcription factor code necessary for dedifferentiation of somatic cells to an embryonic-like state. Thomson's group implemented a lentiviral transduction system to introduce genes encoding these four transcription factors into human mesenchymal cells. While Nanog could be removed from the system and iPS clones successfully derived, this transcription factor was shown to improve clonal recovery, resulting in an over 200-fold increase in reprogramming efficiency (Figure 1.18) (Thomson et al., 1998). iPS induction utilizing these four transcription factors was also performed on human adult somatic cell fibroblasts with similar results.

Case Study 1.2: Induction of pluripotent stem cells from mouse embryonic and adult fibroblast cultures by defined factors

Kazutoshi Takahashi and Shinya Yamanaka

In 2006 Shinya Yamanaka and colleagues at the Institute for Frontier Medical Sciences focused their efforts on the introduction of four key transcription factors via retroviral transduction methods into adult mouse fibroblasts to drive reprogramming of these cells into an embryonic stem cell-like state (iPS). The factors Oct 3/4, Sox2, c-Myc and Klf4 were demonstrated to be necessary and sufficient for fibroblast dedifferentiation and reprogramming. Interestingly, the transcription factor Nanog, which has been suggested by many researchers to be required for the pluripotent properties of embryonic stem cells, was dispensable in this study. The researchers confirmed the pluripotency of reprogrammed, GFP tagged iPS cells by introducing them into host blastocyst stage embryos and monitoring iPS cell contribution to the three primary germ layers of the host embryos (Figure 1.16) (Takashi and Yamanaka, 2006). The key transcription factors noted in this study will be discussed in more detail in Chapter 5.

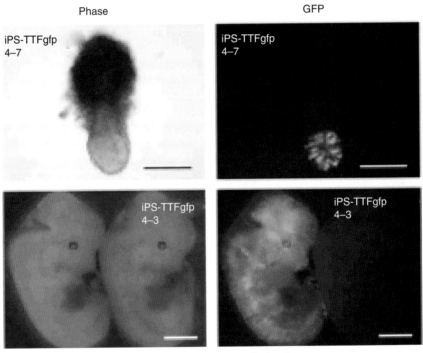

Figure 1.16 Contribution of iPS cells to mouse embryonic development.

iPS cells tagged with green fluorescent protein (GFP) were microinjected into mouse blastocysts. Embryos were characterized at either E7.5 (upper panels) or E13.5 (lower panels) for iPS cell contribution to the embryo proper. (Photos courtesy Shinya Yamanaka and *Cell* (Takashi and Yamanaka, 2006); reprinted with permission.)

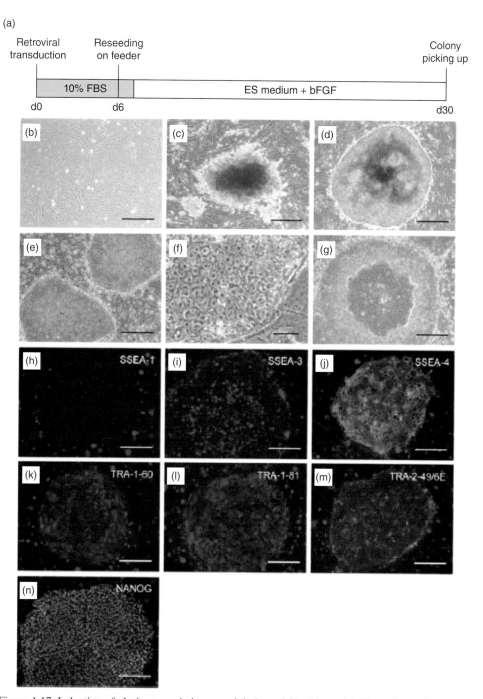

Figure 1.17 Induction of pluripotency in human adult dermal fibroblasts. (a) Chronology of induction strategy; (b) HDF morphology before induction; (c) HDF colony before induction; (d) Example of human ES cell colony; (e) P6 iPS HDF colony; (f) Same under higher magnification; (g) Spontaneous differentiation in the center of the iPS HDF colony; (h–n) Immunocytochemistry for the noted markers demonstrating a similar expression patter in iPS HDFs to that of human embryonic stem cells. (Photos courtesy Shinya Yamanaka and *Cell* (Takashi and Yamanaka, 2006); reprinted with permission.)

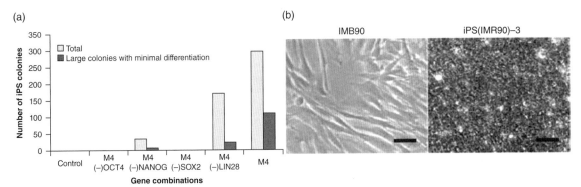

<figure>
Figure 1.18 Gene combinations driving induced pluripotency reprogramming in human adult somatic cells. (a) Comparison of gene combinations for generating iPS colony numbers and sizes. When individual genes are removed from the mixture (M4) colony numbers and sizes drop. (b) Brightfield images of p18 human somatic cell fibroblasts before induction (left) and p18 iPS cells after induction of pluripotency. (Figure courtesy James Thomson and *Science* (Thomson et al., 1998); reprinted with permission.)
</figure>

iPS Cells Derived from Keratinocytes

The ability to utilize somatic cells that are readily available as a source for the generation of iPS cells is a critical factor for ultimately generating banks of autologous cell lines to be used for therapy. While skin fibroblasts are accessible, there is perhaps no more readily available somatic cell source than that of a human hair. Painless collection of samples and a virtually endless supply for almost every individual make hair follicles an ideal adult somatic cell source for iPS cell generation. In October 2008, a team of scientists led by Juan Carlos Izpisúa Belmonte, a professor in the Gene Expression Laboratory at the Salk Institute for Biological Studies in La Jolla, California, successfully reprogrammed keratinocytes isolated from human foreskin into iPS cells. In a manner similar to that for the Yamanaka studies mentioned above, iPS was accomplished via retroviral transduction of genes encoding the key transcription factors Oct4, Sox2 Klf4, and c-Myc. Interestingly, the researchers demonstrated that iPS reprogramming of keratinocytes was 100-fold more efficient and at least 2-fold faster than that for fibroblast conversion. The resulting iPS cell population was dubbed **KiPS cells,** and exhibited striking morphological and molecular similarities to hESCs. The researchers also confirmed the cells' ability to differentiate into lineages representing the three primary germ layers. This study was followed by the induction of KiPS cells from a single plucked human hair. To retrieve keratinocytes needed for iPS induction, the hair was cultured in mouse embryonic fibroblast conditioned hES cell medium promoting the proliferation of keratinocytes out of the hair outer root sheath. Following expansion, retroviral transduction was performed in a manner similar to that for foreskin keratinocytes. The resulting KiPS cells resembled those derived from human foreskin keratinocytes and expressed the classical hES cell pluripotency marker **alkaline phosphatase** (Figure 1.19) (Aasen et al., 2008).

iPS Induction Without the Use of Viruses

The utilization of viruses to induce iPS in cells to be used in regenerative medicine applications results in the inherent risk of transformation of these cells toward a cancerous

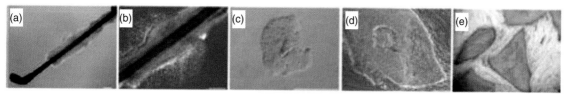

Figure 1.19 Induction of pluripotency in keratinocytes isolated from a single plucked human hair. (a) Portion of follicle cultured in hES medium. (b) Keratinocyte outgrowth from the outer root sheath 5 days after initiation of culture. (c) Colony of hair keratinocytes after reprogramming exhibiting typical hES cell morphology. (d) P1 iPS colony 10 days after picking. (e) High magnification of alkaline phosphatase -stained KiPS colony. (Photos courtesy Juan Carlos Izpisúa Belmonte and *Nature Biotechnology* (Aasen et al., 2008); reprinted with permission.)

phenotype. This is due to the mechanism of viral transduction whereby viral genomic material randomly integrates into the host cell's genome. These random integrations could occur in proto-oncogenes and thus transform the cells. Therefore mechanisms for induction of pluripotency that do not modify the host genome would be of great value and possibly safer to implement. The following sections outline two methods for accomplishing non-viral-mediated induction of pluripotency.

Transposon-Mediated iPS As of 2009 most efforts at inducing adult somatic cells to dedifferentiate and take on pluripotent characteristics had been focused on invasive intra-cellular viral or plasmid-based introduction of the four key transcription factors: Oct4, Sox2, Klf4, and c-Myc. Strategies involving retroviral, lentiviral, adenoviral, and plasmid transfection had the desired effect of resulting in high levels of genetic material and corresponding protein products present within the cells to promote pluripotency. The utilization of retroviral and lentiviral methods to introduce the key genes relies on stable incorporation into the host cell's genome. This is a major issue and consideration as integration into the wrong loci could promote tumorigenesis. It is only adenoviral and plasmid transfection that represented transitional, non-stable introduction of key genes into the cells. However, an obvious diminished capacity to drive and maintain iPS long-term is a key issue with these methods. This is due to the fact that over time and through multiple cell divisions the concentration of genetic material encoded by the transient presence of adenoviral or plasmid vectors becomes low or even non-existent. In addition, there are no guarantees of transient presence, and molecular incorporation into the host genome of viral genetic products may occur in rare instances. In April 2009, researchers at the Samuel Lunenfeld Research Institute, Mount Sinai Hospital, and Department of Molecular Genetics, University of Toronto, Toronto, Canada implemented a novel non-integrative approach to drive gene expression and the formation of key proteins necessary for iPS induction in mouse somatic cell fibroblasts. They employed a transposon/transposase-based system known as ***piggyback*** (**PB**) which allows for the stable but transient integration of genetic material into the host cell genome and drives the maintenance of its expression long-term. *Piggyback* is host factor-independent, making it suitable for transduction of a variety of somatic cell types. The genomic insertion may be seamlessly, efficiently, and precisely excised at will, thus leaving the iPS cell genetically unaltered following excision. The researchers implemented the *piggyback* system to introduce genetic material encoding Oct4, Sox2, Klf4, and c-Myc into mouse fibroblasts. They subsequently demonstrated clean excision of the integrants and confirmed iPS in whole embryos and adult chimeras (Figure 1.20) (Woltjen et al., 2009).

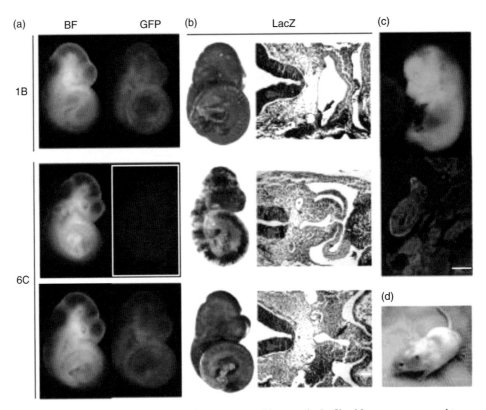

Figure 1.20 *Piggyback* (PB)-mediated factor transposition results in fibroblast reprogrammed to an iPS state. (a) 10.5dpc demonstrating high-percentage contribution of PB GFP labeled iPS cells to the embryo proper. (b) PB iPS cells contribute to all three primary germ layers as assayed by LacZ staining. (c) Tetraploid complementation using iPS cells results in complete derivation of the embryo from iPS cells as assayed by GFP presence. (d) Adult mouse chimera generated by co-culture of PB iPS cells with diploid 8-cell stage albino embryo. (Photos courtesy Knut Woltjen and *Nature* (Woltjen et al., 2009); reprinted with permission.)

Protein-Based iPS Up to 2009 all studies on induced-pluripotency reprogramming involved either the stable or transient introduction of genetic material encoding key factors required for dedifferentiation of somatic cells and promotion of the pluripotent phenotype. The application, for example, of viral-based systems often result in multiple viral integrants within the genome of the host cell, the location of which can, for the most part, not be controlled. As such, one cannot rule out the effect of insertional mutagenesis resulting in tumor-promoting integrants and other unpredictable genetic dysfunctions. From a therapeutic perspective, a system is therefore needed that allows for the safe production of patient-specific stem cells without genetic alteration to produce such cells. In June 2009, a technique was perfected for the introduction of key transcription factor proteins directly into human somatic cell newborn fibroblasts (HNFs). Led by Kwang-Soo Kim, Associate Professor of Psychiatry and Neuroscience at the McLean Hospital, Harvard Medical School, Harvard Stem Cell Institute, Boston, Massachusetts and CHA Stem Cell Institute in Seoul, South Korea, researchers focused

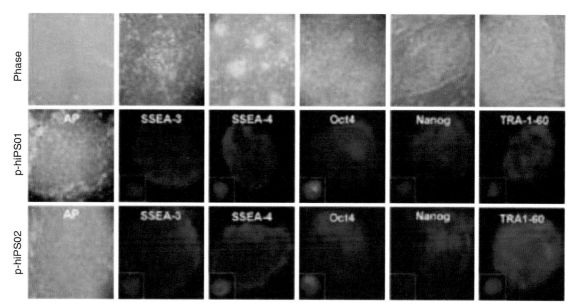

Figure 1.21 Marker expression of protein-induced human iPS cells. Two independent lines were analyzed by immunofluorescence for the classical markers of pluripotency after cycled introduction of CPP-anchored reprogramming factors. Both morphological and marker analysis reveal similarities to iPS lines generated by genetics-based approaches. (Photos courtesy Kwang-Soo Kim and *Cell Stem Cell* (Kim et al., 2009); reprinted with permission.)

on devising a strategy to efficiently introduce the four key reprogramming proteins Oct4, Sox2, Klf4, and c-Myc directly into cells. Proteins and other macromolecules have only a limited ability to cross the cell membrane, thus a technique was needed to drive active transport across the lipid bilayer and into the cytoplasm. Certain proteins and short peptides, referred to as cell penetrating peptides (CPP), have been shown to actively cross the cell membrane, and can carry other macromolecules along with them during this process. Kim's team anchored each of the reprogramming proteins to CPP. After several failures performing single introduction experiments, the researchers attempted repeated protein treatment cycles and observed iPS-like morphology and alkaline phosphatase expression, a key marker of the embryonic stem cell-like phenotype. Dubbed protein-induced human iPS (p-hiPS) cells, they were demonstrated to express all the classical markers of pluripotency and could contribute to the three primary germ layers in murine teratoma studies (Figure 1.21) (Kim et al., 2009). The technique resulted in the induction of pluripotency without the need for viral transduction or plasmid transfection of genetic material encoding these proteins. The authors noted that the process of CPP-anchored protein transduction was inefficient and required further optimization, but this study is an important first step in developing a non-genetics-based approach to pluripotency induction in somatic cells. It opens the door to a safer non-genetics-based alternative to viral transduction or plasmid introduction of iPS induction factors.

THE DISCOVERY OF HUMAN AMNIOTIC STEM CELLS

The derivation of embryonic stem cell lines from human embryos and the induction of pluripotency in adult cells are both technologies that hold much promise in the field of stem cell therapeutics. By creating a source for the generation of terminally differentiated cells of various types, many medical disorders such as diabetes and cardiac hypertrophy could theoretically be addressed. However, a more readily available source of pluripotent stem cells that eliminates the controversy or technical challenges of hES cell derivation/embryo destruction and iPS technology would be advantageous for the treatment of a number of disorders. Sources such as cord blood embryonic-like cells mentioned above eliminate the need to destroy embryos for ES cell derivation or induce pluripotency in tissue culture. They also may provide a readily abundant autologous cell source for some specific desired lineages. In early 2007, a research group led by Anthony Atala, the W.H. Boyce Professor and Director of the Wake Forest Institute for Regenerative Medicine at Wake Forest University School of Medicine in Winston-Salem, North Carolina successfully isolated both human and rodent **amniotic fluid-derived stem (AFS) cells** and demonstrated striking properties inherent in these cells that could make them a valuable source for stem cell therapeutics initiatives. To efficiently isolate the cells from other cell types in the amniotic fluid, the researchers employed **immunoselection**—the isolation of an antigen using antibody specificity—and magnetic isolation to separate cells expressing c-Kit from others in the heterogeneous population. **C-Kit**, also referred to as **CD117**, is a tyrosine kinase cell surface receptor known to be a marker for progenitor and stem cell lineages such as those of the prostate, thymus, and of hematopoietic origin. The isolated cells exhibited striking properties that could make them a valuable cell-based therapeutics source. For example, after more than 250 population doublings, the stem cell lines maintained a normal karyotype and retained long telomeres, a sign of genomic stability. In addition, the cell lines could be directed to differentiate into multiple adult lineages including those of the adipogenic, osteogenic, myogenic, endothelial, neuronal, and hepatic phenotypes; thus they were classified as multipotent in nature (Figure 1.22) (De Coppi et al., 2007) (See Chapter 2).

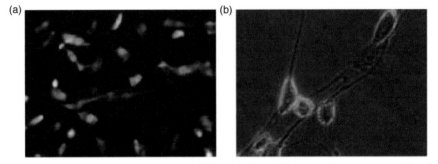

Figure 1.22 Amniotic fluid-derived stem cell differentiation. (a) Immunofluorescence staining for nestin; (b) phase contrast microscopy of dopaminergic neurons. Both were directed to differentiate from AFS cells. (Photos courtesy Anthony Atala and *Nature Biotechnology* (De Coppi et al., 2007); reprinted with permission.)

HUMAN EMBRYONIC STEM CELLS GENERATED WITHOUT EMBRYO DESTRUCTION

The destruction of human embryos for the derivation of embryonic stem cells is perhaps the most controversial aspect of hES cell research. Although embryonic stem cell isolation typically occurs at the very earliest stages of embryonic development, given that it is carried out post fertilization, many feel that this is truly a destruction of human life. This poses ethical concerns and in fact is formally outlawed in a number of countries. Specifically in the United States, federal funding may not be utilized for research that involves the destruction of human embryos. As such, a great deal of effort has been focused on deriving embryonic stem cells or creating embryonic-like stem cells without the need for embryo destruction. Examples of this include the application of iPS technology, which generates embryonic-like stem cells as discussed above, but these are not true ES cells. For the derivation of authentic hESCs, to avoid controversy and, some would say, for the preservation of human life, a method is needed that does not destroy nor harm the developing human embryo. In January of 2008, scientists at Advanced Cell Technology in Worcester, Massachusetts published a scientific article outlining that they had accomplished this by implementing a technique similar to **preimplantation genetic diagnosis (PGD)**, which is an embryonic manipulation procedure employed to profile the genetics of an embryo prior to implantation. In PGD, controlled ovarian stimulation is employed for the release of oocytes, which are subsequently fertilized, most often by intracytoplasmic sperm injection. Cells from embryos are isolated and biopsied at specific stages of development and assessed for genetic defects that might result in a termination of pregnancy. The stage and mechanism for cell isolation in most cases does not damage the developing embryo. Researchers at Advanced Cell Technology isolated **blastomeres**, which are defined as cells resulting from the cleavage of a fertilized ovum during early embryonic development, from fertilized eggs (8 cell stage; 1 or 2 blastomeres isolated per egg) and cultured them via a modified PGD approach which was designed to reconstitute the "inner cell mass niche". To accomplish this goal, a microdrop co-culture system was implemented in the presence of hESCs, which are thought to provide support and secrete factors promoting pluripotency. To distinguish between hES cells used for co-culture and blastomeres, the hES cells were labeled by stable transfection of a gene encoding green fluorescent protein (GFP). Medium was supplemented with laminin and fibronectin and the cultures transferred to a mouse embryonic fibroblast layer for expansion and characterization. This allowed for an improvement in successful ES cell derivation rates, which were comparable to that of derivations utilizing whole embryos. Five independent lines were derived that maintained normal karyotypes and marker expression after more than 50 passages. In addition, it was shown that the lines could differentiate into lineages representing all three germ layers (Figure 1.23) (Chung et al., 2008).

HUMAN CLONING

In the realm of cell biology, **cloning** can be defined as the identical reproduction of another organism at the molecular and cellular level. From the perspective of humans, it is highly controversial and raises ethical issues on individual identity and even on religious grounds. In addition, serious safety concerns are associated with human cloning. In 2008, despite the fact that significant progress had been made with respect to both iPS reprogramming and

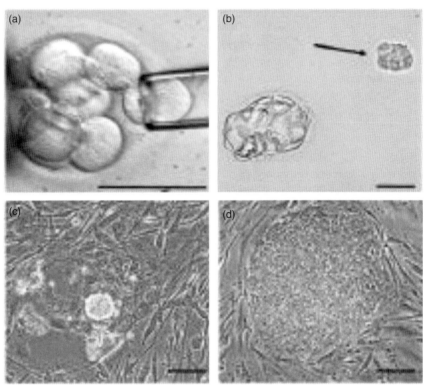

Figure 1.23 Derivation of hES cells without embryo destruction. Brightfield microscopy of (a) Actual biopsy of a blastomere from a human embryo. (b) Biopsied blastomere (denoted by arrow) growing alongside the embryo from which it was derived. (c) Outgrowth of a blastomere culture on mouse embryonic fibroblasts (MEFs). (d) Colony of blastomere-derived human embryonic stem cells. (Photos courtesy Robert Lanza and *Cell Stem Cell* (Chung et al., 2008); reprinted with permission.)

hESC isolation without damage to the embryo, many research groups remained focused on SCNT, also referred to as cloning, as an alternative way to generate hESCs. Andrew J. French and colleagues at Stemagen Corporation and the Reproductive Sciences Center in La Jolla, California published a research article in the journal *Stem Cells Express* that delineated the first recorded example of successful cloning of human blastocysts through SCNT of nuclei from adult fibroblasts into enucleated oocytes. In this study, mature oocytes obtained from donors were enucleated either via extrusion or aspiration and nuclei transferred into these oocytes from adult male fibroblasts exhibiting normal karyotypes. The team observed high rates of **pronucleus** formation (66%), cleavage (47%), and blastocyst development (23%) following SCNT. In addition, the morphology of the embryos during early development post SCNT appeared normal (Figure 1.24) (French et al., 2008). Perhaps in order to deflect from the controversial field of embryonic stem cell research, French's team coined the term **nuclear transfer stem cells (NTSC)**, which are defined as stem cells derived from a SCNT-generated embryo, to distinguish stem cells created and derived through SCNT procedures from other sources. The cloning of humans remains a highly controversial topic and raises many ethical, religious, and moral issues that are beyond the scope of this text.

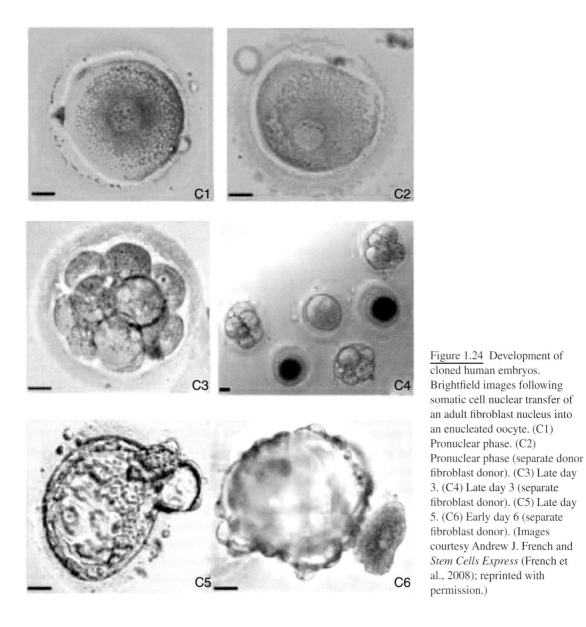

Figure 1.24 Development of cloned human embryos. Brightfield images following somatic cell nuclear transfer of an adult fibroblast nucleus into an enucleated oocyte. (C1) Pronuclear phase. (C2) Pronuclear phase (separate donor fibroblast donor). (C3) Late day 3. (C4) Late day 3 (separate fibroblast donor). (C5) Late day 5. (C6) Early day 6 (separate fibroblast donor). (Images courtesy Andrew J. French and *Stem Cells Express* (French et al., 2008); reprinted with permission.)

MESENCHYMAL STEM CELL-DERIVED HUMAN KNEE CARTILAGE

Regenerative medicine can be defined as the ability to regenerate tissue rather than surgically extracting or altering that tissue. It can be classified into three unique categories, including platelet augmentation, recombinant growth factor amplification, and stem cell isolates. As platelet augmentation methods, including platelet-rich plasma (PRP) implantation, and growth factor injection are beyond the scope of this text, the focus of this initial overview section will be on the application of stem cell isolates for cartilage regeneration as it is in this area where stem cell-based regenerative medicine has had the largest impact clinically. For example, mesenchymal stem cells (MSCs)

TABLE 1.1 Cartilage volume analysis after mesenchymal stem cell injection. Measurements were taken in mm^3.

Image	Area of measurement	Volume ($n = 3$)	STDEV	SE	Change from pre-injection (%)
Pre-injection	Cartilage surface	4,020	12.1	6.99	
	Meniscus	5,178	164.57	95.13	
1 month	Cartilage surface	4,924	149.01	86.13	22.49
	Meniscus	5,647	453.57	262.18	9.06
3 month	Cartilage surface	4,795	113.5	65.61	19.28
	Meniscus	6,661	146.47	84.67	28.64

Source: Centeno et al., 2008. Reproduced with permission from C. Centeno.

(defined in Chapter 4) have been demonstrated to differentiate into bone marrow, as well as synovial and adipose tissues. In June 2008, researchers at Regenerative Sciences, Inc. and Centeno-Schultz Clinic in Westminster, Colorado conducted a clinical study in which mesenchymal stem cells were isolated from the iliac crest of patients, cultured, and subsequently reintroduced **autologously** (transplant from one part of the body into another of the same patient) into the subject's knee to combat degenerative joint disease previously diagnosed by magnetic resonance imaging (MRI). The goal of the study was to drive the growth and development of new cartilage in the degenerated area. To prepare the MSCs for implantation, whole bone marrow was isolated and centrifuged for removal of red blood cells. Plasma was subsequently removed and nucleated cells were cultured for expansion purposes through five passages. **Percutaneous** (under the skin) implantation of cultured, expanded MSCs was performed in an autologous fashion in combination with dexamethasone to induce differentiation of the stem cells into cartilage precursors and ultimately functional cartilage. Table 1.1 outlines the results of this study, revealing considerable new cartilage surface and meniscus growth 1 month and 3 months post injection. This is the first published study of successful cartilage regeneration in a human knee via autologous stem cell therapy (Centeno et al., 2008).

THE FIRST CLINICAL TRIAL USING HUMAN EMBRYONIC STEM CELLS

2010 was a pivotal year for embryonic stem cell research at the clinical level with the launch of a clinical trial utilizing an embryonic stem cell derived therapy termed GRNOPC1 by **Geron Corporation**. Geron is a biotechnology  company based in Menlo Park, California that has developed proprietary technologies for the growth, maintenance, and scaling of hESCs and other cell types for therapeutic purposes. Spinal cord injury is among the most devastating medical problems, and affects over 250,000 individuals in the United States alone. It is both painful and debilitative, causing immeasurable suffering for those unfortunate enough to experience it. As of this publication, Geron has spent a total of $170M USD developing a stem cell treatment platform for spinal cord injury. The company utilized feeder- and serum-free fully defined growth conditions to culture and propagate a variety of stem cell-based therapeutics. The defined growth environment coupled with an absence of any animal components has enabled Geron to obtain approval by the Food and Drug Administration (FDA) for the initiation of a clinical trial of an embryonic stem cell-based platform to treat spinal cord injuries. Initial findings from the clinical study, which focused on the application of ES cell-derived oligodendrocytes progenitor cells (the OPC in GRNOPC1), demonstrated high tolerance without any serious side effects. Unfortunately, the company halted the clinical trial as of November 2011, citing the need to conserve funds. At the time of this publication, Geron was actively seeking partners who "have the technical and financial resources to advance its stem cell programs." It is clear that studies like this must proceed if stem cell-based therapeutics are to become a reality. This and other clinical trials implementation a variety of different stem cell types are discussed in more detail in Chapter 8.

MITOCHONDRIAL DNA: A BARRIER TO AUTOLOGOUS CELL THERAPEUTICS

Much of the research on the generation of embryonic stem cells from SCNT or the production of iPS cells has been based on the premise that by using host nuclear material immune system rejection of the resulting cells would not be an issue post transplantation. Yet it must be noted that these cells are not 100% identical to that of the donor host. Specifically, the **mitochondria**, which are defined as spherical or elongated organelles in the cytoplasm of mostly eukaryotic cells containing genetic material and many enzymes important for cell metabolism, are derived from donor ES cells or oocytes in SCNT. The genetic material of mitochondria (mtDNA) has accumulated mutations over the life of the cell, which results in a unique and potentially lethal difference in the cellular makeup compared to that of the host. In the case of iPS cells generated from host donor cells also have accumulated mitochondrial DNA mutations throughout the aging process. These mutations result in desired differentiated lineages that are different from the host donor, which may contribute to immune rejection by the host recipient of any derived cell therapy. Researchers at the Graduate School of Life and Environmental Sciences, University of Tsukuba, Tsukuba, Japan studied the effects of **allogeneically -** (taken from different individuals of the same species) introduced mtDNA into various mouse strains. These cells shared the same nuclear genetic material and background, but differed in mtDNA makeup. Theoretically, since the nuclear material from these cells was the same as the host strain, no immune response should have been observed. However, the researchers observed that transplants with mtDNA from the same murine strain as the host strain were rejected, most likely due to inherent mutations (Figure 1.25) (Ishikawa et al., 2010). Interestingly, it was confirmed that the rejection was due to innate rather than acquired immune response. **Innate immunity** is naturally or inherently present and is not due to sensitization of the immune system by an antigen. Thus, differences in mtDNA between a donor cell and host recipient for either SCNT- or iPS-based cell therapeutics will be a factor in immunorejection, and must be addressed for purposes of safety.

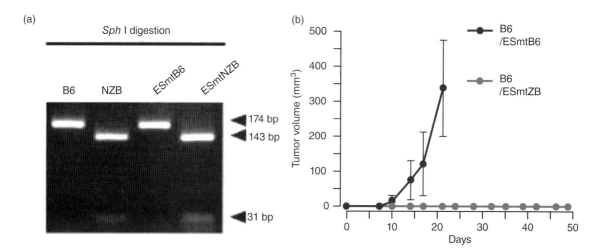

Figure 1.25 Rejection of cells containing allogeneic mtDNA by host mice. (a) Confirmation of mtDNA genotypes by PCR. (b) Analysis of tumor formation size after ES inoculation into host mice. The blue line demonstrates significant tumor growth in allogeneic transplants of B6 mtDNA with a B6 host. The red line reveals suppressed tumor growth in allogeneic transplants of NZB strain mitochondrial DNA with a B6 host suggesting mutations in the mtDNA trigger host rejection due to the innate immune system. (Figures courtesy Jun-Ichi Hayashi and the *Journal of Experimental Medicine* (Ishikawa et al., 2010); reprinted with permission.)

INDUCED PLURIPOTENCY AND THE POTENTIAL TO SAVE ENDANGERED SPECIES

The development and application of technologies for the production of pluripotent stem cells is not just useful in the therapeutic realm. Manufacturing cells capable of producing multiple differentiated lineages or even entire organisms has enormous

implications in zoology. Species preservation has been a top priority for zoologists and conservation biologists for literally hundreds of years. In many examples of highly endangered species there are simply too few animals capable of the reproductive capacity necessary to maintain species numbers. In other cases the species has been declared officially extinct, with no known surviving examples. Yet in both cases SCNT and iPS technologies may allow for a rescue and perhaps even a reintroduction of the species into the ecosystem. In September 2011, Jeanne Loring, Professor and Director of the Center  for Regenerative Medicine at the Scripps Research Institute in La Jolla, California, and her team focused their studies on the induction of pluripotency in adult somatic cells to address the problem of species endangerment and extinction. Specifically, they focused on two endangered species: a primate, the drill, *Mandrillus leucophaeus* and the nearly extinct northern white rhinoceros (NWR), *Ceratotherium simum cottoni*. The drill is considered one of the most endangered species on the African continent and its numbers have drastically declined over the past 20 years due to both the destruction of its native habitat and illegal poaching. Current captive populations of drills are sustained by small reproductive colonies, which poses a considerable risk of inbreeding and therefore genetic issues for the entire population. The horns of the northern white rhinoceros make it a target for hunting and illegal poaching—only seven confirmed living rhinos exist today. The goal of these studies was to generate iPS cells from frozen somatic cell fibroblasts corresponding to each of these species, which might later be utilized to generate fully mature, adults capable of reproduction. To accomplish this, the Loring team retrovirally transduced drill and northern white rhinoceros fibroblasts with human genes encoding Oct4, Klf4, Sox2, and c-Myc. Frozen fibroblast samples were obtained from the Frozen Zoo of the San Diego Zoo Institute for Conservation Research. Following thaw, expansion, and transduction of the fibroblasts, putative reprogrammed lines were initially selected based upon morphological similarities to embryonic stem cells and iPS cells obtained from other species. Four and three independent, clonal iPS lines were derived from the drill and northern white rhinoceros, respectively, and further characterized for pluripotency properties. Figure 1.26 outlines the characterization of the northern white rhinoceros iPS cells illustrating ES cell-like morphology, normal karyotypes, expression of markers and the ability to generate lineages representative of the three primary germ layers (Ben-Nun et al., 2011)). In addition, the researchers performed **glycomic profiling**, which is the analysis of the complete repertoire of glycans and glycoconjugates that cells produce under specified conditions, of each iPS line. The profiling revealed a clustering and close correlation of the drill and NWR iPS lines with those from other species. While complementing SCNT, these studies set the stage for the possibility of a reintroduction of genetic material into endangered species breeding populations for purposes of species survival and the elimination of extinction risks. In addition, it may provide an avenue for the rescue of already extinct species should viable cell samples for these species exist.

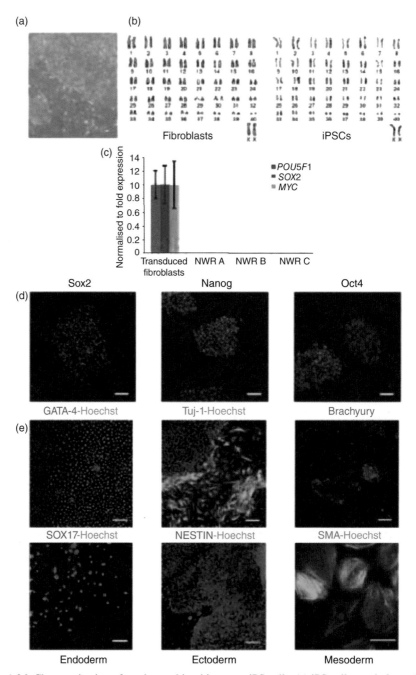

Figure 1.26 Characterization of northern white rhinoceros iPS cells. (a) iPS cell morphology. (b) Normal karyotype of pre-programmed fibroblasts and resulting iPS cells. (c) Quantitative RT-PCR of three of the four key genes involved in reprogramming, (d) Immunocytochemistry of pluripotency markers. (e) Immunocytochemistry of markers for the three primary germ layers. (Data and photos courtesy Jeanne Loring and *Nature Methods* (Ben-Nun et al., 2011); reprinted with permission.)

CHAPTER SUMMARY

Early Studies

1. The existence of stem cells has been contemplated for over 100 years, their presence first recognized by Alexander A. Maximow in his theory of hematopoiesis.
2. Ernest McCulloch and James Till demonstrated the existence of bone marrow-derived stem cells.
3. Joseph Altman and Gopal Das were the first to demonstrate neurogenesis in the adult brain, going against the "no new neurons" central dogma.
4. Robert Alan Good performed the first successful bone marrow transplant.

Hematopoietic Stem Cell Discovery

1. Gregor Prindull and his colleagues discovered the presence of HSCs (HSCs) in human umbilical cord blood.
2. HSCs can form myelocytic CFUs in tissue culture.

Mouse Embryonic Stem Cells

1. Gail R. Martin and Sir Martin Evans simultaneously and independently discovered mouse embryonic stem cells.
2. Mouse ES cells have been pivotal in the development of rodent-based gene targeting technologies.
3. Gail Martin used conditioned medium to derive and culture mouse ES cells.

Successful Neurosphere Culture

1. Brent A. Reynolds and Samuel Weiss were the first to isolate neural stem cells from the adult brain.
2. Constance Cepko generated a v-myc transformed neural stem cell line that could propagate in tissue culture and colonize the adult brain.
3. LacZ staining allows for tracing of individual cell fate.

The Discovery of Cancer Stem Cells

1. Cancer cells proliferate almost indefinitely.
2. Multiple mutations result in stem cell-like properties in CSCs.
3. John Dick identified a subpopulation of cells in a human AML sample that originated from a primitive HSC.

Human Embryonic Stem Cell Discovery

1. James A. Thomson discovered hESCs.
2. The H9 hES cell line is perhaps the most popular and widely studied of the original hES cell isolates given its stable karyotype over extended passages and long periods of cell culture.

3. Telomere lengths are an indication of a cell's lifespan.

4. hES cells may secrete factors that keep neighboring hES cells undifferentiated.

5. Mechanical dissociation of hES cell clumps is preferable to enzymatic digestion for reducing stress and differentiation of the cells.

6. Serious ethical and moral issues have surrounded the derivation and study of hESCs.

Stem Cells and Cloning

1. Sir Ian Wilmut successfully cloned the first mammal, Dolly the sheep.

2. Cloning utilizing embryonic stem cells could allow for the generation of a virtually limitless supply of stem cells that are genetically identical, thus providing an extremely valuable therapeutic or drug discovery platform, as the population of cells would be genetically and phenotypically of an identical origin.

3. Cloned adult somatic cells still have a composition of genetic material that has undergone the rigors of environmental influence and the aging process that so significantly damages that of adult cells.

4. Jose B. Cibelli and Advanced Cell Technology, Inc. created the world's first human cloned eggs via parthenogenesis and SCNT.

Cord Blood Embryonic-Like Stem Cells—An Alternative to ES and Adult Stem Cells

1. Colin P. McGuckin isolated a population of stem cells, termed CBEs, from human umbilical cord blood that bore striking resemblance to hES cells and possessed the capacity to differentiate into a variety of lineages.

2. CBEs may act as a possible autologous stem cell-based therapeutic platform for a number of anomalies.

Breakthrough in Spinal Cord Injury Repair

1. The adult CNS has a very limited capacity to regenerate itself, and, as a consequence, injuries to the spinal cord often result in partial or complete irreversible paralysis.

2. Hans S. Keirstead demonstrated an enhancement in myelination and improvement of locomotor recovery in injured rats upon intraspinal cord transplantation of hESC-derived oligodendrocyte progenitor cells (OPCs).

The Generation of iPS Cells

1. Adult differentiated cells may be driven to lose their morphological and molecular identities and transform into other cell types.

2. In an artificial setting, differentiated cells can be reprogrammed to an embryonic-like state.

3. Shinya Yamanaka and James Thomson were the first to induce adult mouse fibroblast cells to dedifferentiate into stem cells and become pluripotent.

4. The factors Oct 3/4, Sox2, c-Myc, and Klf4 were demonstrated to be necessary and sufficient for fibroblast dedifferentiation and reprogramming.
5. iPS cells can contribute to the three primary germ layers of the host embryos.

The Discovery of Human Amniotic Stem Cells

1. Anthony Atala successfully isolated both human and rodent amniotic fluid-derived stem (AFS) cells and demonstrated striking properties inherent in these cells that could make them a valuable source for stem cell therapeutics initiatives.
2. To efficiently isolate the cells from other cell types in the amniotic fluid the researchers employed immunoselection for CD117.
3. Human amniotic stem cells can maintain a normal karyotype over many population doublings and differentiate into a variety of lineages.

Generation of Human iPS Cells

1. Human iPS cells could provide a valuable source for autologous cell therapy to treat a variety of diseases and disorders.
2. Shinya Yamanaka and James Thomson were the first to generate human iPS cells, but utilized genes encoding different transcription factors to accomplish this.
3. The transcription factor Nanog was shown to improve iPS clonal recovery, resulting in an over 200-fold increase in reprogramming efficiency.

Human Embryonic Stem Cells Generated Without Embryo Destruction

1. The destruction of human embryos for the derivation of embryonic stem cells is perhaps the most controversial aspect of hES cell research.
2. The company Advanced Cell Technology generated embryonic stem cells without embryonic destruction using a modified version of preimplantation genetic diagnosis (PGD) to remove single blastomeres.

Human Cloning

1. Andrew J. French at Stemagen Corporation reported the first example of successful cloning of human blastocysts through SCNT of nuclei from adult fibroblasts into enucleated oocytes.
2. The cloning of humans remains a highly controversial topic and raises many ethical, religious and moral issues that are beyond the scope of this text.

Mesenchymal Stem Cell-Derived Human Knee Cartilage

1. Regenerative medicine can be classified into three unique categories including platelet augmentation, recombinant growth factor amplification and stem cell isolates.
2. Mesenchymal stem cells (MSCs) have been demonstrated to differentiate into bone marrow as well as synovial and adipose tissues.

3. Researchers at Regenerative Sciences, Inc. and Centeno-Schultz Clinic conducted the first successful study of cartilage regeneration in a human knee via autologous stem cell therapy.

iPS Cells Derived from Keratinocytes

1. Painless collection of samples and a virtually endless supply for almost every individual make hair follicles an ideal adult somatic cell source for iPS cell generation.
2. Juan Carlos Izpisúa Belmonte successfully reprogrammed keratinocytes isolated from both human foreskin and hair follicles into iPS cells.

iPS Induction Without the Use of Viruses

1. The utilization of retroviral and lentiviral methods to introduce the key genes relies on stable incorporation into the host cell's genome.
2. Integration into the wrong loci could promote tumorigenesis.
3. Researchers at the Samuel Lunenfeld Research Institute implemented the *piggyback* system to introduce genetic material encoding Oct4, Sox2, Klf4 and c-Myc into mouse fibroblasts that could later be cleanly and precisely excised.

Protein-Based iPS Reprogramming

1. A system is needed that allows for the safe production of patient-specific stem cells without genetic alteration to produce such cells.
2. Kwang Soo Kim and colleagues at the Harvard Stem Cell Institute devised a strategy to efficiently introduce the four key reprogramming proteins Oct4, Sox2, Klf4, and c-Myc directly into cells using cell penetrating peptides (CPP).
3. Protein-induced human iPS (p-hiPS) cells were confirmed to be pluripotent and had no genetic alterations due to protein introduction.

The First Clinical Trial Using Human Embryonic Stem Cells

1. Geron Corporation developed a stem cell therapy called GRNOPC1 to aid in spinal cord injury.
2. Geron used feeder- and serum-free fully defined growth conditions to culture and propagate a variety of stem cell-based therapeutics.
3. Despite solid initial clinical trial findings, Geron halted the trials in November 2011.

Mitochondrial DNA: A Barrier to Autologous Cell Therapeutics

1. Mitochondria are derived from donor ES cells or oocytes in SCNT, carrying with them unique genetic material that has acquired mutations over time.
2. Age-related mitochondrial mutations may contribute to immune rejection by the host recipient of any derived cell therapeutic.
3. Researchers at the University of Tsukuba observed that transplants with mtDNA from the same murine strain as the host strain were rejected—most likely due to inherent mutations—and this rejection was due to innate immunity.

Induced Pluripotency and the Potential to Save Endangered Species

1. The ability to produce cells that may be capable of producing multiple differentiated lineages or perhaps even entire organisms has enormous implications for species preservation.

2. Jeanne Loring and her team at the Scripps Research Institute generated drill and northern white rhinoceros iPS cells through retroviral transduction of the human genes encoding Oct4, Klf4, Sox2, and c-Myc into corresponding fibroblasts from each species.

3. iPS technology may provide an avenue for the rescue of already extinct species should viable cell samples for these species exist.

KEY TERMS

(Key terms are in the order as they appear in the text.)

- **Stem cells**—biological cells capable of self-renewal and that have the capacity to differentiate into a variety of cell types; present within most, if not all, multi-cell organisms.
- **Theory of hematopoiesis**—all blood cellular components are derived from a common precursor stem cell.
- **Spleen colonies**—visible lumps present within irradiated immune-deficient mice due to the injection of bone marrow cells.
- **Neurogenesis**—the generation of neurons and glial cells.
- **Microneurons**—small-caliber mature interneurons.
- **"No new neurons" central dogma**—Ramon and Cajal's hypothesis that mammals are born with a preset number of neurons and that there is no growth or generation of new neurons after birth.
- **Severe combined immunodeficiency syndrome (SCID)**—a genetic disorder in which both B and T cells of the immune system are severely compromised due to a defect in one of several possible genes.
- **Colony forming units (CFUs)**—cells that have the ability to divide and form a clonal colony in tissue culture.
- **Myelocytes**—cells of granulocytic origin present in bone marrow that proliferate during fetal development and accumulate in the cord blood of newborn infants.
- **Conditioned medium**—cell culture media prepared in the presence of live cells that secrete key growth factors needed for cell survival.
- **Subventricular zone**—a paired brain structure situated throughout the lateral walls of the lateral ventricles.
- **v-myc—the viral homolog of c-myc which is capable of cellular transformation.**
- **Process (in neural cells)**—either axon or dendrite-like protrusions.
- **β-galactosidase**—an enzyme that catalyzes the hydrolysis of β-galactosides into monosaccharides.
- **Cancer**—abnormal growth of cells that tend to proliferate in an uncontrolled way and, in some cases, to metastasize.
- **Metastasis**—the spread of cancer cells to other parts of the body.
- **Cleavage-stage**—2-, 4-, 8-, and 16-cell embryonic stages containing blastomeres.
- **Karyotype**—the number and appearance of chromosomes in the nucleus of a cell.

- **H9 cell line**—the most popular and widely studied of the original hES cell isolates given its stable karyotype over extended passages and long periods of cell culture.
- **Telomerase**—an enzymatic ribonucleoprotein which functions to add telomeric repeats to chromosomal ends and plays a critical role in extending the lifespan of a cell.
- **Cloning**—the process of creating genetically identical individuals from a single donor.
- **Somatic cells**—any biological cell forming the body of an organism other than a germ cell, gamete, gametocyte, or undifferentiated stem cell.
- **Somatic cell nuclear transfer (SCNT)**—a technique for creating a clonal embryo by combining an ovum (egg) with a donor nucleus.
- **Reproductive cloning**—propagation in tissue culture of an SCNT-derived cell or its allowance to develop into an embryo and be transplanted into a surrogate mother.
- **Therapeutic cloning**—reproduction of a genetically identical cell for purposes of cell therapy such as cellular replacement.
- **Dolly the sheep**—the first mammal and sheep to be cloned by combining the nucleus of a mammary cell from a Fin-Dorset sheep with an enucleated egg from a Scottish Black-face ewe.
- **Parthenogenesis**—a form of asexual reproduction in females where no fertilization from a male is required in order to reproduce.
- **Pronucleus**—the nucleus of an egg cell during fertilization.
- **Blastocoele**—a cleavage cavity or segmentation cavity present in a developing embryo.
- **Cumulus cells**—specialized granulosa cells that surround and nourish a developing egg.
- **Demyelination**—loss of the myelin sheath insulting the nerves.
- **Oligodendrocyte progenitor cells (OPCs)**—immature myelin-producing stem cells.
- **Transdifferentiation**—a non-stem cell transforming into a different cell type, or when a differentiated stem cell generates other cell types outside of its normal realm of multipotency.
- **Dedifferentiation**—reprogramming of cells to an embryonic-like state.
- **KiPS Cells** - keratinocyte-derived induced pluripotency (iPS) cells.
- **Alkaline phosphatase**—an enzyme that is considered a classical pluripotency marker for embryonic stem cells derived from a variety of species.
- *piggyback* **(PB)**—a host factor-independent transposon/transposase system that allows for the stable but transient integration of genetic material into the host cell genome and drives the maintenance of its expression long-term.
- **Cell penetrating peptides (CPP)**—short peptides that actively cross the cell membrane and can carry other macromolecules along with them during this process.
- **Protein-induced human iPS (p-hiPS)**—iPS cells generated using repeated protein treatment cycles.
- **Amniotic fluid stem (AFS) cells**—multipotent stem cells derived from amniotic fluid.
- **Immunoselection**—the isolation of an antigen using antibody specificity.
- **C-Kit (CD117)**—a tyrosine kinase cell surface receptor known to be a marker for progenitor and stem cell lineages such as those of the prostate, thymus, and of hematopoietic origin.
- **Preimplantation genetic diagnosis (PGD)**—an embryonic manipulation procedure employed to profile the genetics of an embryo prior to implantation.
- **Blastomeres**—cells resulting from the cleavage of a fertilized ovum during early embryonic development, from fertilized eggs (8 cell stage; 1 or 2 blastomeres isolated per egg).

- **Cloning**—the identical reproduction of another organism at the molecular and cellular level.
- **Pronucleus**—the nucleus of a sperm or egg cell during fertilization.
- **Nuclear transfer stem cells (NTSC)**—stem cells derived from a SCNT-generated embryo.
- **Regenerative medicine**—the ability to regenerate tissue rather than surgically extracting or altering that tissue.
- **Autologous**—transplant from one part of the body into another of the same patient.
- **Percutaneous**—under the skin.
- **Geron Corporation**—a biotechnology company based in Menlo Park, California that has developed proprietary technologies for the growth, maintenance, and scaling of human embryonic stem cells and other cell types for therapeutic purposes.
- **Mitochondria**—spherical or elongated organelles in the cytoplasm of nearly all eukaryotic cells containing genetic material and many enzymes important for cell metabolism.
- **Allogeneic**—taken from different individuals of the same species.
- **Innate immunity**—naturally or inherently present and is not due to sensitization of the immune system by an antigen.
- *Mandrillus leucophaeus*—an endangered primate.
- *Ceratotherium simum cottoni*—the northern white rhinoceros.
- **Glycomic profiling**—the analysis of the complete repertoire of glycans and glycoconjugates that cells produce under specified conditions.

REVIEW QUESTIONS

(Answers to select review questions can be found at www.stemcelltextbook.com.)

1. Who first coined the term "stem cell" and what was his background?
2. What happened in 1963 that changed stem cell research forever?
3. What contribution did Joseph Altman and Gopal Das make to the field of stem cell research?
4. Describe the transplant that occurred in 1968 that set the stage for stem cell therapeutics.
5. How were hematopoietic stem cells discovered?
6. How were mouse embryonic stem cells derived and what researchers independently contributed to this discovery?
7. From what region of the brain were the first neural progenitor and stem cells isolated and who accomplished this?
8. Describe a method for tagging stem cells to monitor their *in vivo* presence?
9. What led to the hypothesis that cancer stem cells exist?
10. Who discovered human embryonic stem cells and what changes in cell culture methodology over that of mES cells were necessary to accomplish this?
11. Describe the process of somatic cell nuclear transfer.
12. What is the difference between therapeutic and reproductive cloning?
13. How was Dolly the sheep cloned?

14. Describe the two methods used by Jose Cibelli's group to produce autologous embryos.

15. How did Colin McGuckin's group isolate cord blood stem cells?

16. What hepatic-specific antigens were expressed in differentiated cord blood stem cells?

17. What two factors aggravate spinal cord injury repair?

18. How did Hans Keirstead and colleagues partially restore locomotion in rat models of spinal cord injury?

19. What is the difference between transdifferentiation and dedifferentiation?

20. How did Shinya Yamanaka and colleagues induce pluripotency in adult mouse fibroblasts?

21. What method did Anthony Atala's group employ to isolate human amniotic stem cells?

22. What is the difference in transcription factor identity for human iPS cell creation by Yamanaka's vs. Thomson's groups?

23. Describe preimplantation genetic diagnosis.

24. How did Andrew French's group successfully clone human embryos?

25. How were mesenchymal stem cells prepared for autologous replacement therapy?

26. Why is *piggyback* considered a superior way to introduce genes encoding transcription factors into cells for reprogramming purposes?

27. How did Kwang-Soo Kim and his group efficiently introduce transcription factor proteins into fibroblasts for induction of pluripotency?

28. What was Geron Corporation's strategy for treating spinal cord-injured patients?

29. Why does the mitochondria drive innate immunity rejection?

30. From what two endangered species did Jeanne Loring's group produce iPS cells?

THOUGHT QUESTION

All chapters throughout this book include at least one thought-provoking question designed to test your knowledge of the material and also to give you the opportunity to think critically about the chapter's content and how it might be applied in a real-world setting. Note that there are not necessarily any definitive or correct answers to these questions. They are merely meant to prod your intellect.

How would you go about deriving human embryonic stem cells without embryo destruction and how might you optimize both technique and equipment to increase clonal yield?

SUGGESTED READINGS

Books, Compilations, and Lectures

Altman, J. (2012). *MENTAL EVOLUTION: Origins of the Human Body, Brain, Behavior, Consciousness, and Culture*. In Press, 2012. Independent Publisher.

Evans, MJ. (2007). Embryonic Stem Cells: The Mouse Source–Vehicle for Mammalian Genetics and Beyond. Nobel Lecture, December 7, 2007.

Morgan, S. (2008). *From Microscopes to Stem Cell Research: Discovering Regenerative Medicine*. January 31, 2008. Heinemann-Raintree, Chicago, IL.

Cited Research Articles

Aasen, T., A. Raya, et al. (2008). "Efficient and rapid generation of induced pluripotent stem cells from human keratinocytes." *Nat Biotechnol* **26**(11): 1276–1284.

Altman, J. and G. D. Das (1967). "Postnatal neurogenesis in the guinea-pig." *Nature* **214**(5093): 1098–1101.

Becker, A. J., C. E. Mc, et al. (1963). "Cytological demonstration of the clonal nature of spleen colonies derived from transplanted mouse marrow cells." *Nature* **197**: 452–454.

Ben-Nun, I. F., S. C. Montague, et al. (2011). "Induced pluripotent stem cells from highly endangered species." *Nat Methods* **8**(10): 829–831.

Bonnet, D. and J. E. Dick (1997). "Human acute myeloid leukemia is organized as a hierarchy that originates from a primitive hematopoietic cell." *Nat Med* **3**(7): 730–737.

Centeno, C. J., D. Busse, et al. (2008). "Increased knee cartilage volume in degenerative joint disease using percutaneously implanted, autologous mesenchymal stem cells." *Pain Physician* **11**(3): 343–353.

Chung, Y., I. Klimanskaya, et al. (2008). "Human embryonic stem cell lines generated without embryo destruction." *Cell Stem Cell* **2**(2): 113–117.

Cibelli, J. B., R. P. Lanza, et al. (2002). "The first human cloned embryo." *Sci Am* **286**(1): 44–51.

De Coppi, P., G. Bartsch, Jr., et al. (2007). "Isolation of amniotic stem cell lines with potential for therapy." *Nat Biotechnol* **25**(1): 100–106.

French, A. J., C. A. Adams, et al. (2008). "Development of human cloned blastocysts following somatic cell nuclear transfer with adult fibroblasts." *Stem Cells* **26**(2): 485–493.

Garrison, F. H. (1929). "Ramon y Cajal." *Bull N Y Acad Med* **5**(6): 482–508.

Ishikawa, K., N. Toyama-Sorimachi, et al. (2010). "The innate immune system in host mice targets cells with allogenic mitochondrial DNA." *J Exp Med* **207**(11): 2297–2305.

Keirstead, H. S., G. Nistor, et al. (2005). "Human embryonic stem cell-derived oligodendrocyte progenitor cell transplants remyelinate and restore locomotion after spinal cord injury." *J Neurosci* **25**(19): 4694–4705.

Kim, D., C. H. Kim, et al. (2009). "Generation of human induced pluripotent stem cells by direct delivery of reprogramming proteins." *Cell Stem Cell* **4**(6): 472–476.

Martin, G. R. (1981). "Isolation of a pluripotent cell line from early mouse embryos cultured in medium conditioned by teratocarcinoma stem cells." *Proc Natl Acad Sci U S A* **78**(12): 7634–7638.

McGuckin, C. P., N. Forraz, et al. (2005). "Production of stem cells with embryonic characteristics from human umbilical cord blood." *Cell Prolif* **38**(4): 245–255.

Prindull, G., B. Prindull, et al. (1978). "Haematopoietic stem cells (CFUc) in human cord blood." *Acta Paediatr Scand* **67**(4): 413–416.

Shamblott, M. J., J. Axelman, et al. (1998). "Derivation of pluripotent stem cells from cultured human primordial germ cells." *Proc Natl Acad Sci U S A* **95**(23): 13726–13731.

Snyder, E. Y., D. L. Deitcher, et al. (1992). "Multipotent neural cell lines can engraft and participate in development of mouse cerebellum." *Cell* **68**(1): 33–51.

Takahashi, K., K. Tanabe, et al. (2007). "Induction of pluripotent stem cells from adult human fibroblasts by defined factors." *Cell* **131**(5): 861–872.

Takahashi, K. and S. Yamanaka (2006). "Induction of pluripotent stem cells from mouse embryonic and adult fibroblast cultures by defined factors." *Cell* **126**(4): 663–676.

Thomson, J. A., J. Itskovitz-Eldor, et al. (1998). "Embryonic stem cell lines derived from human blastocysts." *Science* **282**(5391): 1145–1147.

Woltjen, K., I. P. Michael, et al. (2009). "*piggyBac* transposition reprograms fibroblasts to induced pluripotent stem cells." *Nature* **458**(7239): 766–770.

Yu, J., M. A. Vodyanik, et al. (2007). "Induced pluripotent stem cell lines derived from human somatic cells." *Science* **318**(5858): 1917–1920.

Chapter 2

FUNDAMENTALS OF STEM CELLS

Chapter 2 explores the basic biology of stem cells—from genetic and biochemical makeup to the different types of cells and their capabilities for generating therapeutically valuable differentiated phenotypes. In addition, the basic concepts behind the growth and manipulation of cells in tissue culture are explored to give a firm understanding of stem cell research. Finally, a general overview of the potential for stem cells to impact diagnostics, drug screening, and therapeutics is outlined, with more on each of these topics to be discussed in detail in later chapters.

BASIC *IN VITRO* CELL CULTURE—A HISTORICAL PERSPECTIVE

Cell culture may be defined as the *in vitro* growth and manipulation of cells under controlled conditions outside of their natural *in vivo* environment. The term cell culture is often utilized interchangeably with **tissue culture**, a broader term used to describe the growth and manipulation of either tissues or cells under controlled conditions outside of their normal environment. While the vast majority of cell and tissue culture experimental protocols involve eukaryotic animal cells, techniques and methods have also been optimized and are routinely implemented for the propagation and study of plant cells, fungi, and bacteria.

The history of cell culture can be traced back as early as 1885, when the German zoologist Wilhelm Roux (Figure 2.1) successfully maintained a section of avian neural plate, known as the medullary plate, in saline for a period of several days (Carrel and Lindbergh, 1938). This single experiment is considered as the birth of tissue culture experimentation and set the stage for a revolution in the study of cells and tissues in the laboratory.

It was not until over 20 years later, in 1907, that American biologist Ross Granville Harrison cultured amphibian neuroblasts in a specially prepared lymph medium (Abercrombie, 1961). This is widely believed to be the first real experimentation on stem cells or stem cell precursors, and would set the stage for a revolution in stem cell research. These studies would later be expanded upon by Harrison and Austrian biologist Paul Alfred Weiss, who optimized cell culture conditions and demonstrated that cellular growth patterns are directly influenced by their microenvironment and, specifically, the growth substrate utilized for

Stem Cells: A Short Course, First Edition. Rob Burgess.
© 2016 John Wiley & Sons, Inc. Published 2016 by John Wiley & Sons, Inc.

Figure 2.1
Photograph of
Wilhelm Roux
(1850–1924).
(Courtesy
Wikimedia
Commons;
reprinted with
permission.)

their culture. In 1912, the French surgeon and biologist Alexis Carrel initiated a 20-year long experiment culturing avian embryonic cardiac tissue in a controlled, aseptic environment. The tissue remained viable for over 20 years, much longer than the lifespan of a chicken, and confirmed the importance of a controlled, sterile technique (Sanford et al., 1948). Much later, in 1938, Carrel would collaborate on a book with the famous aviator Charles Lindberg titled "The Culture of Organs," which described a jointly developed Pyrex-based glass chamber that could perfuse organs and keep them alive *in vitro* for long periods. It was the heat-resistant aspect of the glass which enabled thorough sterilization of the perfusion apparatus, eliminating troublesome contamination issues that had dogged Carrel over the years (Figure 2.2) (Enders et al., 1949). It should be noted that, in 1912, Carrel won the Nobel Prize in Physiology or Medicine for his work on vascular suturing and blood vessel transplantation.

It was perhaps 1948 when the most significant advancement regarding clonal cell culture was made. It was in this year that W.R. Earle and his colleagues at the National Cancer Institute in Bethesda, Maryland derived the clonal L-929 cell line from his original L strain isolated 8 years prior. The **L strain** was derived from subcutaneous areolar and adipose tissue of a 100-day-old male C3H/An mouse and was one of the most widely studied cell isolates at the time, with clonal line 929 derived from the 95th subculture of the parental strain (Skloot, 2010). These studies solidified the concept that clonal isolates, which propagate indefinitely, could be successfully isolated from a heterogeneous population of cells. Also in 1948, John Enders, an American medical scientist at Children's Hospital in Boston, along with colleagues American virologists Thomas Huckle Weller, and Frederick Chapman Robbins successfully cultured polio vaccine in human tissue, a feat for which they were awarded the 1954 Nobel Prize in Physiology or Medicine (Darnell et al., 1970). In 1952, George Otto Gey and his colleagues at Johns Hopkins University in Baltimore, Maryland isolated a line of human epithelial cells from cervical carcinoma that exhibited robust growth properties. The cells were taken from Henrietta Lacks, a cancer patient, without his/her consent and referred to as **Hela cells**. This was the first human cell line to be propagated successfully for many passages *in vitro,* and the line has since been widely

Figure 2.2
Charles
Lindbergh and
Alexis Carrel
with their
famous
PYREX® glass
pulsating
perfusion pump
on the cover of
Time in 1938.
(Courtesy
Corning and
Time magazine;
reprinted with
permission.)

used worldwide in the study of cancer and as a research tool for basic biological research. Jonas Salk utilized Hela cells as a model system for his research on the polio vaccine in the 1950s, and, by some estimates, there are as many as 60,000 publications implementing the use or study of Hela cells (Figure 2.3) (Littlefield, 1964). In Gey's studies, he realized that planar culture did not adequately mimic the natural environment of a living organism in which exposure to nutrients and waste removal fluctuates. Thus he created a more *in vivo*-like microenvironment, plating cells on the sides of test tubes and rotating the tubes to alternate exposure to medium and air.

Despite early successes in the culture of various cell types, it remained a mystery which growth factors and other agents were keys to cellular survival *in vitro*.

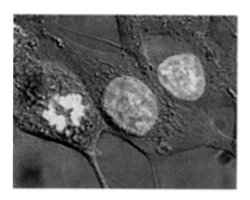

Figure 2.3 Hela cells stained with Hoechst 33258. (Courtesy Wikimedia Commons; reprinted with permission.)

Finally, in 1955, the American physician and pathologist Harry Eagle meticulously dissected the minimum essential components of media that allow for cellular propagation and inhibition of senescence. The formulation was dubbed **Eagle's Minimal Essential Medium (EMEM)** and contained various amino acids, glucose, salts, and vitamins such as folic acid and B12.

Cell lines that could be grown in tissue culture, especially those of human origin, were soon realized to be valuable tools for studying genetics and genomics involved in human physiology and disease. In 1964, J.W. Littlefield developed a medium that allowed for the clonal selection of hybrid cell lines. Termed **hypoxanthine-aminopterin-thymidine (HAT) medium**, it is based on *de novo* DNA synthesis requirements of mitotic cells. In Littlefield's selection strategy, the drug aminopterin acts as a blocking agent, preventing DNA synthesis that can only be overcome by the presence of hypoxanthine when hypoxanthine-guanine phosphoribosyltransferase (HGPRT) is present in the cells. This is referred to as **positive selection**, which is defined as selection for the presence of a survival or resistance trait. The second aspect of HAT selection involves the presence of thymidine in the medium, another necessary precursor required for *de novo* DNA synthesis. The enzyme **thymidine kinase (TK)** phosphorylates thymidine thus generating the thymidine monophosphate precursor also crucial for DNA synthesis. HAT selection is a dual positive selection strategy allowing for the survival of cells containing both HGPRT (HPRT) and TK, allowing for efficient identification of clonal hybrid cell lines whereby unfused cells die due to a lack of the essential enzymes required for *de novo* DNA synthesis (Figure 2.4) (Ham, 1965).

Up to this point, cell culture applications had been performed in the presence of pure or dialyzed animal serum, purified serum proteins, or unknown factors dialyzed into the medium from serum. This was necessary to provide key nutritional components needed for cell survival, expansion, and propagation over multiple **passages**, which is the replating of cells upon reaching confluence. In 1965, a major leap forward in the area of serum-free cell culture was made when Richard G. Ham in the department of Biophysics at the University of Colorado Medical Center in Denver developed a chemically defined, fully synthetic medium for the culture of mammalian cells. Ham studied Chinese Hamster Ovary (CHO) cells, performing serial cultures of these cells in the presence or absence of a combination of over 40 defined factors. Various characteristics of cell growth and viability—including colony formation, colony morphology, long-term passaging, and plating efficiencies—were analyzed. The final version of the fully synthetic, defined medium was termed "**F12 medium**" and is still widely used today (Figure 2.5) (Wigler et al., 1978).

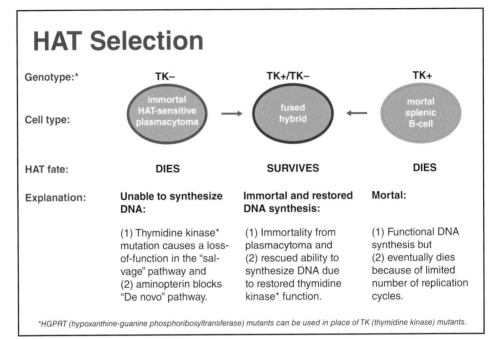

Figure 2.4
HAT selection
for hybrid cell
lines. (Courtesy
Wikipedia.org;
reprinted with
permission part
of Wikimedia
Commons.)

The ability to use cell lines as manufacturing or production vehicles has long been a staple of cell culture applications not only for purposes of producing research-grade reagents but also therapeutics. This phenomenon began in 1975 when the Argentine biochemist Cesar Milstein and German biologist Georges J. F. Köhler developed a technique for the application of cell lines to produce monoclonal antibodies. Referred to as **hybridoma technology**, which is defined as the fusing of antibody-producing cells with myeloma cancer cells for the generation of antibody-producing hybridomas, the methodology paved the way for the high-throughput production of monoclonal antibodies (Figure 2.6) (Evans and Kaufman, 1981; Martin, 2011). Interestingly, the hybridoma technology was published in the journal *Nature* before the filing of a patent, thus precluding any intellectual property ownership by the inventors or the British government. Milstein and Kohler were awarded the 1984 Nobel Prize in Physiology or Medicine for this work.

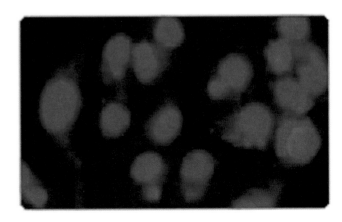

Figure 2.5 CHO cells grown in Life Technologies, Inc. F12 medium. (Courtesy Life Technologies; reprinted with permission.)

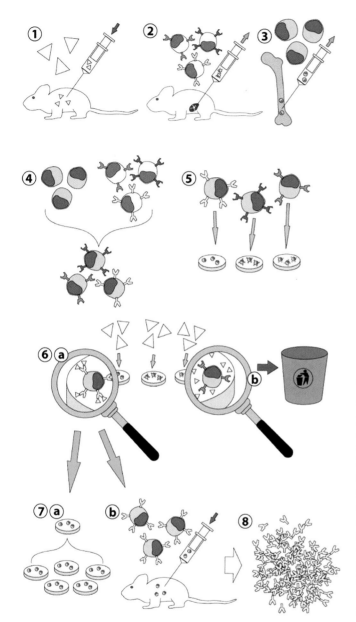

Figure 2.6 Diagrammatic illustration of hybridoma technology for the production of monoclonal antibodies. (1) Immunization of a mouse. (2) Isolation of B cells from the spleen. (3) Cultivation of myeloma cells. (4) Fusion of myeloma and B cells. (5) Separation of cell lines. (6) Screening of suitable cell lines. (7) *In vitro* (a) or *in vivo* (b) multiplication. (8) Harvesting. (Courtesy Wikimedia Commons; reprinted with permission.)

In 1978, 2 years after the introduction of hybridoma technology to the arsenal of cell culture platforms, a method was developed for the stable introduction of single genes into cells by American microbiologist Michael Wigler, American neuroscientist Richard Axel, and American microbiologist Saul Silverstein. In these studies Wigler and colleagues utilized murine L cells (described above) deficient in the enzyme TK, introduced herpes simplex virus (HSV) TK by **transfection**, which is defined as the introduction of exogenous nucleic acid material into cells, and confirmed its presence via both serological and biochemical techniques (see Figure 2.4). It was confirmed that TK had become stably integrated into the genome of the host cell and its presence—as well as corresponding enzymatic activity—was observed to be stable over hundreds of cell generations. Cells were con-

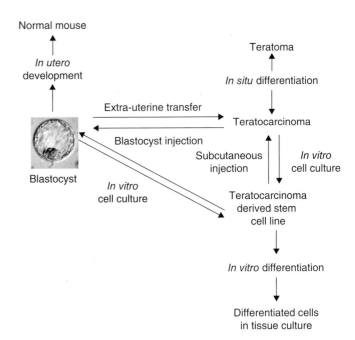

Figure 2.7 Relationship
between normal murine
embryos and
teratocarcinoma cells.

firmed to have only a single copy of the TK gene and genomic DNA from these cells was isolated and transformed into a second cell population, also deficient in TK activity. Single copy integrants were confirmed as being present in this second cell population.

From the first successful *in vitro* culture of living tissue it was quickly realized that different types of tissues and cells most often require different microenvironments to effectively maintain viability, mitotic activity, and even differentiation capacity. Some cell types thrive in a simple planar environment supplemented with minimal media, while others require unique three-dimensional characteristics and key growth factors for survival. As described in Chapter 1, this was evident in the successful derivation and culture of mouse embryonic stem cells, performed independently by Gail Martin and colleagues at UCSF and Martin Evans and colleagues at the University of Cambridge in 1981. The combination of a three-dimensional microenvironment created by the use of a fibroblast feeder layer coupled with media conditioned in the presence of teratocarcinoma stem cells allowed for the isolation and propagation of embryonic stem cells from the inner cell mass (ICM) of murine blastocysts (Figure 2.7) (Martin, 1981). Martin's, Evans', and Kaufman's work set the stage for the utilization of conditioned medium and feeder cells for the derivation and *in vitro* manipulation of a multitude of stem cell types.

STEM CELL CULTURE—OPTIMAL CONDITIONS AND TECHNIQUES

The following section focuses on current procedures and methodologies for optimizing the culture of stem cells. As these technologies are unique for each type of stem cell, the section has been broken down accordingly for the most widely studied types of cells including embryonic stem, hematopoietic, and adipose-derived lineages. The focus is on lines of human origin, but in many cases these techniques can be applied to similar cell types derived from other species. The culture of cell types other than stem cells is beyond the scope of this book and will not be discussed.

Embryonic Stem Cell Culture

As mentioned above and in Chapter 1, embryonic stem cells are derived from the inner cell mass (ICM) of mammalian blastocysts. Cell culture techniques utilizing embryonal carcinoma cells as a model system were optimized for the isolation and culture of embryonic stem cells from the ICM of blastocysts. This has now been successfully accomplished for a variety of species including mouse, primate, and human. Although there are some differences in the microenvironment required for successful derivation, propagation, and maintenance of pluripotency for these cells, many commonalities exist. For example, the first derivation of human embryonic stem cells employed **mouse embryonic fibroblast (MEF) feeders** to maintain the cells in a pluripotent, undifferentiated state (Nur et al., 2006). Yet the underlying ideal cell culture environment for embryonic stem cells of human origin has always been in a xenobiotic-free environment, eliminating the presence of animal components or animal by-products which would prevent any realistic therapeutic use of the cell platform. Thus, feeder cells were established from numerous human sources, including fetal foreskin, muscle, bone marrow (BM), amniotic epithelium, and even fallopian tube epithelium. However, in order to reduce variability in feeder cell culture conditions and quality, progress has been made in both utilizing media conditioned with feeders in the absence of a feeder layer, often employing a fibronectin or **Matrigel** (a gelatinous protein mixture secreted by Engelbreth-Holm-Swarm (EHS) mouse sarcoma cells and marketed by BD Biosciences) solid support. Ironically, hES cells cultured in the absence of feeders often yield a layer of stromal- and fibroblast-like cells that act as a support matrix and secrete factors driving cellular viability and pluripotency. What are the factors secreted by feeder cells that maintain hES cells in an undifferentiated state? Basic FGF, noggin, activin A, and TGF-β1 have been defined as crucial defined factors needed for hES cell culture (Figure 2.8) (Wong et al., 2011).

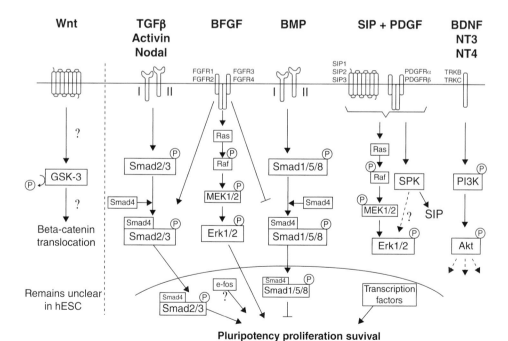

Figure 2.8 Pathways involved in maintaining human embryonic stem cell pluripotency. (Courtesy Wong, Pebay and Nova Publishers (Wong et al., 2011); reprinted with permission.)

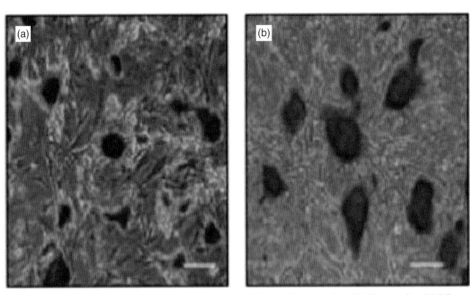

Figure 2.9 Mouse embryonic stem cell culture on a 3D nanofiber matrix. (A) No matrix. (B) Ultra-web nanofiber matrix. Both populations were stained for the pluripotency marker alkaline phosphatase. (Images courtesy Alam Nur-E-Kamal and *Stem Cells* (Nur et al., 2006); reprinted with permission.)

Although the vast majority of cell culture has, for both research and therapeutic applications, been performed utilizing a planar two-dimensional environment, it is clear that recapitulating the three-dimensional properties of a living organism provides a much more *in vivo*-like niche for cultured cells to thrive. Parameters such as cell-to-cell signaling and nutrient availability are considerably affected by the dimensional nature of the cell culture platform. Indeed, it has been shown that murine embryonic stem cells cultured on a nano-fibrillar three-dimensional matrix exhibited increased viability and self-renewal properties in comparison to classical 2D cell culture formats. The increased proliferative activity correlated directly with increased expression of the pluripotency marker and transcription factor Nanog and the pluripotency marker alkaline phosphatase (Figure 2.9) (Nur et al., 2006).

Human embryonic stem cells almost always grow in clumps known as **colonies** (Figure 2.10). Proper expansion of these colonies requires their dissociation and replating. There are two primary modes for dissociating embryonic stem cell colonies without damage to individual cells: (1) Mechanical dissection/dissociation; and (2) Enzymatic digestion. In mechanical dissection, a pipette is utilized to break apart the colony through repeated, albeit gentle, pipetting of the colony, often against the edge of the tissue culture flask or plate. Researchers will often pipette the same ES cell colony 30 times or more to ensure proper dissociation of cells from one another. Interestingly, this mechanical dissociation and physical separation does not overly stress the cells. Enzymatic digestion often involves the use of a **collagenase**, an enzyme that breaks down peptides present in collagen. Enzymes successfully utilized for the dissociation of human embryonic stem cells and those of other species include trypsin, dispase, and collagenase IV, and collagenase I. While the implementation of enzymatic digestion is certainly less labor intensive than mechanical dissociation, enzymes often stress the cells far more than mechanical dissociation and can lead to unwanted genetic abnormalities, resulting in cellular senescence and differentiation.

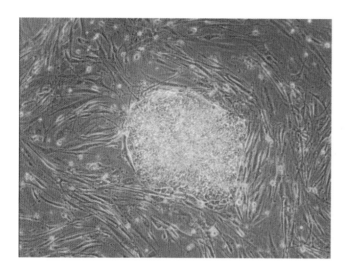

Figure 2.10 A colony of embryonic stem cells, from the H9 cell line (NIH code: WA09) grown on a fibroblast feeder layer viewed at 10 X with Carl Zeiss Axiovert scope. (Courtesy Wikimedia Commons; reprinted with permission.)

While the vast majority of experimentation involving embryonic stem cell culture has occurred in the presence of feeders or in a feeder-free environment supplemented with key growth factors, either technique can be considered two-dimensional in nature. It is the 2D environment that is thought by many researchers to limit the "embryonic"-like state of the cells and prevent them from realizing their true pluripotency and differentiation capacities. Thus it has long been a goal of many embryonic stem cell researchers to more effectively recapitulate the 3D nature of the *in vivo* environment of a developing embryo. Sally Meiners' group in the Department of Pharmacology, Robert Wood Johnson Medical School-University of Medicine and Dentistry, Piscataway, New Jersey utilized a synthetic polyamide matrix referred to as **Ultra-Web**[R], recapitulating the basement membrane / extracellular matrix 3D geography found in the developing embryo. Mouse ES cells grown in this environment exhibited enhanced proliferation and self-renewal properties in comparison to those cultured under 2D conditions. These properties were directly attributed to the activation of the GTPase Rac, the PI3 kinase pathway, and corresponding upregulation of Nanog expression (Figure 2.11). It was speculated that the three-dimensionality of the nanofibrillar surfaces provided by Ultra-Web yielded a more *in vivo*-like environment promoting pathway activation that drove the expression of Nanog (Nur et al., 2006).

Specific types of **scaffolds**, which in this context are defined as three-dimensional matrices for cell culture, that have shown promise for embryonic stem cell propagation include:

- Polyamide matrix (Ultra-Web)
- Cellulose acetate
- Poly-glycerol-sebacate-acrylate (PGSA) elastomers
- Hyaluronic acid (HA) hydrogel

Each of these platforms has unique molecular features which affect the growth and behavior of cells differently. For example, Ultra-Web might be ideal for use in the propagation of embryonic stem cells while PGSA elastomers may be better suited for neural stem cells. Finally, a combination of 3D scaffolding along with growth factors may prove to be the most efficient yet at promoting pluripotency properties and inhibiting differentiation. Conversely, the addition of differentiation inducing factors to a 3D cell culture platform may allow for more efficiently directed cell differentiation into desired mature lineages.

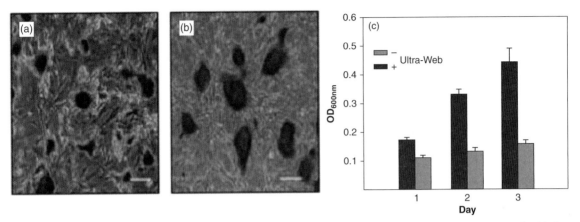

Figure 2.11 UltraWeb's effects on mouse ES cell morphology and proliferation. Top panels: (a) the absence of UltraWeb and (b) the presence of UltraWeb. Lower panel (c): the presence of UltraWeb increases cell proliferation rates. (Courtesy Sally Meiners and *Stem Cells* (Nur et al., 2006); reprinted with permission.)

Hematopoietic Stem Cell Culture

Hematopoietic stem cells are often described as HSCs and hematopoietic stem and progenitor cells (HSPCs). For the purposes of this text they will be referred to as HSCs. Early studies on HSC or HSPC expansion and *ex vivo* culture implemented the use of specific, water-soluble cytokines. These growth factors, while actually promoting survival of specific, terminally differentiated lineages, were thought to also support the proliferation and viability of HSCs. Behind this convoluted speculation was the idea that cell lineage determination involved stochastic events, with random access to various factors driving multipotency and cell division as well as commitment to particular phenotypes. Yet it is now widely accepted that specific, defined factors and pathways such as Notch signaling represent a more deterministic process for HSC survival and expansion in a cell culture setting. It should be noted that BM and peripheral blood stem cells (PBSCs) have been routinely cultured successfully. Yet as the vast majority of recent success in HSC culture has utilized cord blood (CB) cells as a model system, this section will focus on the processes and methodologies used to culture this cell type.

Notch Regulation of HSC Proliferation As the 1990s saw a revolution in stem cell culture techniques primarily directed at human embryonic stem cell manipulation in an *ex vivo* setting, it became clear that both differentiation capacity and cell viability depended primarily on intercellular interactions. The physical and chemical connections between adjacent and neighboring cells were denoted to provide crucial signaling cues, both for defining potency as well as cell division capabilities. In HSCs it was discovered that the Notch signaling cascade plays a significant role in their propagation. Four **Notch** homologs (1–4) and five corresponding ligands (Jagged 1 and 2 and Delta 1–3) have been discovered in mammals. It is the interaction between Notch's extracellular domain with one of the five ligands on adjacent cells which activates receptor proteolysis, driving an intracellular domain into the nucleus for direct interaction with the transcription factor CSL (CBF1/RBPJκ) which switches from a transcrip-

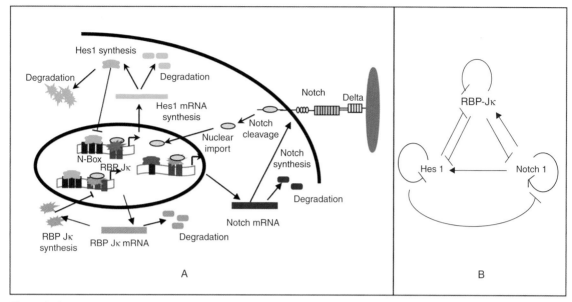

Figure 2.12 Diagrammatic illustration of the Notch signaling pathway. A. Notch signaling at the cellular level. B. Notch signaling at the transcriptional level. (Courtesy Wikimedia Commons; reprinted with permission.)

tional repressor to a transcriptional activator. CSL subsequently activates the HES1 and HES5 repressors, thereby inhibiting expression of genes essential for driving B-cell and myeloid cell terminal differentiation. For a general overview of the Notch signaling cascade see Figure 2.12.

Pioneering researchers in the Clinical Division of Pediatric Oncology at the Fred Hutchison Cancer Research Center in Seattle, Washington have manipulated the Notch signaling pathway to culture HSCs *ex vivo*. Irwin Bernstein and colleagues chose to activate endogenous Notch signaling in HSCs by introducing engineered ligands Jagged1 and Delta1. The presence of these ligands activated endogenous receptors and drove murine HSC stem cell progenitor expansion. The researchers extended these studies to human HSPCs, combining the delivery of ligand with fibronectin fragments and various cytokines including Flt3 ligand, IL-3, IL-6, SCF, and TPO. This resulted in an over 200-fold increase in the number of CD34+ HSPCs (Dromard et al., 2011).

Other Drivers of HSC Proliferation In addition to direct Notch pathway manipulation via the provision of ligands in culture, other approaches have been successful at driving HSPC or HSC expansion. Co-culture of HSCs with stromal cells and growth factors/cytokines has been demonstrated to result in a 40-fold expansion of CD34+ cells. Prostaglandin E2 (PGE2) has also been shown to drive HSC formation—specifically in zebrafish—presumably through the Wnt pathway, thereby activating β-catenin expression (Figure 2.13). Finally, it has long been postulated that copper plays a role in the regulation of proliferation and differentiation of HSPCs, with higher levels driving cellular senescence and differentiation into mature lineages. Addition of the copper chelator tetraethylenepentamine (TEPA) combined with cytokines TPO, IL-6, SCF, and Flt3 ligand resulted in a 160-fold increase of CD34+ cells after 7 weeks of culture.

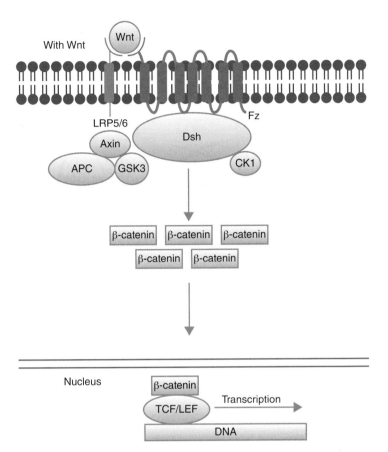

Figure 2.13
Diagram of the Wnt
pathway. (Diagram
courtesy Wikimedia
Commons; reprinted
with permission.)

Adipose-Derived Stem Cell Culture

Adipose-derived stem cells (ASCs) are thought by many researchers to be the next type of stem cell with potential to significantly impact cell-based therapeutics and regenerative medicine. ASCs are often referred to as multipotent adipose derived stem (MADS) cells, or, if of human origin, hMADS. These cells differentiate into a variety of lineages and may be isolated from fat tissue without serious risk to the patient (Figure 2.14) (Pisani et al., 2011).

In fact, ASCs exhibit many of the characteristics and properties inherent in mesenchymal stem cells (MSCs) isolated from the BM, yet they are easier to harvest. Adipose tissue has long been known to provide an abundant supply of both MSCs and ASCs, and large numbers of cells can be harvested during routine procedures with low chances of morbidity. Successful isolation and purification of MADS cells is largely due in part to the well-established marker expression patterns inherent in these cells allowing for efficient flow cytometric characterization and purification. Table 2.1 outlines the standard MADS cell marker phenotype.

A number of clinical trials have been conducted utilizing hMADS in the surgical field with promising results in plastic, oral, and cardiac surgery. This is the result of not only successful harvesting and isolation but optimization of *in vitro* cell culture conditions. **Lipoaspirates**, most often a disposable byproduct of routine and specialized

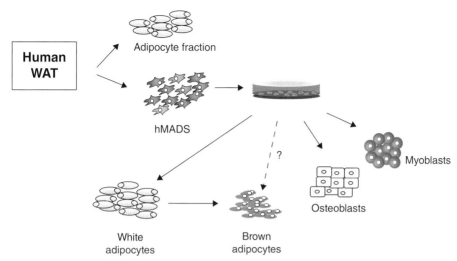

Figure 2.14 Diagrammatic illustration of the isolation of human multipotent adipose derived stem cells (hMADS) from white adipose tissue (WAT), culture and differentiation into terminal lineages. (Courtesy D. Pisani and *Front. Endocrin.* (Pisani et al., 2011); reprinted with permission.)

cosmetic surgery, are the most widely used source of hMADS for reasons mentioned earlier. Typical initial isolation of hMADS from lipoaspirates involves selecting cells positive for CD44, CD105, and CD271 (LNGFR) and negative for CD45. Given the extensive heterogeneous population of cells harvested from lipoaspirates, some laboratories have used magnetic beads conjugated with specific antibodies for precise isolation of hMADS. Miltenyi Biotec, for example, provides a CD271 MicroBead kit that allows for the efficient isolation of CD271$^+$ cells via flow cytometry. Flow also allows for an exclusion of CD45$^-$ cells, thus resulting in a pure population of primary hMADS that can subsequently be expanded and directed to differentiate into numerous mature lineages.

V. Planat-Benard's research team at the University of Toulouse in France developed a methodology for culturing hMADS in serum-free media as floating spheres. Specifically, a 3D cell culture system was developed that allowed for the direct plating of cells isolated from tissues in a non-adherent system without preliminary monolayer propagation or the use of serum. Spheroid formation efficiency averaged 0.8%, indicating that only the rarest of cells survived the cell culture conditions (Figure 2.15). These cells expressed the classical markers of hMADS, could be expanded over several passages, and retained their differentiation capacities (Dromard et al., 2011).

TABLE 2.1 Marker phenotype of MADS cells.

Positive markers	Negative markers
CD10, CD13, CD29, CD34, CD44, CD54, CD71, CD90, CD105, CD106, CD117, CD271, and STRO-1	CD45, CD14, CD16, CD45, CD56, CD61, CD62E, CD104, CD106, CD31, CD144, and von Willebrand factor

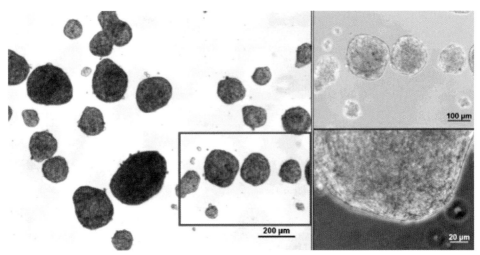

Figure 2.15 Morphology of hMADS cell spheres 7 days after seeding. (Courtesy V. Planat-Benard and *Experimental Cell Research* (Dromard et al., 2011); reprinted with permission.)

THE STUDY OF EMBRYONIC DEVELOPMENT

Embryonic Development and the Origin of Stem Cells

The past century has seen a wealth of information gleaned from developmental biology studies on how embryos develop from fertilized ova to maturity and birth. These findings originally focused on the morphological transitions behind embryonic maturation, but as research techniques were refined and the importance of molecular and biochemical signaling mechanisms realized particular internal and environmental cues and chronological transitions began to be outlined in great detail. In the past 30 years or so, morphological transitions and organ development have now been intricately tied to internal molecular events, signaling molecule gradients and environmental influences. As the stem cell niche is perhaps the most crucial component of embryonic development for much of gestation, each of these influences must be understood as to how they impact the variety of stem cells present in the developing embryo. The following sections review key regulatory pathways essential to mammalian embryonic development, from biochemical signaling to transcriptional modulation, as they pertain to the stem cell niche.

Early Events in Embryogenesis **Fertilization** is defined as the process of sperm and egg fusion. Perhaps one of the most crucial events in all of embryonic development occurs soon after fertilization and is known as **cleavage**. During the cleavage process the single fertilized egg cell divides into equal two counterparts that subsequently divide in a **symmetric** fashion to produce cells known as **blastomeres**. Blastomeres that result from symmetric division are essentially equal in morphology and genetic potential. They continue to divide, generating daughter cells whose progeny will eventually become specialized and lose differentiation potential. The embryo is now referred to as a **blastocyst**. The blastocyst is cylindrical and contains three primary groups of cells: the trophectoderm (TE), the primitive ectoderm, and a lineage known as **epiblasts**. Epiblasts are key cells that differentiate into the three primary germ layers—endoderm, ectoderm, and mesoderm —and give rise to the embryo proper during a process known as gastrulation (defined and described in

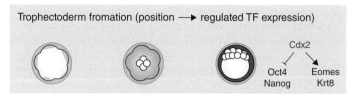

Figure 2.16 Diagrammatic illustration of trophectoderm and trophoblast maturation. (Courtesy Janet Rossant and *Developmental Biology* (Ralston and Rossant, 2008); reprinted with permission.)

more detail below). During gastrulation, the single layered blastula becomes reorganized through an invagination of cells to form a trilaminar **gastrula**. The transcriptional control of gastrulation and organization of the gastrula, represented by the three primary germ layers, is critical to early embryo formation and proper organization of cell types that will later become determined and commit to specific lineages. The **homeobox** transcription factor Cdx2 is perhaps one of the main players in the establishment of early cell lineage commitment during gastrulation, specifically trophoblast cells. Janet Rossant and Amy Ralston in the Department of Molecular Genetics at the University of Toronto demonstrated through a series of elegant transfection studies in murine embryonic stem cells that overexpression of Cdx2 drives these cells toward a trophoblast fate. Cdx2 acts to repress transcription of the transcription factors Sox2, Nanog, and Oct4 which normally confer stem cell pluripotency. Thus, negative regulation by Cdx2 of pluripotency markers drives trophoblast lineage commitment and maturation. Cdx2 is also thought to activate the transcription of Eomesodermin (Eomes), which has been shown through gene targeting studies in mice (Chapter 7) to be required for TE formation (Figure 2.16) (Ralston and Rossant, 2008).

The **inner cell mass (ICM)** is the mass of cells inside the blastocoele of a blastocyst - stage embryo which will eventually give rise to the embryo proper. It is also known as the pluriblast or embryoblast (Figure 2.17).

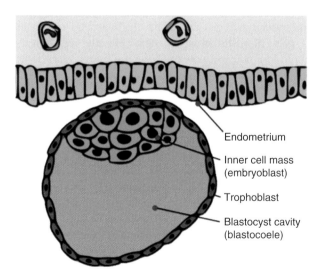

Figure 2.17 Diagrammatic illustration of a blastocyst containing a definitive inner cell mass. (Courtesy Wikimedia Commons; reprinted with permission.)

The signaling and transcriptional regulatory mechanisms underlying ICM formation and development have been well characterized. For example, the pluripotency defining transcription factors Nanog and Oct4 are expressed widely at the earliest stages of embryonic development, yet quickly become restricted to the ICM upon its appearance in the blastocyst due to Cdx2's active repression in trophoblast cells. As the ICM expands and matures, Oct4, Sox2, and Nanog act to repress the trophoblastic phenotype and promote pluripotent characteristics in ICM cells. Oct4, Sox2, and Nanog each act in a **positive autoregulatory feedback loop**, thereby maintaining expression to drive pluripotency and repress trophoblast marker expression.

The first evidence of germ layer commitment becomes apparent with the expression of the growth factor receptor bound protein, Grb2, which plays a dual role in both inhibiting Cdx2 and Nanog expression while activating Gata6 expression in a subpopulation of ICM cells destined to become the primitive endoderm. These transcriptional events are perhaps the most crucial with respect to the transition from toti- to pluripotency (discussed in detail in the next section), the establishment of intra- versus extraembryonic lineages, and the formation of the three primary germ layers (Figure 2.18) (Rossant and Tam, 2009).

Interestingly, by the time the blastocyst has fully formed, ICM cells have already lost potency, having transitioned from a toti- to pluripotent phenotype. Thus, ICM cells are destined to become either epiblasts (demarcated by Nanog expression) or primitive endoderm (demarcated by Gata6 expression). This has been confirmed by Rossant's group in a series of elegant expression studies outlining the activity of the key transcription factors present in the ICM. The random, mutually exclusive expression pattern of Nanog and Gata6 has been referred to by Rossant's group as the "**salt and pepper**" pattern (see Figure 2.16) (Burgess et al., 1995). The salt and pepper mosaic expression of Nanog and Gata6 sets the stage for and immediately precedes **gastrulation**, which is defined as the process by which three primary germ layers are acquired. It is the point during early embryonic development whereby a morphological transition occurs with specific cells of the ICM forming a visible invagination to set up germ layer specification: endoderm, ectoderm, and mesoderm (Figure 2.19).

Each layer of the gastrula will give rise to different facets of the developing embryo. The **endoderm** yields the internal organs, the **ectoderm** will produce neural tissue including the brain, and the **mesoderm** will give rise to muscle, bone, and vasculature. It is important to note that, at the point of gastrulation, the stem cells that reside in the resulting germ layers rapidly begin to lose potency, now transitioning from a pluri- to a multipotent phenotype. This loss of potency is known as **determination**.

As above with respect to blastocyst formation and gastrulation, to fully understand the stem cell component and contribution to embryonic development, it is important to have a firm grasp of the transcriptional regulatory mechanisms that govern germ layer determination. Within the ICM, a subset of cells express the homeodomain proteins Gata4, Gata6 and the low density lipoprotein receptor-related protein Lrp2. These factors have been shown to drive the endodermal phenotype through the activation of endoderm - specific target genes. **Basic helix-loop-helix (bHLH) transcription factors** are transcription factors characterized by two "α" helices linked by a loop juxtaposed to a basic DNA binding domain. These factors both homo- and heterodimerize to form powerful transcriptional complexes on promoter, enhancer, and silencing elements of target genes. They are known as potent mediators of lineage-specific target gene transcription during and post-embryonic development. As the mesodermal layer begins to mature, specific **bHLH** transcription factors that have been demonstrated to be crucial for the development of mesodermal lineages. **Twist**, for example, is a bHLH transcription factor expressed very early during mesoderm

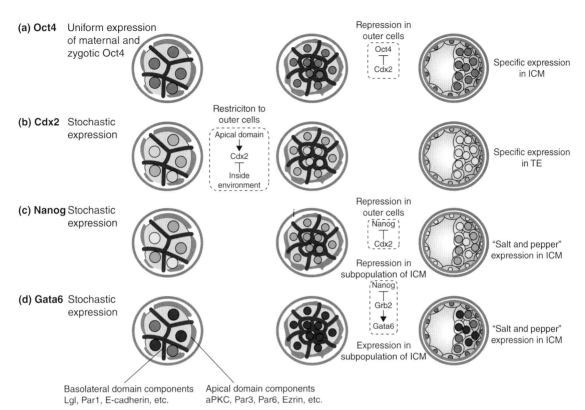

Figure 2.18 Molecular players in the formation of early embryonic lineages. Four lineage-specific transcription factors, Oct4, Cdx2, Nanog, and Gata6 are important for the generation of the first three lineages in the blastocyst. The initial expression of these transcription factors is not restricted to specific cell populations. Lineage-specific expression is gradually established in association with the maturation of cellular structures (such as apical–basolateral cell membrane domains and intercellular junctions) and of positive and negative interactions among the transcription factors themselves. (a) Oct4: Oct4 protein is observed in all blastomeres throughout early cleavage stages. At the eight-cell stage, all blastomeres contain Oct4. At the blastocyst stage, Oct4 is gradually downregulated in the outer trophectoderm (TE) cells by Cdx2 through direct physical interaction and transcriptional regulation. (b) Cdx2: Cdx2 protein is detected beginning at the 8- to 16-cell stage; its initial expression appears to be stochastic. By the early morula to early blastocyst stages, Cdx2 expression is ubiquitous but higher in outer, apically polarized cells. Restricted expression in outer TE cells is established by the blastocyst stage. (c) Nanog and (d) Gata6: Nanog and Gata6 are detected from the eight-cell stage. Both proteins are expressed uniformly in all cells until the early blastocyst stage. Nanog expression is downregulated in outer cells by Cdx2 and in a subpopulation of the ICM by Grb2-dependent signaling. By contrast, Gata6 expression is maintained by Grb2-dependent signaling. By the late blastocyst stage, ICM cells express either Nanog or Gata6 exclusively. (Courtesy Janet Rossant, Patrick P.L. Tam and *Development* (Rossant and Tam, 2009); reproduced/adapted with permission from *Development*.)

development and has been shown by Patrick Tam's group in the Embryology Unit at the Children's Medical Research Institute in Westmead, Australia to play a role in the maintenance of mesoderm-specific progenitor cell proliferation and viability (Ota et al., 2004). A second bHLH transcription factor, **paraxis**, is expressed at much later stages of embryonic development, in the developing paraxial **somites**, which are defined as masses of paraxial

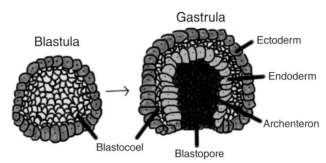

Figure 2.19 Diagrammatic illustration of the process of gastrulation. Gastrulation occurs when a blastula, made up of one layer, folds inward and enlarges to create a gastrula. A gastrula has three germ layers—the ectoderm, the mesoderm, and the endoderm. Some of the ectoderm cells from the blastula collapse inward and form the endoderm—the blastopore is the hole created in this action. Ectoderm, blue. Endoderm, green. Blastocoele (the yolk sack), yellow. Archenteron (the gut), purple. (Courtesy Wikimedia Commons; reprinted with permission.)

mesoderm on either side of the neural tube destined to form mesodermal compartments— the **dermatome**, **myotome** and **sclerotome** —compartments that give rise to the dermis, axial skeletal muscle, and axial skeleton, respectively (Burgess et al., 1995). It is the multipotent nature of the stem cells in both the uncompartmentalized somites and later the specific compartments, which drive mature mesodermal lineage specification. Eric Olson and colleagues, then in the Department of Biochemistry at the University of Texas M.D. Anderson Cancer Center in Houston, demonstrated in mouse knockout experiments (see Chapter 7) that paraxis is required for somite formation and musculoskeletal patterning. Paraxis null embryos demonstrate a failure of the somites to form epithelial spheres, thereby disrupting axio-skeletal patterning. Interestingly, the failure of somites to epithelialize did not affect rostral-to-caudal axis segmentation or lineage specification (Figure 2.20). These studies suggest that the "stemness" nature of the stem cells present in both uncompartmentalized somites and the three compartments was not affected by the loss of paraxis function (Burgess et al., 1996).

The ectodermal layer also expresses specific transcription factors that function primarily to drive neural development. Pax3 is a homeodomain transcription factor expressed in a narrow band of ectoderm at early stages of germ layer development. It has been postulated that Pax3 represses Pax7 expression in the ectodermally derived neural tube, and experiments utilizing antisense oligonucleotides to inhibit Pax3 function have demonstrated

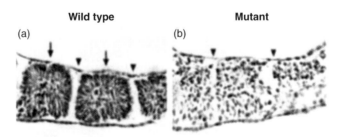

Figure 2.20 Histological examination of somite development in wild-type and paraxis -null embryos. Large arrows denotes somites. Small arrows denote somitic segmentation or lack thereof. (Courtesy Rob Burgess and *Nature* (Burgess et al., 1996); reprinted with permission.)

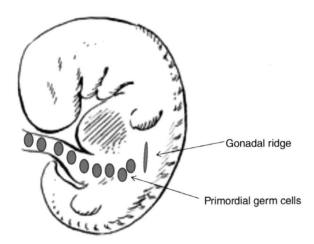

Figure 2.21 Diagrammatic illustration showing the location of primordial germ cells in the developing human embryo during the 4th week of gestation.

neural tube defects. Thus, for all three germ layers, ultimately it is the signaling and transcriptional regulatory hierarchies that govern the transition from pluri- to multi-potency of cells destined to commit to specific lineages during embryonic development.

Germ Cell Development As this text is dedicated to the study of stem cells and their contribution to transient and mature lineages during both embryonic and postnatal development, a discussion on primordial germ cell (PGCs) specification during embryogenesis and in the adult is warranted. **Germinal (germ) cells**, also known as PGCs, are defined as biological cells that give rise to the gametes of an organism that reproduces sexually. It was realized as early as 1970 that they are totipotent (described in detail below), with the ability to give rise to all embryonic and extra-embryonic cell types and tissues. The embryo's genital (gonadal) ridge will become the future gonads, and it is populated by germinal cells that will commit to either a male or female phenotype (Figure 2.21) (Kousta, et al., 2010).

It is the process of **meiosis**, defined as the production of gametes (sperm and egg cells), through cellular division, in which each gamete retains half the original chromosome number and is referred to as **haploid**. Gametes provide the full complement of chromosomes upon successful fertilization between a sperm and egg cell, with sex determination defined by the presence of either the XX or XY complement. Thus, germ cells are a unique stem cell population that is present both in the early developing embryonic genital ridge and in the adult. The totipotent nature of germ cells also predisposes them to a tumorigenic phenotype, with studies in the 1970s demonstrating this basic property via the production of **teratocarcinomas** (malignant teratomas possessing numerous extra-embryonic and embryonic cell types) in mice. What drives germ cell specification? Recent studies have defined three events that must occur to drive development of the germ cell lineage: full repression of the somatic cell phenotype; reacquisition of pluripotency; and epigenetic reprogramming. These three processes are driven primarily by specific transcriptional regulatory cascades. One of the earliest transcriptional markers of the PGC lineage is the transcription factor BLIMP1/PRDM1, which is expressed in epiblast cells that will later become committed PGCs. PRDM14 is also expressed in both pre-committed PGC epiblasts and committed PGCs. Expression here is exclusive, and gene-targeting studies in mice have demonstrated crucial roles for both PRDM proteins in mammalian germ cell maturation. Hox genes are also speculated to play a considerable role in both early and late PGC development. Prior to PGC commitment, epiblast cells express a multitude of Hox genes such as

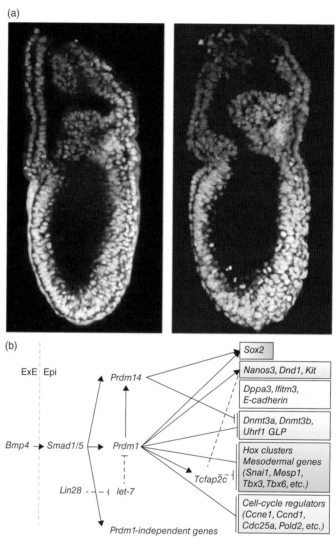

Figure 2.22 A regulatory model of primordial germ cell specification. (A) Expression of Prdm1 (left) and Prdm14 (right) in the LS stage embryo visualized by the Prdm1-mVenus and Prdm14-mVenus reporters, respectively. Prdm1 is expressed in the nascent PGC precursors emerging from the most proximal part of the posterior epiblast as well as in the visceral endoderm. Prdm14 is exclusively expressed in the germ cell lineage and pluripotent cell lines. (B) A summary of genetic pathways for PGC specification. See text for details. Black arrows and black lines with terminal bars indicate genetic pathways for activation and for repression, respectively, as demonstrated by in vivo experiments. Dotted arrows and dotted lines with terminal bars indicate genetic pathways for activation and for repression, respectively, as proposed based on *in vitro* experiments. (*Source:* Saitou & Yamaji, 2010. Reproduced with permission from BioScientifica Ltd.)

Nanog, Sox2, and Zic3, which not only drive pluripotency but are known to specify meso-dermal cell types later in development. The influences that each of these factors have upon each other's expression and even the upstream signaling molecules such as BMP4 have now been elucidated. These transcriptional networks, activated by growth factors, work to drive PGC commitment and maturation (Figure 2.22) (Saitou and Yamaji, 2010).

It has been speculated that early PGCs become transiently committed to a mesodermal fate, but regain pluripotency at later timepoints. Interestingly, but not surprisingly, many of the same cascades that promote embryonic stem cell development are conserved in germ cell development. As mentioned above, Oct4 plays an essential role in conferring ES cell pluripotency. It is also expressed in and becomes restricted to PGCs immediately after gastrulation, and Oct4 loss-of-function experiments in mice have confirmed it to be necessary for PGC survival. In addition, Oct4 is expressed in pluripotent cells present in human germ cell tumors.

BASIC PROPERTIES OF STEM CELLS

The following sections outline and describe the two primary characteristics of stem cells: long-term self-renewal and the ability to differentiate into other cell types. These characteristics will be discussed in detail for individual types of stem cells in later sections.

Long-Term Self-Renewal

Long-term self-renewal is defined as a cell's ability to replicate itself over extended periods and through multiple passages while maintaining an undifferentiated state. While replicative capacity is key to stem cell number expansion in preparation for populating a particular tissue or organ with differentiated cell types, it varies with each stem cell phenotype and even varies within the same cell type depending upon the environment. Cancer stem cells, discussed in more detail in Chapter 6, are thought to lose their ability to regulate mitotic activity causing them to divide in an uncontrolled manner. Stem cell self-renewal can be classified as either symmetric, in which each parent stem cell gives rise to two daughter stem cells, or asymmetric, in which each parent stem cell gives rise to one daughter, undifferentiated stem cell and one differentiated cell specific for the potency of the parent stem cell. This is also known as **obligatory asymmetric replication** and has been studied extensively in the nematode *Caenorhabditis elegans*. The function of asymmetric cell division is to contribute both to the expansion of the undifferentiated stem cell population while also populating the developing embryo (or in some cases even mature post-natal tissues) with much needed differentiated lineages. In rare cases, cells may divide asymmetrically to produce two entirely unique progeny. This occurs in plants in which uncommitted cells divide to produce semi-differentiated guard cells that can subsequently divide through very limited replicative capacity to produce more guard cells. **Symmetric cell division** — more prominent during the course of embryonic development than after birth— drives the expansion of an undifferentiated stem cell population for use in an embryo's tissue and organ generation. In symmetric cell division, each daughter cell appears genetically and morphologically identical to one another and to the parent cell. It is more prominent in pluripotent embryonic stem cells than in somatic stem cells. It should be noted that symmetric and asymmetric stem cell division are not mutually exclusive, and can occur stochastically within the progeny resulting from a single undifferentiated stem cell (Figure 2.23).

Different Potency Capabilities

The ability of a stem cell to differentiate into different lineages is known as **potency**. It can take on a variety of degrees and defines perhaps the stem cell's most important property,

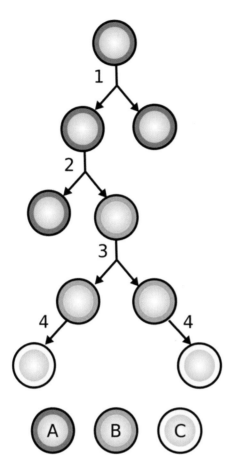

Figure 2.23 Diagrammatic illustration of stochastic stem cell replication. Steps 1 and 3 illustrate symmetric stem cell division; step 2 illustrates asymmetric stem cell division and step 4 illustrates final differentiation. (Courtesy Wikimedia Commons; reprinted with permission.)

which is the ability to provide mature, functional cell types in tissues and organs needed for survival of the organism. The following sections describe the various degrees of stem cell potency. Figure 2.24 gives a diagrammatic overview of some the stages of potency described below.

Totipotency **Totipotency** is defined as the ability to differentiate into both embryonic and extra-embryonic tissues. Totipotency is sometimes referred to as **omnipotency**. Totipotent stem cells are produced immediately after fusion of the egg and sperm and, in mammals, are generated within the first few cell divisions post-fertilization. In the context of the developing human embryo, the first totipotent cell is known as the **zygote**. The zygote undergoes symmetric cell division to give rise to additional totipotent cells, all of which have the capacity to differentiate into extra-embryonic cells and tissues such as trophoblasts or into any of the three primary germ layers. In human embryos, only morula stage stem cells are classified as totipotent (Figure 2.24). It is at the 16-cell stage when totipotent stem cells begin to commit to either a trophoblastic or epiblastic lineage, with trophoblasts later contributing to extra-embryonic tissues such as the layers of the placenta and epiblasts driving inner cell mass (ICM) expansion (discussed above).

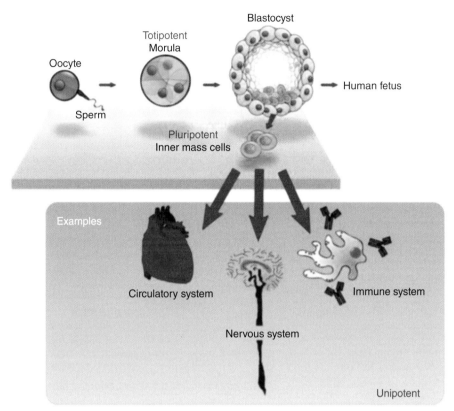

Figure 2.24 Diagrammatic illustration of the different stages of stem cell potency exhibited during human embryonic development. (Courtesy Wikimedia Commons; reprinted with permission.)

Pluripotency Cells are **pluripotent** if they have the capacity to differentiate into all of the three primary germ layers but cannot contribute to extra-embryonic tissues. This definition has now been expanded in the context of embryonic stem or embryonic germ cells to include the following generic criteria:

- Originate from a pluripotent cell population
- Maintain a normal karyotype
- Immortal
- Can be propagated indefinitely in the embryonic state
- Can be clonally derived and are capable of spontaneous differentiation into somatic cells representative of all three embryonic germ layers in teratomas or *in vitro*

It is generally accepted that pluripotent stem cells have the capacity to give rise to any lineage of the embryo proper and the adult organism. As outlined below and in Figure 2.25, these germ layers will give rise to specific organs and tissues in the adult:

Endoderm—lungs, gastrointestinal tract, interior lining of the stomach
Mesoderm—urogenital tract, muscle, blood, portions of the dermis, bone
Ectoderm—nervous system and epidermal tissues

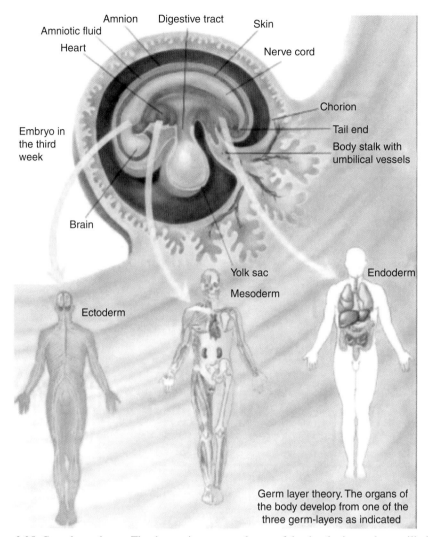

Figure 2.25 Germ layer theory. The three primary germ layers of the developing embryo will give rise to specific organs and tissues in the adult human. (From Van de Graaff and Fox (1989); reprinted with permission.)

Over the past 20 years a massive effort to define the signaling and transcriptional regulatory mechanisms governing pluripotency and the transition from this state to germ layer commitment has been undertaken by researchers worldwide. The result of these efforts has been a fairly complete breakdown of the transcription factors and pathways both necessary and sufficient to maintain the pluripotent phenotype or initiate the earliest stages of germ layer commitment. These signaling molecules and transcription factors will come up time and again throughout the text of this book as they are critical to virtually all aspects of stem cells. Interestingly, transcriptional networks reorganize in response to external stimuli during the transition from pluripotency to multi-potency and germ layer commitment. Researchers in the Department of Stem Cell and Regenerative Biology at Harvard University in Cambridge, Massachusetts, led by Sharad Ramanathan, used murine embryonic stem cells as a model system to define the roles of key individual transcription factors during this process (Thomson et al., 2011). Their exceptional research is summarized in Case Study 2.1.

Case Study 2.1: Pluripotency factors in embryonic stem cells regulate differentiation into germ layers

Matt Thomson et al.

The question of how stem cells regulate their fate has long dogged developmental biologists. It is clear that complex regulatory networks promote certain cellular phenotypes, but how these networks are reorganized during germ layer commitment is poorly understood. Sharad Ramanathan and colleagues at Harvard asked how murine embryonic stem cells transition for a pluripotent phenotype to germ layer commitment. They specifically analyzed and characterized the dynamic transcriptional circuitry present in mESCs by assessing published gene expression microarray data and applying the findings in an elegant *in vitro* directed differentiation assay. Upstream growth factor signaling was shown to activate the transcription factor Brachyury and, surprisingly, they found that Oct4 and Sox2, well-known pluripotency-defining transcription factors, also have unique roles in promoting germ layer specification. Oct4 was demonstrated to suppress ectoderm while promoting mesoderm formation. Sox2 was shown to play a converse role, promoting ectoderm while simultaneously inhibiting mesoderm induction. The researchers also found that differentiation-inducing signals dynamically modulate both Oct4 and Sox2 protein levels thus promoting cell fate direction by altering each transcription factor's genome-binding properties. It was noted that differential expression of Oct4 and Sox2 and downregulation of Nanog was required prior to cell fate selection (Figure 2.26) (Thomson et al., 2011).

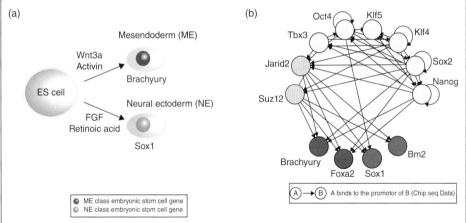

Figure 2.26 Transcriptional networks of mES cells during differentiation. (A) ESCs lose pluripotency and differentiate into ME progenitor cells to express Brachyury in response to Wnt3a or Wnt agonist, CHIR. They differentiate into NE progenitors to express Sox1 in response to FGF signals or retinoic acid. (B) Diagram of interactions between the pluripotency factors (yellow), key epigenetic regulators (gray), and regulators of the ME (red) and NE (green) lineage inferred from ChIP-seq data in the literature. (Images courtesy Sharad Ramanathan and *Cell* (Thomson et al., 2011); reprinted with permission.)

Multipotency **Multipotent** cells can be defined as cells that have the capacity to give rise to multiple, but not all, cell lineages in the developing embryo or adult. In some cases, multipotent cells have a limited ability to divide, and will senesce after a certain number of cell divisions. Multipotent stem cells have been determined to commit to specific lineages upon exposure to the appropriate environmental conditions conducive to lineage development. Interestingly, multipotent cells are prevalent in the adult organism, as they are often needed to replenish various types of differentiated cells that diminish over time. Examples of multipotent stem cells include HSCs—which give rise to various mature blood cell types—MSCs—which provide a rich source for differentiated bone (chondrocytes and osteocytes) and fat (adipocytes)—and adult neural stem cells, which give rise to glial cells and neurons. Mesenchymal cardiac-specific stem cells also give rise to cardiomyocytes, smooth muscle, and endothelial cells (Figure 2.27).

Oligopotency **Oligopotency** refers to a stem cell's ability to differentiate into a limited number of cell types, usually two. Oligopotent cells are often found in adult organisms as they specialize in the replacement of two or more differentiated lineages that exhibit high turnover rates. Vascular stem cells, for example, have the capacity to differentiate into either smooth muscle or endothelial cells, each of which must be replaced often due

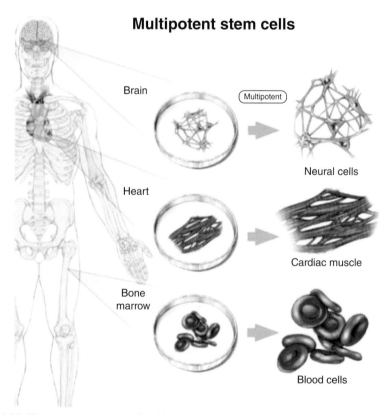

Figure 2.27 Three organ systems defined by the existence of multipotent stem cells. (Courtesy *CancerResearchJournal.com*; reprinted with permission.)

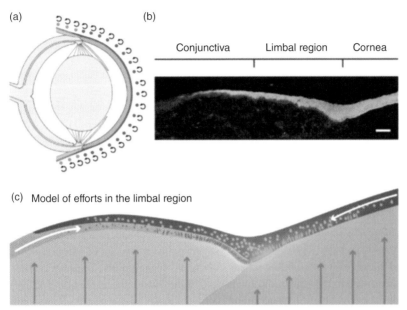

Figure 2.28 Model of ocular stem cell oligopotency. (a) Epithelial stem cells of equal potency are distributed throughout the entire ocular surface. (b) Suprabasal cells at the limbal region of the pig express K3$^+$ corneal-type differentiation whereas basal cells are negative. Differentiated corneal cells appear to slide over limbal basal cells, as suggested by their orientation. Scale bar, 100 µm. (c) Schematic representation of the same region in b. (Courtesy Yann Barrandon and *Nature* (Majo et al., 2008); reprinted with permission.)

to new blood vessel growth. Other examples of oligopotent stem cells include those residing on the mammalian ocular surface. The **limbus**, which is defined as the edge of the cornea where it joins the sclera, has been postulated to act as a prime niche for oligopotent stem cells that repopulate the corneal squamous epithelium with new corneal and conjunctival cells. Through *in vitro* culture and transplant experiments, researchers at the Laboratory of Stem Cell Dynamics, École polytechnique fédérale de Lausanne in Switzerland demonstrated the oligopotency of stem cells isolated from both mice and pigs (Figure 2.28) (Majo et al., 2008).

Unipotency **Unipotent** stem cells, also commonly referred to as **precursor cells**, are cells which have the capacity to differentiate into only one cell type. These cells tend to be differentiated to the extent that they have limited mitotic activity and have committed to become, or produce progeny that will become, a single differentiated lineage. Unipotent stem cells typically arise from multipotent stem cells and will either divide to increase stem cell numbers or differentiate into a mature phenotype. Unipotent cell presence and activity is most often present in specific tissue compartments of the adult organism, the purpose of which is to provide structural or functional components of that particular system. One example of unipotent stem cells is those cells present in the **epithelium**, which is the outermost layer of skin tissue in mammals (not including the dead layer of squamous cells overlying it). It is the unipotent nature of stem cells in the epithelium that allow for it to be used for the generation of new skin *in vitro* for transplantation purposes. A second example of unipotent stem cells is hepatocytes. Hepatocytes

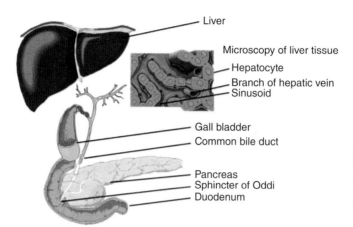

Figure 2.29
Diagrammatic illustration
of hepatocyte location
within the liver. (Courtesy
Wikimedia Commons;
reprinted with permission.)

represent the majority (70–80%) of the liver mass in mammals, completely encompassing and surrounding the sinusoidal liver regions, and allow for near-complete liver regeneration in the event of cell loss due to damage or disease (Figure 2.29). They also exist to a certain degree in the pancreas. Hepatocytes are unique in that they are differentiated stem cells that maintain the capacity to divide and regenerate more differentiated stem cells. They not only repopulate the liver with additional hepatocytes but also play crucial roles in liver function such as undertaking protein, cholesterol, bile salts and phospholipid synthesis, protein and carbohydrate storage, and metabolism. Hepatocytes can be induced to divide *in vitro* via the addition of hepatocyte growth factor (HGF), which activates a tyrosine kinase-signaling cascade through the HGF receptor.

TYPES OF STEM CELLS

The following sections give a brief description of the main individual types of stem cells currently studied with later chapters providing more detailed descriptions of each cell type and their potential in diagnostic or therapeutic applications.

Embryonic Stem Cells

In mammals, **embryonic stem (ES) cells** can be defined as pluripotent cells present in or derived from the inner cell mass (ICM) of blastocyst -stage embryos. Embryonic stem cells are defined by two basic properties:

1. Pluripotency
2. Indefinite replication

Fetal Stem Cells

Fetal stem cells can be defined as primitive cells present within the organs and tissues of a developing fetus than can give rise to differentiated phenotypes. The isolation of fetal stem cells and their potential use in therapeutic applications is highly controversial because they are typically derived from aborted fetal tissue.

Amniotic Stem Cells

Amniotic stem cells are mesenchymal in origin, multipotent, and extracted from amniotic fluid. They may be considered a subset of fetal stem cells if they are isolated from the amniotic fluid prior to birth. Their use in research and therapy is not controversial as their isolation does not require the destruction of human embryos.

Adult Stem Cells

Adult stem cells are defined as undifferentiated cells present in organs and tissues after birth that are capable of self-renewal and can give rise to at least one differentiated cell type.

Induced Pluripotency (iPS) Cells

Induced pluripotency stem (iPS) cells are cells artificially derived from non-pluripotent stem cells that exhibit pluripotent differentiation capabilities.

Cancer Stem Cells

Cancer stem cells are defined as stem cells present in various types of tumors or cancers that are capable of self-renewal and can give rise to a cell type or cell types present within that particular tumor or cancer.

All of the above types of stem cells will be broken down in more detail in later chapters.

THE POTENTIAL OF STEM CELLS IN MEDICINE AND MEDICAL RESEARCH

Therapeutics

The multitude of stem cell types mentioned above has vast potential for therapeutic applications, including tissue engineering and cell replacement therapeutics. Below is a brief description of each therapeutic platform. These will be expanded upon in detail in later chapters for specific cell types.

Tissue Engineering As defined in Chapter 1, regenerative medicine regenerates tissue rather than surgically extracting or altering that tissue. It overlaps closely with cell therapy and often the two terms are used interchangeably. **Tissue engineering**, however, has more often than not been used to describe the *ex vivo* production of living tissue for transplant using cells, methods, and materials to create said tissue. Cell therapy tends to refer to the introduction of individual cells in the body to replace or repopulate a diminished supply of a particular cell type. In tissue engineering applications, doctors often cultivate, grow, and generate whole tissues or components of organs for transplant to replace that damaged by injury or disease. It is performed primarily when the body is unable to repair and heal on its own, and typically is applied when the ultimate goal is to repair or replace the bulk or the entirety of a particular tissue organ. Examples of different types of human tissues and organs that may be candidates for tissue engineering include:

- Bladder
- Skin
- Vasculature

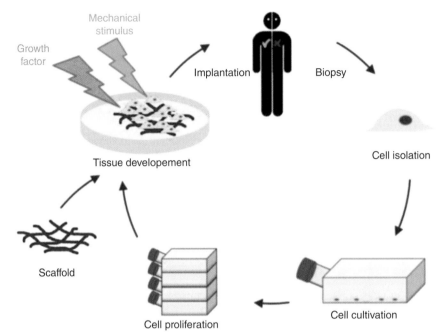

Figure 2.30 Diagrammatic illustration of the process for autologous tissue engineering therapeutics. See text for details. (Courtesy Wikimedia Commons; reprinted with permission.)

- Cardiac Tissue
- Smooth and skeletal muscle
- Bone and cartilage

The process of tissue engineering often involves the use of the patient's own cells or tissues as the source for newly generated tissue and/or organ components. This is typically referred to as **autologous transplantation therapy**. In autologous transplantation therapy a patient's own cells or tissue is biopsied, cultivated for expansion, and infiltrated with a three-dimensional scaffold for stability. The scaffold is often made of biodegradable materials that are resorbed in the body over time. Examples of these include hyaluronic acid (HA) and poly (L-lactide-*co*-glycolide) (PLLG) nanofibers. After culture in the presence of growth factors or mechanical stimuli to prepare and

---

Focus Box 2.1: James Blundell and the first successful blood transfusion

James Blundell (1791–1878) was a British physician and obstetrician who originated the concept that postpartum hemorrhaging could be treated with a transfusion of new blood into the patient. He determined that the rate of transfusion and removal of extraneous air in the syringe was key to success. He conducted 10 documented transfusions during his career and invented numerous instruments for the procedure that are still in use today. (Photo courtesy Wikimedia Commons; reprinted with permission.)

prime the tissue/matrix for both function and survival *in vivo,* it is implanted into the patient (Figure 2.30).

Cell Therapy **Cell therapy** can be defined as the introduction of cells into a tissue or organ to allow for the growth or repair of that tissue or organ or for the treatment of disease. Cell replacement therapeutic platforms and the idea of directly implanting new cells into the body are not a new concept and in fact became routine after British physician James Blundell performed the first successful human-to-human blood transfusion in 1829 (See Focus Box 2.1).

Cell therapy can be broken down into two primary platforms with each ultimately providing a unique therapeutic benefit. These include:

1. **Stem cell transplantation**—The transplant of progenitor or stem cells that will act as a source for mature, functional lineages.
2. **Differentiated cell transplantation**—The transplant of mature, functional, differentiated lineages.

In addition, each of these transplantation platforms can be modified to introduce a genetic component, for example, a gene that produces a particular desired therapeutic benefit. This is often referred to as **cell-based gene therapy**. This technique, however, is cumbersome and, with the exception of adenovirus associated virus (AAV) technology, requires the stable integration of a transcriptionally active gene into the candidate cell's genome. This results in considerable safety concerns, as genomic integration could at some point activate an oncogenic pathway and lead to a tumorigenic or metastatic phenotype. Typical modes of genetic information introduction into cells include:

1. Plasmid DNA transfection —Host genome integration
2. Adeno-associated virus (AAV) transduction—Host genome integration or cytoplasmic episomal maintenance and replication
3. Retroviral vector transduction—Host genome integration

An example of stem cell-based gene therapy includes the modification of a patient's own HSCs with vectors providing desired genomic constructs, followed by reintroduction of the genetically modified cells into patients (Figure 2.31) (Bagnis and Mannoni, 1997).

The typical anomalies treated with cell replacement therapy, whether gene- or non-gene based, often include inherited disorders in which the patient simply does not produce enough of the differentiated functional cells or a desired functional gene product. Examples of these disorders include:

• Batten disease
• Alzheimer's and Parkinson's disease
• Hurler syndrome
• Blood and immune disorders
• Diabetes

Chapter 8 will provide more detail into the use of stem cells in cell replacement therapeutics with a focus on specific physiological disorders and diseases.

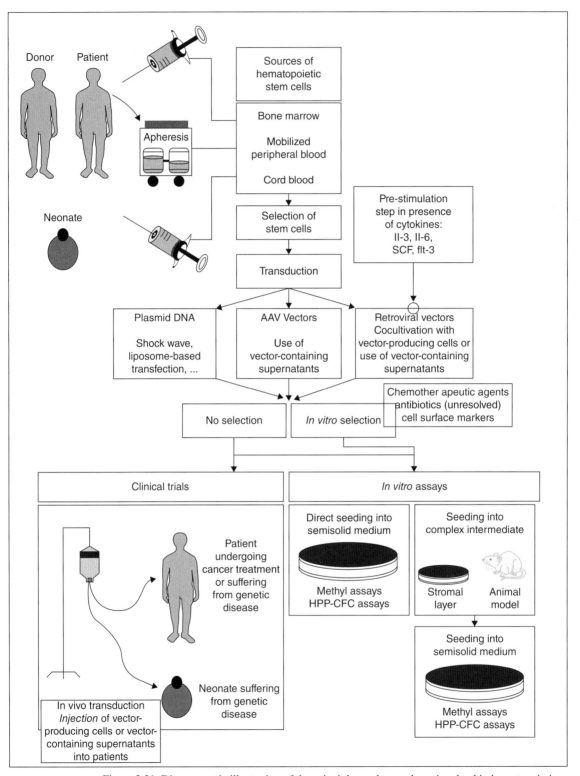

Figure 2.31 Diagrammatic illustration of the principles and procedures involved in hematopoietic stem cell based gene therapy. (*Source:* Bagnis & Mannoni, 1997. Reproduced with permission from Alphamed Press.)

Cell-Based Drug Screening

Cell-based drug screening can be defined as the use and application of cells to identify compounds of therapeutic benefit or to assess specific drug safety and toxicity properties (ADME-Tox). The application of stem cells in drug screening initiatives is a relatively new concept, but holds considerable promise in the drug development process. Over the past decade, technologies implemented for the use of cells as combinatorial small molecule library screening platforms, also known as **high-content screening (HCS)**, have become optimized, and in many cases may eliminate the need for a great deal of *in vitro* or animal screening efforts. Cell-based drug screening is most often carried out in a microtiter plate on a monolayer of cells, allowing for a semi-high-throughput implementation of the screening process. HCS libraries or individual drug candidates are added to the wells and a particular response is monitored. In each assay, most often a specific readout is designed to provide information as to the drug's effect on cellular viability or function. For example, particular readouts might include:

- Intracellular cAMP levels
- Intracellular calcium levels
- G Protein-Coupled Receptor (GPCR) activity
- Gene expression

Cell-based drug screening technologies are constantly evolving. For each technology, however, the method for detecting a cellular response to a compound or small molecule is key to the sensitivity, reproducibility, and success of the system. These may include:

- Fluorescence
- Luminescence
- Colorimetry
- Radiologic readouts

In addition, output data may occur in mass at the end of the screen or in real time, for example, in the analysis of intracellular calcium concentrations upon exposure to external stimuli. Cell-based drug screening for the identification not only of drug candidates but also of druggable targets has now become commonly accepted in both primary and secondary drug screening initiatives. Drug screening using more conventional biochemical assays, while efficient and amendable to high-throughput screening platforms, often yields false positives (or negatives) and does not truly mimic a cell's physiological response to a drug candidate. The testing of compounds against living cells has the clear advantage of providing a more accurate therapeutic analysis. In addition, it allows for an opportunity to dissect the regulatory cascades that may be affected by agonistic and antagonistic compounds. There are numerous examples where cell-based screening technologies and efforts have resulted in the dissection of a regulatory cascade, the identification of an unexpected cellular function or the characterization of drug efficacy, toxicity, or safety concerns. Perhaps one of the most widely used in the area of cardiotoxicity is that of Ether-a-go-go Related Gene (hERG) ion channel activity monitoring in cardiomyocytes for assessing cardiotoxicity. The human **hERG** gene encodes a voltage-gated potassium channel expressed in cardiomyocytes. It primarily acts as an inward rectifying potassium channel, allowing potassium inflow into the cell's cytoplasm. hERG, in effect, acts as a molecular break for the action potential exhibited by cardiomyocytes, and any blockage of its activity results in a potentially lethal hyperpolarization phenomenon known as long QT. Many drugs are known to interact with hERG and induce long QT, thus a cell-based screen assessing drug hERG activity is a valuable tool for safety and toxicology assessment.

Gold Standard Assay: Whole Cell Voltage Clamp

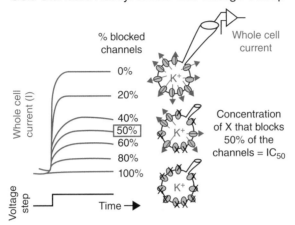

Figure 2.32 Overview of the cell-based screening whole cell patch clamp assay for monitoring ionic channel activity. (Courtesy LabAutopedia.com; reprinted with permission.)

Figure 2.32 outlines the hERG cell-based screening assay. It is based on the application of **patch clamp** monitoring of intra- versus extracellular ionic concentrations.

How do stem cells impact cell-based drug screening? Through directed differentiation they provide an unlimited source for a variety of mature, often functional cell types that can be assessed for activity and response in the presence of small molecules or other types of agonists and antagonists. Induced pluripotent (iPS) cells, discussed in detail in Chapter 5, have gleaned most of the focus for HCS cell-based applications over the past several years. This is most likely due to their unlimited supply, ability to differentiate into a multitude of mature, functional lineages, and the elimination of any controversy surrounding the application of embryonic stem cells. Figure 2.33 outlines the stages at which either iPS or embryonic stem cell-based screening may impact the drug discovery and development process (Sartipy and Bjorquist, 2011).

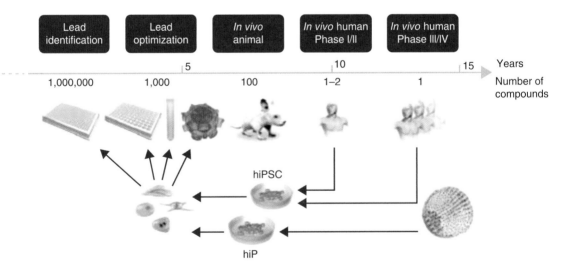

Figure 2.33 Diagrammatic overview of the drug discovery and validation phases affected by the application of iPS and embryonic stem cell directed differentiation technology. Terminally differentiated, functional lineages derived from iPS or ES cells provide opportunities for HCS implementation at both the lead identification and lead optimization preclinical phases of drug discovery and development. (Courtesy Peter Sartipy, Peter Bjorquist and *Stem Cells* (Sartipy and Bjorquist, 2011); reprinted with permission.)

CHAPTER SUMMARY

Basic *In Vitro* Cell Culture—A Historical Perspective

1. Cell culture and tissue culture are terms often used interchangeably.
2. Cell culture goes back as far as the late 1800s.
3. The first real experiment on stem cells or stem cell precursors was performed by Ross Granville Harrison in 1907.
4. French surgeon Alexis Carrel cultured avian cardiac tissue for over 20 years.
5. Clonal cell culture was developed first by W.R. Earle via generation of the L-929 line.
6. The Hela cell line was the first human cell line to be propagated successfully for many passages *in vitro* and has since been widely used worldwide in the study of cancer.
7. Harry Eagle defined the essential components of media that allow for cellular propagation and inhibition of senescence.
8. J.W. Littlefield developed HAT medium for clonal selection of hybrid lines.
9. Richard Ham developed the first fully synthetic medium, F12, for the culture of CHO cells.
10. Cesar Milstein and George J.F. Kohler developed hybridoma technology.
11. Michael Wigler and Richard Axel developed a method for the stable introduction of single genes into cells.
12. Gail Martin and Martin Evans are independently credited with the isolation and propagation of embryonic stem cells from the ICM of murine blastocysts.

Stem Cell Culture—Optimal Conditions and Techniques

Embryonic Stem Cell Culture

1. Many commonalities exist for the successful culture of murine and human ES cells.
2. Basic FGF, noggin, activin A, and TGF-β1 have been defined as crucial defined factors needed for hES cell culture.
3. Recapitulating the three-dimensional properties of a living organism provides a much more *in vivo*-like niche for cultured cells to thrive.
4. Human ES cells tend to grow as colonies and can be dissociated via mechanical or enzymatic digestion.
5. Feeder cells are often used as a monolayer for ES cell culture. Synthetic matrices may also be used.
6. 3D scaffolds provide a more *in vivo*-like environment than monolayer cells for ES cell culture.

Hematopoietic Stem Cell Culture

1. The Notch signaling cascade plays a significant role in HSC propagation.
2. In HSCs Notch works through the transcription factor CSL to activate HES repressor expression inhibiting expression of genes essential for driving B-cell and myeloid cell terminal differentiation.
3. Coculture of HSCs with stromal cells improves the efficiency of expansion of CD34+ cells.
4. Copper chelators improve the efficiency of expansion of CD34+ cells from HSCs.

Adipose-Derived Stem Cell Culture

1. ASCs exhibit many of the characteristics and properties inherent in MSCs.
2. Large numbers of ASCs can be harvested without risk to the patient.
3. A number of clinical trials have been conducted utilizing hMADS in the surgical area with promising results in plastic, oral, and cardiac surgery.

The Study of Embryonic Development

Embryonic Development and the Origin of Stem Cells

1. Embryonic morphological transitions and organ development have now been intricately tied to internal molecular events, signaling molecule gradients and environmental influences.
2. The stem cell niche is perhaps the most crucial component of embryonic development.

Early Events in Embryogenesis

1. The transcriptional control of gastrulation and organization of the gastrula, represented by the three primary germ layers, is critical to early embryo formation and proper organization of cell types which will later become determined and commit to specific lineages.
2. Cdx2 acts to repress transcription of the transcription factors Sox2, Nanog, and Oct4 which normally confer stem cell pluripotency.
3. Oct 4, Sox2 and Nanog play key roles in driving ICM pluripotency and repressing trophoblast development.
4. Toti- versus pluripotency ICM phenotypes are regulated by dual Grb2 activation and repression functions.
5. After gastrulation and upon formation of the three primary germ layers cells begin to become determined, transitioning from pluri- to multipotent capacities.
6. Twist and paraxis drive mesoderm development both early and late in embryogenesis.
7. Pax3 drives ectoderm development.

Germ Cell Development

1. Germ cells are a unique totipotent stem cell population that is present both in the early developing embryonic genital ridge and in the adult.
2. Germ cells may easily become tumorigenic due to their totipotent nature.
3. Germ cells commit to a mesodermal fate then revert to pluripotency.

Basic Properties of Stem Cells

1. Long-term self-renewal and potency are the basic properties of stem cells.
2. Self-renewal can be classified as either symmetric or asymmetric.
3. Symmetric cell division drives the expansion of an undifferentiated stem cell population as the basis for tissue and organ generation in a rapidly growing embryo.
4. Symmetric and asymmetric stem cell division are not mutually exclusive.
5. Differing potency capabilities yield stem cell uniqueness and allow for the development of all differentiated cell types.

Totipotency

1. Totipotent stem cells are produced immediately after fusion of the egg and sperm.
2. In human embryos only morula stage stem cells are classified as totipotent.

Pluripotency

1. Pluripotent stem cells have the capacity to give rise to any lineage of the embryo proper and the adult organism.
2. Oct4 and Sox2 antagonize each other's function in germ layer development.

Multipotency

1. Multipotent stem cells play roles in differentiation during both embryogenesis and in the adult.

Oligopotency

1. Oligopotent cells are often found in adult organisms as they specialize in the replacement of two or more differentiated lineages that exhibit high turnover rates.
2. Examples of oligopotent cells include vascular stem cells and those of the ocular limbus.

Unipotency

1. Unipotent stem cells typically arise from multipotent stem cells and are often present in specific tissue compartments of the adult organism.
2. Examples of unipotent stem cells are those in the epithelium that play a direct role in the generation of new skin and hepatocytes which drive liver regeneration.

Type of Stem Cells

Embryonic Stem Cells
1. Embryonic stem cells are defined by pluripotency and indefinite replication.

Fetal Stem Cells
1. The isolation of fetal stem cells and their potential use in therapeutic applications is highly controversial because they are typically derived from aborted fetal tissue.

Amniotic Stem Cells
1. Amniotic fluid contains mesenchymal multipotent stem cells.

Adult Stem Cells
1. Adult stem cells are capable of self-renewal and can give rise to at least one differentiated cell type.

iPS Cells

1. iPS cells can be derived artificially and are pluripotent in differentiation capacity.

Cancer Stem Cells

1. Cancer stem cells are thought to drive tumor growth and have the capacity to differentiate into a variety of cell types representative of the tumor of origin.

The Potential of Stem Cells in Medicine and Medical Research

Therapeutics

1. Tissue engineering and cell replacement therapeutics are the main stem cell-based therapeutic platforms being currently studied.

Tissue Engineering

1. In tissue engineering applications doctors cultivate, grow, and generate whole tissues or components of organs for transplant to replace those damaged by injury or disease.
2. Examples of different types of human tissues and organs which may be candidates for tissue engineering include bladder and skin.
3. In autologous transplantation therapy a patient's own cells or tissue is biopsied, cultivated for expansion, and infiltrated with a three-dimensional scaffold for stability.
4. James Blundell performed the first successful blood transfusion.

Cell Therapy

1. Cell therapy can be broken down into stem cell transplantation and differentiated cell transplantation.
2. Stem cell-based gene therapy may allow for the long-term or permanent incorporation of a gene into a patient's body for therapeutic purposes.
3. Examples of disorders that could be treated with cell replacement therapy include Batten disease and diabetes.

Cell-Based Drug Screening

1. HCS has allowed for the elimination of a great deal of *in vitro* or animal screening efforts.
2. HCS readout examples include intracellular cAMP levels and GPCR activity.
3. The testing of compounds against living cells has the clear advantage of providing a more accurate therapeutic analysis.
4. Stem cells provide an unlimited source for a variety of mature, often functional cell types which can be assessed for activity and response in the presence of small molecules or other types of agonists and antagonists.

KEY TERMS

(Key terms are listed in order as described in the text.)

- **Cell culture**—the *in vitro* growth and manipulation of cells under controlled conditions outside of their natural *in vivo* environment.
- **Tissue culture**—the growth and manipulation of either tissues or cells under controlled conditions outside of their normal environment.
- **L strain**—a strain of cells derived from subcutaneous areolar and adipose tissue of a 100-day-old male C3H/An mouse and one of the most widely studied cell isolates in the late 1940s.

- **Hela cells**—A cervical carcinoma cell line taken from Henrietta Lacks without her consent.
- **Eagle's minimal essential medium (EMEM)**—a media formulation containing the minimum essential components for cellular propagation and inhibition of senescence.
- **HAT medium**—a medium developed by J.W. Littlefield that allowed for the clonal selection of hybrid cell lines.
- **Positive selection**—selection for the presence of a survival or resistance trait.
- **Thymidine kinase (TK)**—an enzyme which phosphorylates thymidine thus generating the thymidine monophosphate precursor also crucial for DNA synthesis.
- **Passages**—the replating of cells upon reaching confluence.
- **F12 medium**—a chemically defined, fully synthetic medium for the culture of mammalian cells.
- **Hybridoma technology**—the fusing of antibody-producing cells with myeloma cancer cells for the generation of hybridomas to produce antibodies.
- **Transfection**—the introduction of exogenous nucleic acid material into cells.
- **Mouse embryonic fibroblast (MEF) feeders**—a cell line used as a support system and to "feed" embryonic stem cells in order to maintain them in a pluripotent state.
- **Matrigel**—a gelatinous protein mixture secreted by Engelbreth-Holm-Swarm (EHS) mouse sarcoma cells and marketed by BD Biosciences.
- **Colonies**—clumps of stem cells growing in cell culture.
- **Collagenase**—an enzyme that breaks down peptides present in collagen.
- **Ultra-Web**—a synthetic polyamide matrix that recapitulates the basement membrane / extracellular matrix 3D geography found in the developing embryo.
- **Scaffolds**—three-dimensional matrices for cell culture.
- **Notch proteins**—a family of transmembrane proteins with repeated extracellular EGF domains and the notch (or DSL) domains involved in embryogenesis.
- **Adipose-derived stem cells**—multipotent stem cells isolated from fat tissue.
- **Lipoaspirates**—a disposable byproduct of routine and specialized cosmetic surgery and the most widely used source of hMADS.
- **Fertilization**—the process of sperm and egg fusion.
- **Cleavage**—the division of a single fertilized egg cell into two equal counterparts which subsequently divide in a symmetric fashion to produce cells known as blastomeres.
- **Symmetric cell division**—cell division which produces two cells with the same fates.
- **Blastomere**—one of the two cells produced by cleavage of a zygote.
- **Blastocyst**—cylinder-stage mammalian embryo.
- **Epiblast**—The epithelial tissue that develops from the ICM and gives rise to the ectoderm, mesoderm, and definitive endoderm during gastrulation.
- **Gastrula**—An embryo at the stage following the blastula, consisting of a hollow, two-layered sac of ectoderm and endoderm surrounding an archenteron that communicates with the exterior through the blastopore.
- **Homeobox**—a 180 base pair nucleotide sequence which codes for a key domain in transcription factors regulating embryonic development.
- **Inner cell mass (ICM)**—the mass of cells inside the blastocoele of a blastocyst -stage embryo which will eventually give rise to the embryo proper. It is also known as the pluriblast or embryoblast.

- **Positive Autoregulatory Feedback Loop**—a transcriptional regulatory cascade that reinforces a particular expression characteristic.
- **"Salt and pepper" pattern**—the random, mutually-exclusive expression pattern of Nanog and Gata6 in the inner cell mass of a developing embryo.
- **Gastrulation**—the process by which three primary germ layers are acquired.
- **Endoderm**—embryonic germ layer that yields the internal organs.
- **Ectoderm**—embryonic germ layer that yields the neural tissue including the brain.
- **Mesoderm**—embryonic germ layer that yields the muscle, bone vascular, and portion of the dermis of the back.
- **Determination**—loss of potency.
- **Basic helix-loop-helix (bHLH) transcription factors**—transcription factors containing two helices separated by a loop and a basic DNA binding domain. They often regulate aspects of embryonic development.
- **Paraxis**—a bHLH transcription factor involved in embryonic mesoderm development.
- **Somites**—masses of paraxial mesoderm on either side of the neural tube destined to form mesodermal compartments.
- **Dermatome**—a compartment of the somite which gives rise to the dermis.
- **Myotome**—a compartment of the somite which gives rise to axial muscle.
- **Sclerotome**—a compartment of the somite which gives rise to axial skeleton.
- **Germinal (germ) cells**—also known as primordial germ cells (PGCs), they are biological cells that give rise to the gametes of an organism that reproduces sexually.
- **Meiosis**—the production of gametes (sperm and egg cells), through cellular division in which each gamete retains half the original chromosome number.
- **Haploid**—a cell which contains half the chromosome number.
- **Teratocarcinoma**—malignant teratomas possessing numerous extra-embryonic and embryonic cell types.
- **Long-term self-renewal**—the ability of a cell to replicate itself over extended periods and through multiple passages while maintaining an undifferentiated state.
- **Obligatory asymmetric replication**—each parent stem cells gives rise to one daughter, undifferentiated stem cell and one differentiated cell specific for the potency of the parent stem cell.
- **Symmetric cell division**—cell division which drives the expansion of an undifferentiated stem cell population for use in an embryo's tissue and organ generation.
- **Potency**—the ability of a stem cell to differentiate into different lineages.
- **Totipotency (omnipotency)**—the ability to differentiate into both embryonic and extra-embryonic tissues, sometimes referred to as omnipotency.
- **Zygote**—the first totipotent cell formed after fertilization.
- **Pluripotency**—the ability to differentiate into all of the three primary germ layers endoderm, ectoderm, and mesoderm but cannot contribute to extra-embryonic tissues.
- **Multipotency**—to give rise to multiple, but not all, lineages of the developing embryo or adult.
- **Oligopotency**—the ability to differentiate into a very limited number of cell types, usually two.
- **Limbus**—the edge of the cornea where it joins the sclera.

- **Unipotency**—the ability to differentiate into only one cell type.
- **Precursor cell**—unipotent cell.
- **Epithelium**—the outermost layer of skin tissue in mammals (not including the dead layer of squamous cells overlying it).
- **Embryonic stem cells**—pluripotent cells present in or derived from the inner cell mass (ICM) of blastocyst -stage embryos.
- **Fetal stem cells**—primitive cells present within the organs and tissues of a developing fetus than can give rise to differentiated phenotypes.
- **Amniotic stem cells**—mesenchymal multipotent stem cells extracted from amniotic fluid.
- **Adult stem cells**—undifferentiated cells present in organs and tissues after birth that are capable of self-renewal and can give rise to at least one differentiated cell type.
- **Induced pluripotency stem (iPS) cells**—artificially derived from non-pluripotent stem cells which exhibit pluripotent differentiation capabilities.
- **Cancer stem cells**—stem cells present in various types of tumors or cancers which are capable of self-renewal and can give rise to a cell type or cell types present within that particular tumor or cancer.
- **Tissue engineering**—the *ex vivo* production of living tissue for transplant using cells, methods, and materials to create said tissue.
- **Autologous transplantation therapy**—the process of tissue engineering often involves the use of the patient's own cells or tissues as the source for newly generated tissue and/ or organ components.
- **Cell therapy**—the introduction of cells into a tissue or organ to allow for the growth or repair of that tissue or organ or for the treatment of disease.
- **Stem cell transplantation**—The transplant of progenitor or stem cells that will act as a source for mature, functional lineages.
- **Differentiated cell transplantation**—The transplant of mature, functional, differentiated lineages.
- **Cell-Based gene therapy**—the introduction of a genetic component, for example, a gene that produces a particular desired therapeutic benefit.
- **Cell-based drug screening**—the use and application of cells to identify compounds of therapeutic benefit or to assess specific drug safety and toxicity properties (ADME-Tox).
- **High-content screening (HCS)**—technologies implemented for the use of cells as combinatorial small molecule library screening platforms.
- **Patch clamp**—a laboratory technique in electrophysiology that allows the study of single or multiple ion channel activity in cells.

REVIEW QUESTIONS

(Answers to select review questions can be found at www.stemcelltextbook.com.)

1. What was the first real experimentation on stem cells and who conducted it?
2. Who won the 1912 Nobel Prize in Physiology or Medicine for his work on vascular suturing and blood vessel transplantation?

3. What was cervical cancer patient Henrietta Lacks' contribution to the study of cancer?

4. Describe the use of HAT medium in positive selection to identify hybrid cell lines.

5. What is hybridoma technology ?

6. Describe the tissue culture conditions optimized by Gail Martin and Martin Evans, which allowed for the successful isolation and propagation of embryonic stem cells.

7. Name four crucial factors required for human embryonic stem cell culture.

8. What are the two primary modes for dissociating embryonic stem cell colonies without damage to individual cells?

9. How did Sally Meiner's group recapitulate the basement membrane /extracellular matrix 3D geography found in the developing embryo?

10. Name four types of cell culture 3D scaffolds.

11. Describe the Notch signaling cascade and how it affects hematopoietic stem cell differentiation.

12. How did V. Planat-Bénard's group successfully culture hMADS in a 3D environment?

13. Describe the early morphological events in embryogenesis, from fertilization to gastrulation.

14. What transcription factors drive inner cell mass formation and ICM cell pluripotency?

15. What is the "salt and pepper" pattern of expression in the ICM?

16. What are the three germ layers of the embryo and what tissues does each contribute during development?

17. Name two basic helix-loop-helix transcription factors that regulate mesoderm development.

18. What transcription factors are involved in germ cell commitment and specification?

19. List and describe the two basic properties of stem cells.

20. List and define the various degrees of potency exhibited by stem cells.

21. How did Sharad Ramanathan characterize the dynamic transcriptional circuitry present in mESCs?

22. List and describe the major types of stem cells.

23. What is the difference between tissue engineering and cell therapy?

24. What was James Blundell's major contribution to medicine?

25. List and describe the two types of cell therapy.

26. What methods are used to introduce genes into cells in cell-based gene therapy ?

27. List three readouts to monitor cellular response in high-content screening.

28. List three detection methodologies to that provide good sensitivity and reproducibility for monitoring cellular response in high-content screening.

29. Why are iPS cells so potentially valuable for high-content screening?

THOUGHT QUESTION

Describe in detail your own ideal cell culture matrix and system for the growth and propagation of stem cells including the properties required, matrices you would use, and conditions you would implement for optimal stem cell culture.

SUGGESTED READINGS

Abercrombie M. (1961). "Ross Granville Harrison. 1870-1959 Biogr." *Mems Fell. R. Soc.* **7** 110–126; 1748–8494.

Bagnis, C. and P. Mannoni (1997). "Stem cell-based gene therapy." *The Oncologist* **2**(3): 196–202.

Burgess, R., P. Cserjesi, et al. (1995). "Paraxis: a basic helix-loop-helix protein expressed in paraxial mesoderm and developing somites." *Dev Biol* **168**(2): 296–306.

Burgess, R., A. Rawls, et al. (1996). "Requirement of the paraxis gene for somite formation and musculoskeletal patterning." *Nature* **384**(6609): 570–573.

Carrel, A. (1912). "On the permanent life of tissues outside of the organism." *J Exp Med* **15**(5): 516–528.

Carrel, A. and C. A. Lindbergh (1938). *The Culture of Organs*. P. B. Hoeber, Inc., New York.

Dahlberg, A., C. Delaney, et al. (2011). "Ex vivo expansion of human hematopoietic stem and progenitor cells." *Blood* **117**(23): 6083–6090.

Darnell, J. E., L. Levintow, et al. (1970). "Harry Eagle." *J Cell Physiol* **76**(3): 241–252.

Dromard, C., P. Bourin, et al. (2011). "Human adipose derived stroma/stem cells grow in serum-free medium as floating spheres." *Exp Cell Res* **317**(6): 770–780.

Enders, J. F., T. H. Weller, et al. (1949). "Cultivation of the lansing strain of poliomyelitis virus in cultures of various human embryonic tissues." *Science* **109**(2822): 85–87.

Evans, M. J. and M. H. Kaufman (1981). "Establishment in culture of pluripotential cells from mouse embryos." *Nature* **292**(5819): 154–156.

Ham, R. G. (1965). "Clonal growth of mammalian cells in a chemically defined, synthetic medium." *Proc Natl Acad Sci U S A* **53**: 288–293.

Kohler, G. and C. Milstein (1975). "Continuous cultures of fused cells secreting antibody of predefined specificity." *Nature* **256**(5517): 495–497.

Koka, P. S. (2008). *Stem Cells Research Compendium*. New York, Nova Science Publishers.

Kousta, E., A. Papathanasiou, et al. (2010). "Sex determination and disorders of sex development according to the revised nomenclature and classification in 46,XX individuals." *Hormones (Athens)* **9**(3): 218–131.

Kurz, H., K. Sandau, et al. (1997). "On the bifurcation of blood vessels–Wilhelm Roux's doctoral thesis (Jena 1878)—a seminal work for biophysical modelling in developmental biology." *Ann Anat* **179**(1): 33–36.

Littlefield, J. W. (1964). "Selection of hybrids from matings of fibroblasts in vitro and their presumed recombinants." *Science* **145**(3633): 709–710.

Majo, F., A. Rochat, et al. (2008). "Oligopotent stem cells are distributed throughout the mammalian ocular surface." *Nature* **456**(7219): 250–254.

Martin, G. R. (1981). "Isolation of a pluripotent cell line from early mouse embryos cultured in medium conditioned by teratocarcinoma stem cells." *Proc Natl Acad Sci U S A* **78**(12): 7634–7638.

Nur, E. K. A., I. Ahmed, et al. (2006). "Three-dimensional nanofibrillar surfaces promote self-renewal in mouse embryonic stem cells." *Stem Cells* **24**(2): 426–433.

Ohinata, Y., H. Ohta, et al. (2009). "A signaling principle for the specification of the germ cell lineage in mice." *Cell* **137**(3): 571–584.

Ota, M. S., D. A. Loebel, et al. (2004). "Twist is required for patterning the cranial nerves and maintaining the viability of mesodermal cells." *Dev Dyn* **230**(2): 216–228.

Pisani, D. F., M. Djedaini, et al. (2011). "Differentiation of human adipose-derived stem cells into "brite" (brown-in-white) adipocytes." *Front Endocrinol (Lausanne)* **2**: 87.

Ralston, A. and J. Rossant (2008). "Cdx2 acts downstream of cell polarization to cell-autonomously promote trophectoderm fate in the early mouse embryo." *Dev Biol* **313**(2): 614–629.

Rossant, J. and P. P. Tam (2009). "Blastocyst lineage formation, early embryonic asymmetries and axis patterning in the mouse." *Development* **136**(5): 701–713.

Saitou, M. and M. Yamaji (2010). "Germ cell specification in mice: signaling, transcription regulation, and epigenetic consequences." *Reproduction* **139**(6): 931–942.

Sanford, K. K., W. R. Earle, et al. (1948). "The growth in vitro of single isolated tissue cells." *J Natl Cancer Inst* **9**(3): 229–246.

Sartipy, P. and P. Bjorquist (2011) "Concise review: Human pluripotent stem cell-based models for cardiac and hepatic toxicity assessment." *Stem Cells* **29**(5): 744–448.

Skloot, R. (2010). *The immortal life of Henrietta Lacks*. Crown Publishers, New York.

Thomson, J. A., J. Itskovitz-Eldor, et al. (1998). "Embryonic stem cell lines derived from human blastocysts." *Science* **282**(5391): 1145–1147.

Thomson, M., S. J. Liu, et al. (2011). "Pluripotency factors in embryonic stem cells regulate differentiation into germ layers." *Cell* **145**(6): 875–889.

Van de Graaff, K.M. and S. I. Fox (1989–2002). *Concepts of Human Anatomy and Physiology*. William C. Brown Publishers.

Wigler, M., A. Pellicer, et al. (1978). "Biochemical transfer of single-copy eucaryotic genes using total cellular DNA as donor." *Cell* **14**(3): 725–731.

Wong, R. C., P. J. Donovan, et al. (2011) "Molecular mechanism involved in the maintenance of pluripotent stem." *J Stem Cells* **6**(4): 213–232.

Chapter 3

EMBRYONIC STEM, FETAL, AND AMNIOTIC STEM CELLS

Developmental biology is perhaps the most intensely studied field of research as it pertains to deciphering the properties of stem cells and their potential in medicine. It is because many of the molecules and pathways that drive cell fate during development can be utilized or manipulated in a laboratory environment for further discovery or translational medicine. This chapter will delve more closely into the basic properties—including potency—of embryonic stem (ES), fetal, and amniotic stem cells. Examples of each cell type will be outlined along with examples of differentiated, mature lineages that may be derived from each with a general overview of specific therapeutic or diagnostic applications.

ES CELLS

Basic Properties

Pluripotency ES cells, no matter the species origin, can be considered to be pluripotent: giving rise to all cell types of the embryo proper, but not extraembryonic tissues. The term "pluripotent" is derived from the Latin roots "plurimus," meaning "very many," and "potens," defined as "having power". As discussed briefly in Chapter 2, the pluripotent nature of ES cells enables them to contribute to all three primary germ layers—the endoderm, ectoderm, and mesoderm—and thus all organ and tissue types in the developing embryo and the adult, represented by over 220 cell types. This is in contrast to earlier state totipotent cells present in the morula, which yield embryonic and extraembryonic tissues, and in contrast to adult stem cells, which tend to be considerably restricted in their ability to differentiate into varying lineages (often multi or unipotent). It should be noted that ES cells, while having the capacity to differentiate into any cell type present in the embryo proper or adult cell type, cannot result in the generation of a fully functioning embryo and/or adult organism, as extraembryonic tissues such as the placenta are also required for support, nutrient supply, and various other functions.

SIGNALING AND TRANSCRIPTIONAL REGULATION OF PLURIPOTENCY
Since the original discovery of ES cells over 30 years ago, an intense effort has been undertaken to understand the signaling and transcriptional control of the pluripotent nature of ES

Stem Cells: A Short Course, First Edition. Rob Burgess.
© 2016 John Wiley & Sons, Inc. Published 2016 by John Wiley & Sons, Inc.

cells. Deciphering the key signaling molecules and transcriptional hierarchies underlying the pluripotent nature of ES cells will, in all likelihood, yield new avenues for manipulating potency for therapeutic purposes. This is clearly evident in the scientific advancements related to induced pluripotency (iPS) technologies discussed in detail in Chapter 5. Mouse ES cells have served as an excellent model system for deciphering the signaling events crucial to the maintenance of the pluripotent phenotype. Over the years, two key extrinsic signaling proteins, Wnt and leukemia inhibitory factor (LIF), have been defined to be necessary for pluripotency in mES cells. **Wnt proteins** were originally identified in *Drosophila melanogaster* through mutational analysis of the Wingless gene via viral integration. In mouse ES cells, Wnt proteins have been shown to act through the β-catenin pathway to promote pluripotency. Interestingly, β-catenin has been demonstrated to interact directly with the pluripotency transcription factor Oct4, in concert with TCF proteins, to drive the activation of Oct4 target genes. This would then suggest a dual role for Wnt and β-catenin signaling on both early pluripotency and later mesendoderm induction during embryonic development. By contrast, FGF signaling promotes neurogenesis but actively inhibits pluripotency in mES cells (Figure 3.1) (Sokol, 2011). LIF is a cytokine of the interleukin 6 class that acts to inhibit differentiation of mES cells. It acts through the LIF receptor via the formation of heterodimers with gp130 subsequently resulting in the activation of two independent kinase cascades, those of the JAK/STAT and MAPK pathways. The activation of these intracellular signaling events results in a repression of target genes representing differentiated cells.

It should be noted that human ES cells do not require LIF for maintenance of pluripotency, but instead require FGF2. It is speculated that human ES cells were isolated at a later time, thus resembling more murine epiblast cells and requiring different signaling effectors to maintain pluripotency.

PLURIPOTENCY MARKER EXPRESSION

A number of markers have now been characterized as defining a truly pluripotent ES cell. These include alkaline phosphatase, the cell surface antigens (stage-specific embryonic antigens, SSEA-3 and -4), and the keratin sulfate antigens, TRA-1-60 and TRA-1-81. As described throughout this book, the transcription factors Sox2, Oct4, and Nanog also play crucial roles in driving ES cell pluripotency and are classical transcriptional regulatory markers of the pluripotent phenotype (Figure 3.2) (Giritharan et al., 2011).

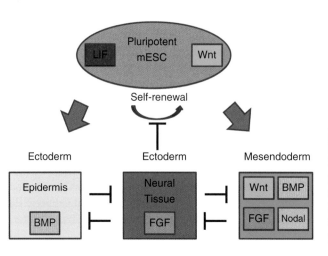

Figure 3.1 Diagrammatic illustration of pluripotency and lineage commitment signaling pathways in mouse embryonic stem cells. (Courtesy Sergei Y. Sokol and *Development* (Sokol, 2011); reprinted with permission.)

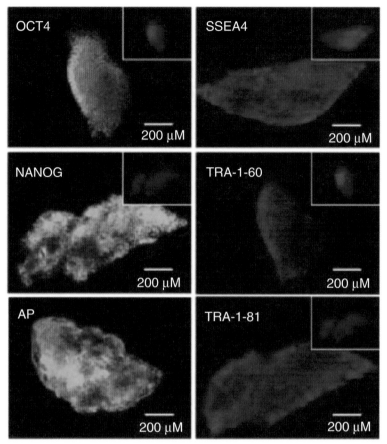

Figure 3.2 Expression of classical markers for ES cell pluripotency at the blastocyst stage of embryonic development as detected by immunofluorescence. Insets represent Hoechst staining of nuclei. (Courtesy Gnanaratnam Giritharan and *PLOS* (Giritharan et al., 2011); reprinted with permission.)

Indefinite Replicative Capacity If cultured under proper conditions, ES cells can be propagated indefinitely as demonstrated by numerous groups of researchers. This allows for the generation of basically an unlimited supply of cells from which to develop cell-based therapeutic or diagnostic platforms. These conditions have been refined and optimized over the years through rigorous characterization of the growth factors and matrices needed to maintain cell viability, mitotic activity, and pluripotency. Specific defined serum- and feeder-free media that are animal component-free have now been developed that allow for ES cell manipulation *in vitro*. Protocols have now been developed for the highly efficient directed differentiation of ES cells into desired mature cell types. This has allowed for the possibility that ES cells may be used in regenerative medicine applications. The capacity for ES cells to divide and produce a multitude of virtually any cell type provides avenues to treat many medical disorders including Parkinson's disease, spinal cord injuries, diabetes, immune disorders, and others (Figure 3.3).

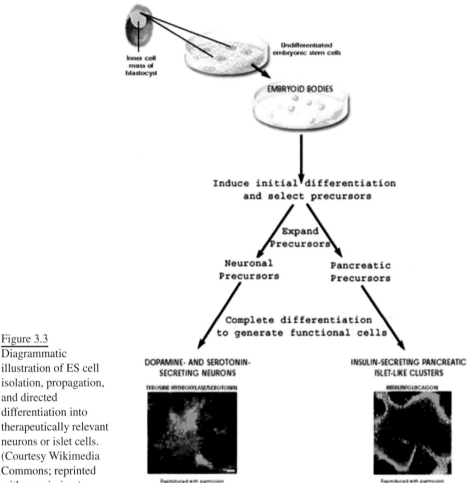

Figure 3.3
Diagrammatic
illustration of ES cell
isolation, propagation,
and directed
differentiation into
therapeutically relevant
neurons or islet cells.
(Courtesy Wikimedia
Commons; reprinted
with permission.)

Signaling and Transcriptional Control of ES Cell Replication Self-renewal is of
crucial importance for cell populations to expand and is a key fundamental requirement for
the classification of a cell as embryonic and stem in nature. True ES cells have been shown
to divide indefinitely, with no observable damage or loss in pluripotent capabilities if cultured under proper conditions. It is the medium, the microenvironment, and even the
mechanical and enzymatic manipulation of stem cells that determine whether they will
remain committed to the cell cycle and divide or exit the cell cycle and undergo differentiation. In mouse ES cells, perhaps the most well-known and studied regulatory cascade
promoting self-renewal is that involving **LIF** activation of STAT3 (Figure 3.4). LIF was
originally demonstrated to inhibit the growth of leukemic cells in tissue culture. Ironically,
it has the opposite effect on murine ES cells, driving self-replicative properties and indefinite expansion capabilities. It has since been confirmed as necessary for the expansion of
most murine ES cell lines derived to date.

There are two primary disadvantages regarding human ES cell therapeutic applications.
First, as mentioned in Chapter 1, it should be noted that there are ethical concerns related
to the use of human ES cells in basic research and diagnosing or treating disease. The

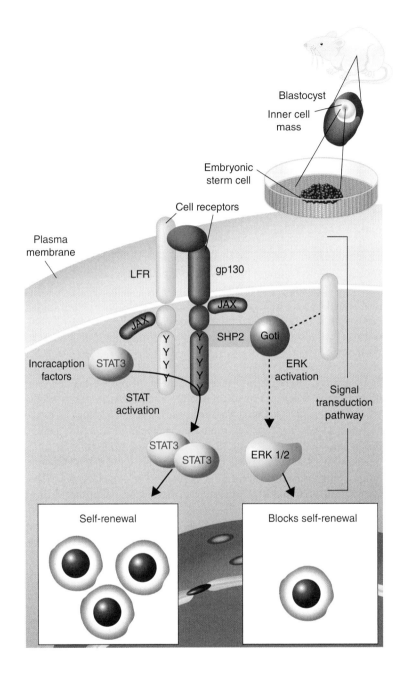

Figure 3.4 Leukemia inhibitory factor (LIF) activation of STAT3 signaling for promotion of self-renewal in mouse embryonic stem cells. (Courtesy National Institutes of Health; reprinted with permission.)

destruction of human embryos for the derivation of ES cell lines has been a major point of contention worldwide for the pursuit of this research discipline. Fortunately the use of adult stem cells or iPS cells (discussed in more detail in Chapters 4 and 5), may make the clinical or diagnostic use of human ES cells a moot point. Second, there are issues related to the immune system compatibility of ES cells and their progeny transplanted **allogeneically** (transplantation from one human to another, genetically nonidentical human). Immunorejection is a real concern for allogeneic transplants of any sort, and is of particular concern in cell transplantation, given its widespread applications. Again, induced pluripotency may come to the rescue, as cells isolated directly from the patient may be utilized to create new therapeutically relevant lineages.

In addition to human ES cells showing promise in therapeutic or diagnostic applications, this cell platform may yield a new model system for the study of human genetic disorders. Specific genetic manipulation of human ES cells or the isolation of cells from embryos exhibiting disease as assessed through prenatal genetic diagnosis has now allowed for the study of numerous disorders which currently have no ideal model system, including cystic fibrosis and fragile X syndrome. ES cells, especially those of human origin, will be discussed in detail throughout this book. For an overview of the history of ES cell discovery, see Chapter 1. For a more detailed description of ES cell discovery, properties, and therapeutic potential continue reading this chapter.

Examples of ES Cells

ES cells have now been derived from a variety of species. Each line is unique. Factors that may play a role in efficiency of derivation, clonal isolation, and pluripotent capabilities include pregnancy status, developmental point of the embryo, cell culture conditions, and general technique. Below is a brief summary of some ES cell lines isolated from higher profile species which have significantly impacted stem cell research and regenerative medicine initiatives along with an overview of the key factors involved in promoting both cellular self-renewal and pluripotency.

Mouse ES Cells The discovery, isolation, and characterization of mouse ES cells has had a profound impact on virtually every aspect of stem cell research. In addition, the existence of mouse ES cells and the corresponding ability to manipulate them in tissue culture and *in vivo* has allowed for the discovery and characterization of many genetic and biochemical cascades involved in embryonic development and adult cellular function. It is widely accepted that the presence of truly pluripotent stem cells in the inner cell mass (ICM) of developing embryos is a mere transient event, and these cells rapidly lose their capacity to differentiate into any cell of the embryo proper. Thus, a great deal of effort was expended to isolate cells from this developmental window of pluripotency and harness their abilities as a tool to study embryogenesis *in vitro* and *in vivo*. Much effort was directed at defining techniques and cell culture conditions that would accomplish just such a task. As mentioned in Chapter 1, murine ES cells were first isolated in 1981 by Gail Martin's group at the University of California, San Francisco, and the Evans/ Kaufman team at the University of Cambridge. These cells were shown to have indefinite replicative capacity and be pluripotent in nature through the formation of teratocarcinomas in mice. Unlike embryonal carcinoma (EC) cells described below, these cells were demonstrated to have normal karyotypes and could contribute to virtually any tissue in mouse chimeras, even colonizing the germ line. These differences are most likely due to the cell's different origins with the restriction of EC cell potentiality and karyotype maintenance thought to be due at least partly to the selective pressures of the teratocarcinoma environment. Since then, mouse ES cells have revolutionized the study of not only developmental biology but also gene function through gene targeting studies discussed in Chapter 7. A variety of mouse ES cell lines have now been derived and characterized from numerous strains, with the 129 strain remaining as the most popularly studied to date. Most of these strains have been isolated from the ICM of 3.5 days post coitum (dpc) mouse blastocysts although some strains have been successfully derived from earlier morula-stage embryos (Figure 3.5).

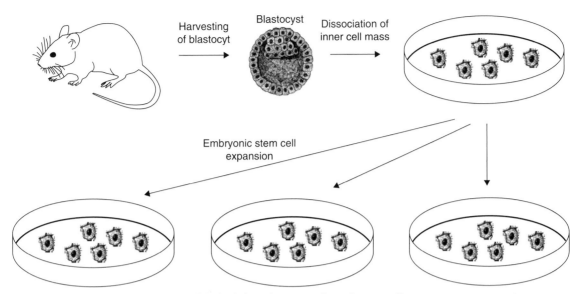

Figure 3.5 Diagrammatic illustration of the isolation of mouse embryonic stem cells.

   It should be noted that indeed genetic background plays a major role in ES cell derivation success and long-term propagation in tissue culture. Derivation of ES cells from the C57Bl/6 strain, for example, is far less efficient in comparison to that of 129SV, yet the 129 strain exhibits the disadvantage of poor breeding efficiencies and low progeny numbers, limiting the implementation of studies such as gene targeting and functional analysis. Some alterations of traditional mouse ES cell derivation methodologies have also allowed for the isolation of ES cells from nonpermissive strains such as CBA.

Murine ES Cell Culture Conditions
What are the cell culture conditions that allow for successful mouse ES cell derivation? Early studies adapted the use and application of feeder cells to support both mouse EC and ES cells. In addition, medium that has been "conditioned," that is, prepared in the presene of co-culture with various cell types, was demonstrated to maintain certain mES cell lines in an undifferentiated state. This was even in the absence of the crucial supplement LIF. LIF interaction with gp130 receptors results in an activation of the Janus-associated tyrosine kinases (JAK) to drive STAT3 activation and subsequent stem cell self-renewal (Figure 3.4). This activation must be balanced with an inhibition of the ERK/MAPK cascade which inhibits self-renewal. It should also be noted that in the absence of serum, LIF alone is not sufficient to drive mouse ES cell self-renewal and it must be combined with **bone morphogenetic protein (BMP)**. BMPs induce the expression of inhibitor of differentiation (Id) by activating the Smad pathway. BMPs may also act through the inhibition of MAPK independent of the Smads to drive mES cell self-renewal, although this alternative pathway may not be essential for this process. In addition, researchers have recently identified several key small molecules that have allowed for the derivation of mouse ES cells from all strains tested including the valuable research strains C57BL/6, DBA/1lacJ, and BalbC. These factors include inhibitors of the GSK3 and FGF/MAPK pathways. Referred to as two-inhibitor (2i) cell culture system, these small molecules have also allowed for derivation of rat ES cells (see below).

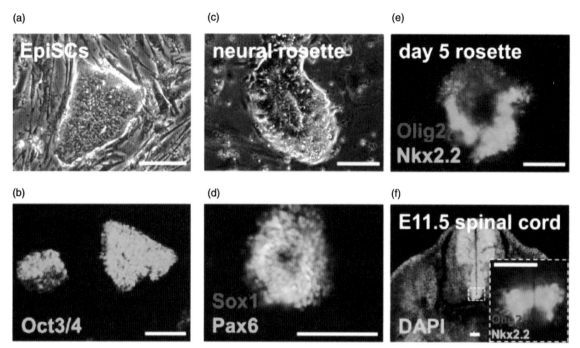

Figure 3.6 Differentiation of mouse EpiSCs into region-specific neuroepithelial cells.
(a) Undifferentiated, (b) undifferentiated demonstrating Oct3/4 expression, (c) neural rosettes,
(d) neural rosettes expression the neural markers Sox1 and Pax6, (e) neural rosettes expressing
spinal cord region-specific markers Olig2 and Nkx2.2, and (f) reference point of spinal cord-
specific expression of Olig2 and Nkx2.2. (Images courtesy Fadi J. Najm and *Nature Methods* (Najm
et al., 2011); reprinted with permission.)

A unique mouse cell line, **epiblast stem cells (EpiSCs)**, warrants mention here, yet
given their similarity to ES cells they do not justify a separate categorization. EpiSCs are
cell lines isolated from later embryonic timepoints than that of more traditional mouse
ES cells. EpiSCs are derived from E5.5 to E6.5 post-implantation mouse embryos and
thus are strikingly different than their ES counterparts. For example, LIF and BMP do
not allow for their derivation, but, in a manner similar to that for human ES cells, both
the activin/nodal and FGF signaling pathways are critical in this respect. In addition,
genome-wide profiling has revealed that EpiSCs are transcriptionally different than ES
cells and resemble the genomic profile of epiblasts. EpiSCs, however, do exhibit both the
capacity to self-renew and differentiate into a variety of lineages (Figure 3.6) (Najm et
al., 2011).

Finally, the morphological and molecular events that transgress during mouse embryonic development differ considerably from that of other rodents and higher organisms
including humans (Figure 3.7) (Thomson and Marshall, 1998), These differences may
influence the properties and differentiation capacity of ES cells depending upon the species
studied. As such, a great deal of effort has been directed at the isolation and propagation of
ES cells from other species for cross-species comparison and in order to more accurately
mimic cellular properties in humans. The sections below illustrate successes and challenges in this respect for both rodents and primates.

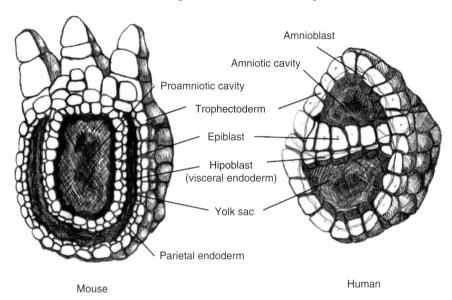

Figure 3.7 Diagrammatic comparison of human and mouse embryos at the gastrulation stage. (Artwork courtesy Connie Zhao.)

Rat ES Cells The next most logical research animal model for which ES cells would have a major impact is the rat. A great deal of effort has been expended attempting to derive rat ES cells for a number of reasons. First, they are much more physiologically similar to humans than mice. In addition, their organ systems are larger, allowing for easier dissection and study. Examples of human diseases preferentially studied in rats as opposed to mice are hypertension and neurological disorders. From a neurological perspective, rats provide much more reliable and reproducible behavioral data than mice. To date, most rat models have been developed seeking specific traits from particular strains with limited genetic modification possible, typically accomplished by using the supermutagen *N*-ethyl-*N*-nitrosurea (ENU) to introduce random mutations. Thus, the derivation of true, pluripotent rat ES cells would be of enormous value to the study of embryology, developmental biology, gene function, and human disease. Classical procedures for mouse ES cell derivation have been attempted with rat embryos but have been unsuccessful. Rat embryonic development differs from that of mice, which may be why difficulties in rat ES cell derivation have been encountered. For example, mouse and rat epiblast cells have different differentiation potential with respect to parietal endoderm. In addition, it has been speculated that different extrinsic signaling stimuli may be responsible for the pluripotency of cells in the rat embryonic inner cell mass. In this respect, a breakthrough in rat ES cell derivation efforts was made in 2009 by researchers at the University of California, San Francisco, studying various kinase signaling pathways in rat cells. Case Study 3.1 summarizes these findings and illustrates the derivation and discovery of the world's first true rat ES cells.

Case Study 3.1: Germline competent embryonic stem cells derived from rat blastocysts

Ping Li et al.

A research team led by Qi-Long Ying at the Eli and Edythe Broad Center for Regenerative Medicine and Stem Cell Research, University of Southern California, demonstrated that manipulation of specific kinase-dependent signaling pathways using small molecule inhibitors referred to as **3i** allowed for the derivation, propagation, and genetic manipulation of rat ES cells. Specifically, the researchers employed small molecules to inhibit the function of GSK3, MEK, and FGF receptor tyrosine kinases in cells isolated from the ICM of rat blastocysts. The resulting cell lines expressed classical pluripotency markers, replicated indefinitely and had the capacity to differentiate into lineages representing the three primary germ layers. The definitive confirmation and true ES cell classification of these cell lines were accomplished when it was demonstrated that the cells contributed to high rates of chimerism in rats and transmitted genetic information through the germ line (Figure 3.8) (Li et al., 2008).

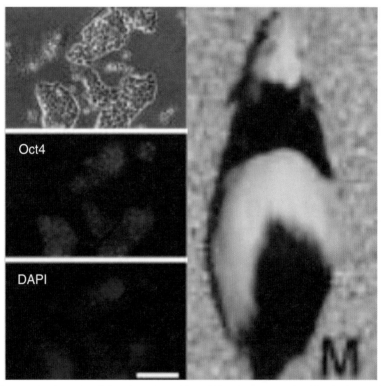

Figure 3.8 Derivation of rat ES cells and resulting chimeric rat. Rat ES cell colonies (left) as observed in brightfield microscopy (upper panel) and stained for the pluripotency marker Oct4 (middle panel) as well as nuclei (lower panel). Chimeras (right) generated from derived rat ES cells contributing to the agouti coat color on an albino background. (Figures courtesy Qi-Long Ying and *Cell* (Li et al., 2008); reprinted with permission.)

Nonhuman Primate ES Cells Although the existence of mouse and rat ES cells has allowed for a wealth of information to be acquired in embryology and stem-cell-based therapeutics, cells from these species differ fundamentally from pluripotent cells present in the ICM of human or nonhuman primates. As discussed above, this is due primarily to the differences in embryonic development between the species. In this respect, various morphological and molecular differences include the chronology of gene expression, the structure of extraembryonic membranes and the placenta, and the morphological transition of early embryos to either an egg cylinder or embryonic disc. In addition, it is clear that human and nonhuman primate embryonic development are extremely similar from a morphological, biochemical, and molecular perspective. These similarities and the obvious restrictions on studying human embryonic development have driven the need for a nonhuman primate ES cell platform.

There are over 200 primate species in existence worldwide yet only a very small number of these are actually used in biomedical research, with the rhesus (*Macaca mulatta*) monkey perhaps leading the way in this respect (Figure 3.9).

Rhesus macaques tend to be widely used in biomedical research due to their similarities with humans in relation to both reproduction and development. Some properties of rhesus monkey reproduction, however, make their study a challenge and include

- Release of a single oocyte during ovulation
- Relatively long gestation period of 5 months
- Sexual maturity age of 4–5 years

Therefore, other species more distantly related to the rhesus have been utilized for studying primate embryology and reproduction. The common marmoset (*Callithrix jacchus*) is another species of monkey often utilized as a model system for these purposes (Figure 3.10). Some characteristics of marmosets that make them suited for embryological research include

- Relatively small size (300–400 grams)
- Routinely have twins or even triplets
- Short time to sexual maturity (~18 months)

Figure 3.9 Photograph of an adult female rhesus (*M. mulatta*) monkey with infant. (Photograph courtesy Wikimedia Commons; reprinted with permission.)

Figure 3.10 Photograph of a common marmoset (*C. jacchus*). (Photograph courtesy Wikimedia Commons; reprinted with permission.)

The reproductive cycle of the marmoset may also be effectively manipulated via the synchronized administration of prostaglandins to enable the collection of developmentally similar embryos. This, coupled with the properties described above, has allowed for significant advances in the understanding of basic primate reproductive mechanisms including embryonic implantation and maternal recognition of pregnancy. These studies and findings correlate well with what is now known to be the reproductive cycle in humans and would not have been possible via the study of mouse embryology. Despite the fact that advances have been made in the study of primate reproduction, development, and embryology using primates as a model system, the events that occur at the cellular level regarding competency, commitment, and differentiation cannot be studied using these systems. Thus, a great deal of effort has been directed at isolating ES cells from various primate species. Broken down by specific species, marmoset ES cells may give insight into the optimization of embryo transfer technologies, while rhesus ES cells might provide valuable information regarding lineage determination and differentiation in humans given their close evolutionary relatedness. What are the basic criteria that would define a primate ES cell? These include:

1. Derivation from the preimplantation or peri-implantation embryo;
2. Prolonged, undifferentiated proliferation; and
3. Stable developmental potential after prolonged culture to form derivatives of all three embryonic germ layers (Ennis et al., 2008).

Long-term maintenance of a normal karyotype would also be of obvious value but is not an absolute prerequisite for classification as a primate ES cell. In 1995, James Thomson's group at the University of Wisconsin–Madison modified the classical approach to the derivation of mouse ES cells to derive ES cells from nonhuman primates, specifically rhesus and marmoset. The procedure for their derivation and propagation was similar to that for mouse ES cells and utilized inactivated mouse embryonic fibroblasts as a feeder layer to support growth, yet when cultured without feeder cells, the presence of LIF in the cell culture media had opposite effects, inhibiting differentiation of mouse ES cells while promoting differentiation of nonhuman primate ES cells. It should be noted that there was no observable difference in cloning efficiency, cellular division rates, or differentiation in the presence of both feeder cells and LIF. Clonally isolated rhesus and marmoset ES cells were confirmed to express all the classical markers for pluripotency and cause tumor formation in SCID mice (Figure 3.11) (Thomson et al., 1995).

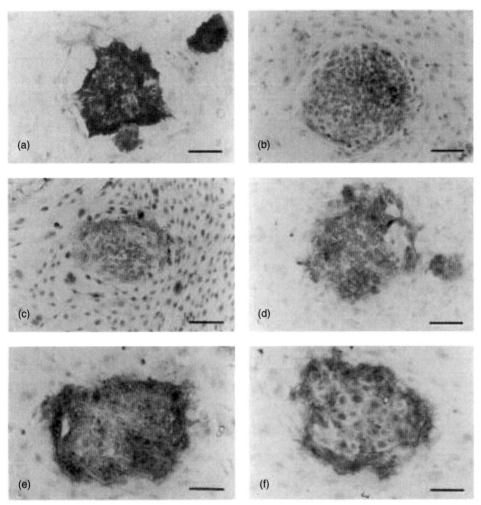

Figure 3.11 Expression of cell surface markers by undifferentiated nonhuman primate (Rhesus) ES cells. (a) Alkaline phosphatase, (b) SSEA-1, (c) SSEA-3, (d) SSEA-4, (e) TRA-1-60, (f) TRA-1-81 (bars = 100 μm). (*Source:* Thomson et al., 1995. Reproduced with permission from National Academy of Sciences, USA.)

Two key differences were noted with respect to nonhuman primate versus murine ES cells. First, these cells were demonstrated to have the ability to differentiate into trophoblast cells as well as the derivatives of the three primary germ layers endoderm, mesoderm, and ectoderm. This suggests that perhaps the cells were isolated from a different developmental point, perhaps earlier, allowing for a semitotipotent phenotype. Second, in the absence of feeder cells but in the presence of LIF, nonhuman primate ES cells exhibited significant differentiation. This is in contrast to murine ES cells and suggests additional as yet unidentified extrinsic factors secreted by feeder cells required to maintain self-replicative and pluripotent capacities. Second, it should be noted that primate ES cells are quite different in morphology compared to that of the mouse and are more susceptible to differentiation when exposed to mechanical dissociation procedures.

Human ES Cells

hES Cell Growth Factor Signaling

As described in Chapter 1, the discovery of hES cells by James Thomson et al., in 1998 has significantly impacted ES cell research and opened countless doors for potential therapeutic applications, as well as drug and drug target screening validation platforms. It is the hES cells' ability to expand almost indefinitely, coupled with their ability to differentiate into a vast array of cell types, that has driven intense study in this area of stem cell research. Yet a period of almost 20 years passed between the first successful derivation of mouse ES cells and that of human cells. It became apparent to embryologists early on that differences in the morphological, molecular, and biochemical makeup of developing mouse and human embryos corresponded to cellular differences as well. Thus, culture media formulations that had been successful with respect to mouse ES cell derivation failed when applied to cells isolated from human embryos. Like nonhuman primate ES cells, human ES cells are unlike those derived from mice in that unique extrinsic stimuli are required to maintain their ability to self-replicate and differentiate into lineages representing the three primary germ layers of the embryo. Specifically, LIF (and related cytokines) is not required for maintaining the undifferentiated state of human ES cells and in fact has been noted by several research groups to actually induce their exit from the cell cycle and differentiation. In fact it has now been confirmed that hES cells do not express or only express at low levels individual components of the LIF signaling pathway including the gp130 receptor and the signaling kinases JAK1 and 2. Consistent with this finding, the need for a fibroblast feeder layer did become apparent in hES cell studies early on. However, original human ES cell lines isolated by James Thomson's group at the University of Wisconsin–Madison revealed that the feeder cells did not act through similar pathways, as the extrinsic stimuli needed to maintain pluripotency are different between mouse and human ES cells. Other factors, such as bone morphogenetic proteins (BMPs), which have been demonstrated to drive self-renewal in mouse ES cells, actually promoted differentiation in human ES cells. What are the pathways that promote the proliferation and pluripotent capacities of hES cells? As outlined in Chapter 2 and Figure 2.8, FGF and TGFβ/Activin/nodal signaling pathways are crucial for promoting both the proliferation and pluripotent properties of human ES cells. It is basic FGF that allows for the clonal isolation and growth of hES cells on fibroblast feeders utilizing commercially designed serum replacement media. At high levels of bFGF, it is also possible to culture the cells in a feeder-free environment, again in the presence of serum replacement medium. Basic FGF is thought to suppress BMP signaling in parallel with Noggin or other BMP inhibitors, thus drastically reducing levels of background differentiation observed in tissue culture. There is some evidence to support that bFGF signals the activation of TGFβ ligands, which drives self-renewal. Interestingly, human ES cells produce FGFs, but not at levels sufficient to allow for cell culture at low density. Activin and TGFβ have also been demonstrated to drive proliferation if utilized in the presence of low-to-moderate FGF concentrations. Thus, there appears to be a reciprocal relationship between FGF and TGFβ/Activin signaling on hES cell proliferative capacities. Other growth factors have also been demonstrated to have a positive effect on the growth of hES cells in culture, including the Wnt family, platelet-derived growth factor (PDGF), pleiotrophin, and insulin-like growth factor 1 (IGF1); yet how these factors and their corresponding cascades interact with those of FGF and TGFβ remains to be determined. It is also clear that other as yet unidentified growth factors will most likely be discovered that have positive effects on hES cell growth.

hES CELL MORPHOLOGY

The human ES cells isolated by Thomson's group in 1998 and others hence display a striking phenotypic resemblance to that of nonhuman primate ES cells when maintained in an undifferentiated state. In addition, they express the same markers for pluripotency (Figure 3.12) (Thomson et al., 1998). Other research groups have since clonally isolated numerous lines of hES cells with over 300 lines isolated as of January, 2006 in over 20 countries (Figure 3.13) (Guhr et al., 2006).

Yet only half of these hES cell lines have been published in scientific journals. Thus it remains to be determined if the remaining 50% are true hES cells and what properties they may exhibit. It should be noted that the vast majority of research publications involve only 13 hES cell lines. The majority of these cells were provided on behalf of the WiCell Research Institute (Table 3.1).

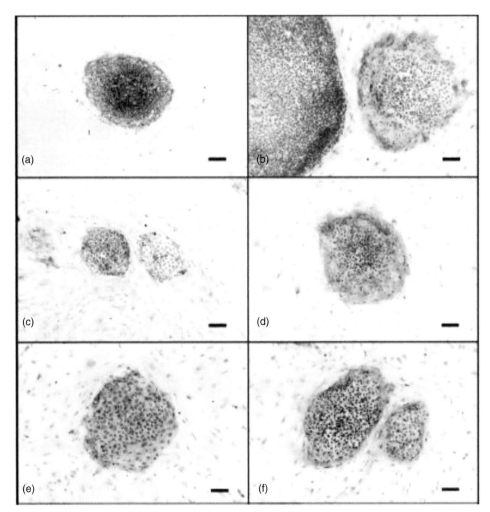

Figure 3.12 Expression of markers by the human embryonic stem cell line H9. (a) Alkaline phosphatase, (b) SSEA-1, (c) SSEA-3, (d) SSEA-4, (e) Tra-1-60, (f) Tra-1-81. (Courtesy James Thomson and *Science* (Thomson et al., 1998); reprinted with permission.)

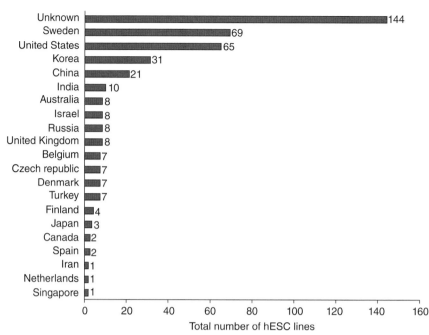

Figure 3.13 Graph of the total number of hES cell lines isolated as of January, 2006. (Graph courtesy Anke Guhr and *Stem Cells* (Guhr et al., 2006); reprinted with permission.)

TABLE 3.1 Top 13 hES cell lines as ranked by numbers of publications cited as of January 2006.

hES Cell line	Number of publication cited	Provider
H9	110	WiCell Research Institute
H1	93	WiCell Research Institute
H7	37	WiCell Research Institute
BG01	26	BresaGen, Inc.
HES-3	23	ES Cell International
HES-2	18	ES Cell International
HSF-6	16	University of California, San Francisco
BG02	16	BresaGen, Inc.
HES-1	13	ES Cell International
Miz	12	MizMedi Hospital, Seoul National University
H13	11	WiCell Research Institute
H14	11	WiCell Research Institute
HES-4	11	ES Cell International

Source: Adapted from Baba et al., 2012.

Figure 3.14 is a comparison of the morphology of some of the more popular human ES cell lines studied to date and cultured under various conditions. The study and application of hES cells for both cell-based therapeutic and diagnostic applications will be discussed in detail throughout many sections of this book with a particular emphasis on these concepts in Chapter 8 (Allegrucci and Young, 2007).

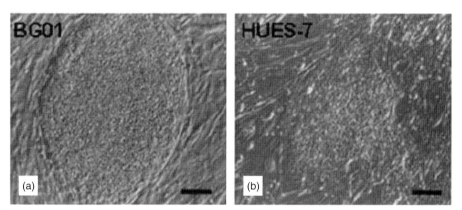

Figure 3.14 Brightfield images of BG01 and HUES-7 human ES cells grown under standardized conditions in the presence of feeder cells. Note the difference in morphology between the two lines. Magnification 160×. (Images courtesy C. Allegrucci and L.E. Young, *Human Reproduction* (Allegrucci and Young, 2007); reprinted with permission.)

It is clear that the study of human ES cells will continue to play a major role in elucidating the key molecular and biochemical events associated with growth and differentiation properties. The derivation of new lines and the continuous optimization of *in vitro* culture conditions for both maintaining pluripotency as well as directing differentiation will have considerable implication on the application of hES cells in therapeutic and diagnostic applications for many years to come. This will be discussed in more detail in Chapters 7 and 8.

EC CELLS

Teratocarcinomas can be defined as malignant germ cell tumors that consist of both an undifferentiated EC component and a differentiated component that may or may not include representative cell types from all three primary germ layers. Teratocarcinomas in humans are considerably rare, but examples of these tumors have now been identified in numerous other species. As the 129 mouse strain exhibits a high rate of testicular teratomas, it has provided numerous cell lines for further study. EC lines isolated from the mouse 129 strain have been shown to be successfully serially transplanted between mice and can contribute to different tumorigenic phenotypes. More importantly, in 1964 it was demonstrated that a single EC cell has the capacity to differentiate into representative cells of the three primary germ layers and also self-renew indefinitely. This is considered the first definitive demonstration of the existence of a cancer stem cell (Ohinata et al., 2009). Stable propagation of mouse EC cells lines was successfully accomplished in the early 1970s and laid the groundwork for the study of differentiation, which would have been impossible utilizing intact embryos. Interestingly, mouse EC cells express many of the same markers present on cells that reside within the inner cell mass ICM. EC cells also have striking resemblance to ES cells in both morphology and differentiation potential, yet in general, most lines characterized to date have limited or restricted developmental potential and do not contribute well to the chimerism of mice when injected into mouse blastocysts.

Focus Box 3.1: Brigid Hogan and the discovery of human EC cells

Brigid Hogan is a British developmental biologist and is one of the pioneers of embryonic cell study. In 1977, Dr. Hogan discovered human EC cells, a finding which laid the groundwork for the delineation of both intracellular cascades involved in potency and proliferation and the optimization of cell culture techniques for cultured stem cells. Dr. Hogan is currently George Barth Geller Professor of Research in Molecular Biology and Chair of the Department of Cell Biology at Duke University as well as the director of the Duke Stem Cell Program. (Photo courtesy Wikipedia; reprinted with permission.)

Human EC cells were first isolated in 1977 and revealed surprising differences from their mouse counterparts (Tam and Loebel, 2009). Many of these differences were noted in the expression of pluripotent markers such as the stage-specific embryonic antigens. An example of this is the abundance of SSEA-1 on the surface of mouse EC cells compared to its absence on the surface of human EC cells. In addition, human EC cells exhibit a relative inability to differentiate into a variety of mature somatic cell types. This is most likely due to the fact that these cells are highly **aneuploid**, which is defined as a cell having an abnormal number of chromosomes. Yet the cells still exhibit a fair degree of morphological similarities to mouse EC cells (Figure 3.15).

Perhaps the most widely studied mouse EC cell line to date is the **P19** clonal isolate (Figure 3.16). P19 EC cells were first isolated in 1982 and have the inherent capacity to differentiate into lineages representing the three primary germ layers. In addition, the P19 line is the most widely studied EC line with respect to the characterization of cardiogenesis, skeletal muscle differentiation, and neurogenesis pathways. Exposure of P19 cells to dimethyl sulfoxide (DMSO) induces the development of cardiac and skeletal muscle while exposure of the cells to retinoic acid induces neuronal and glial cell formation.

EC cells have played a crucial role in defining early events related to lineage commitment and have also allowed for the delineation of pathways related to cardiac and skeletal muscle development as well as neurogenesis. In addition they have paved the way for other more recently discovered cell types such as ES and iPS cells to yield even more insight into lineage commitment and developmental pathways.

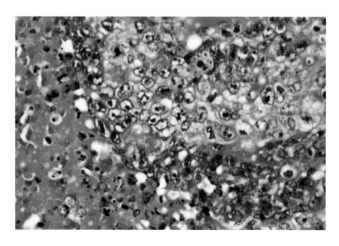

Figure 3.15
Photomicrograph of EC cells. H&E stain. (Photograph courtesy Wikimedia Commons; reprinted with permission.)

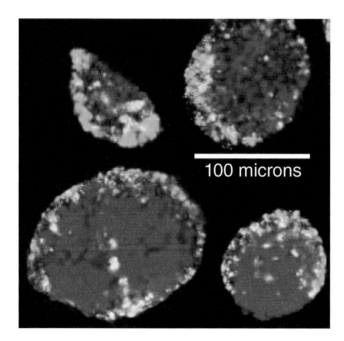

Figure 3.16 Cultured P19 EC cells. Live cells were stained with either DiI (red) or DiO (green). The red cells were genetically altered and express higher levels of E-cadherin than the green cells. The image was captured by scanning confocal microscopy. (Image courtesy J. W. Schmidt and Wikimedia Commons; reprinted with permission.)

EMBRYONAL GERM CELLS

Embryonal (embryonic) germ (EG) cells are defined as cells that have the capacity to give rise to gametes that also take on properties similar to ES cells. They are typically isolated from the primordial germ cells of the gonadal ridge of developing embryos and are highly related to EC cells, yet they have a more restricted potentiality than those of EC origin, being isolated from a narrow developmental window. This restriction may be due to the fact that they have been isolated from different developmental points and tissues. However, like other embryonically derived stem cells, EG cells have the capacity to differentiate into lineages representing all three primary germ layers.

EG Cell Growth Factor Signaling

What signaling molecules play a role in specifying primordial germ cell (PGC) specification *in vivo* and EG cells *in vitro*? Mitinori Saitou's group at the Laboratory for Mammalian Germ Cell Biology, RIKEN Kobe Institute in Kobe, Japan performed an elegant series of experiments to outline the underlying signaling mechanisms behind germ cell specification and the transition from the epiblast state to PGC and correspondingly EG cells. Utilizing extensive *in vivo* genetic analysis combined with serum-free *in vitro* cell culture techniques, the group outlined a signaling activity paradigm for the specification of mammalian germ cell fate. Specifically, an *in vitro* culture protocol using BMP4, BMP8b, and WNT3A allowed for the induction of primordial germ cells from epiblast cells. This resulted in serial expression of specific genes, combined with epigenetic remodeling to induce PGC formation (Figure 3.17) (Tam and Loebel, 2009).

It is the combination of both positive and negative signals that drives the transition from epiblast to PGC/EG cell. This is accomplished through epiblastic response to both BMP

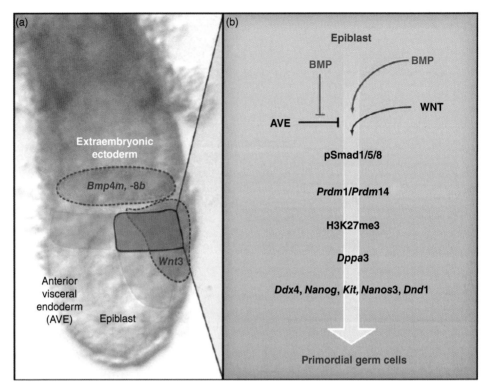

Figure 3.17 Induction of mouse primordial germ cells. A combination of BMP and Wnt signaling induces the serial expression of key genes involved in PGC specification. See text for details. (Figure courtesy Patrick P.L. Tam, David A.F. Loebel and *Cell* (Tam and Loebel, 2009); reprinted with permission.)

and Wnt signaling, thus driving serial expression of germ cell specification-related genes. These studies have laid a solid foundation of knowledge pertaining to signaling molecules that may be used in an *in vitro* format to drive lineage commitment, such as that of pancreatic islet cells, from EG cells for potential therapeutic applications.

COMPARING EMBRYONICALLY DERIVED CELLS

A comparison of embryonically derived pluripotent cells derived from various species including those discussed above reveal differences in morphology and differentiation characteristics under various conditions of growth. Table 3.2 outlines some of these differences.

Cell surface markers also differ to a certain extent for pluripotent stem cell lines isolated from various species. Each cell line expresses characteristic glycolipids as well as high molecular weight glycoproteins and these markers act as signatures of pluripotency. The enzyme alkaline phosphatase is also a well-known marker of pluripotency expressed by numerous EC and ES cell lines. Table 3.3 is a global comparison of pluripotent marker expression for some of the more widely studied pluripotent stem cell lines.

TABLE 3.2 Comparison of pluripotent cell line characteristics.

Cell line	Characteristics
Mouse ES	• High nuclear/cytoplasmic ratio • Prominent nucleoli • Compact colonies • Differentiate rapidly if grown at high density
Rat ES	• Highly refractory in appearance • Many outgrowths
Rhesus ES	• Flat colonies • Individual cells easily visible early • Compact colonies after 1 week of culture • Differentiate rapidly if grown at high density
Marmoset ES	• Intermediate morphology to mouse ES and human EC cells • Flat colonies
Human EC	• Piled up, irregular colonies • Undifferentiated growth promoted at higher densities
Human ES	• High nuclear/cytoplasmic ratio • Prominent nucleoli • Highly refractory in appearance • Flat colonies • Sensitive to mechanical dissociation

TABLE 3.3 Comparison of pluripotent stem cell marker expression.

Marker	Human EC (NT2/D1)	Marmoset ES (CJ11)	Rhesus ES (R278.5)	Mouse ES (D3)	Human ES (H1)
SSEA-1	−	−	−	+	−
SSEA-3	+	+	+	−	+
SSEA-4	+	+	+	−	+
TRA-1-60	+	+	+	−	+
TRA-1-81	+	+	+	−	+
Alkaline phosphatase	+	+	+	+	+

Source: Adapted from *Current Topics in Developmental Biology* (Thomson and Marshall, 1998.)

FETAL STEM CELLS

Fetal stem cells represent a unique and intriguing subset of cells that have the capacity to differentiate into a wide variety of mature lineages. Fetal stem cells not only provide both unique insight into mechanisms underlying embryonic development but may also act as a source for valuable cell-based therapeutics. The following sections outline the basic properties and describe specific examples of various fetal stem cell types.

Basic Properties

Fetal stem cells encompass a broad range of cell types isolated from a variety of tissues either pre- or postnatally. These may include those isolated from the amniotic fluid, amniotic membrane, or placenta. In addition fetal stem cells may be isolated from **Wharton's jelly**, which is defined as a gelatinous substance within the umbilical cord, also present in the vitreous humor of the eyeball, largely made up of mucopolysaccharides (hyaluronic acid and chondroitin sulfate). It also contains both fibroblasts and macrophages. It is derived from extraembryonic mesoderm. Fetal stem cells from any of these sources exhibit "stemness," and therefore have the capacity to restricted multipotency and ultimately a progenitor phenotype, thus becoming unipotent and acting as a source for only one mature differentiated cell type (Figure 3.18) (Pappa and Anagnou, 2009). As there is always an exception to every rule, truly pluripotent fetal stem cells of amniotic origin were recently isolated and are discussed below.

Fetal stem cells exhibit many of the same properties regarded to ES cells including specific markers and the ability to self-renew. In most cases, however, fetal stem cells have a considerably restricted ability to differentiate into various cells types, that is, they are multipotent, in contrast to pluripotent ES cells.

Amniotic Fluid Stem Cells **Amniotic fluid (AF) stem cells**, typically isolated from amniotic fluid, are heterogeneous in nature and are derived from the three primary germ layers endoderm, mesoderm, and ectoderm. AF stem cells exhibit an epithelial morphology given their origins from either the fetus itself or the amniotic membrane (Figure 3.19). The developmental point at which AF stem cells are isolated influences the presence of germ

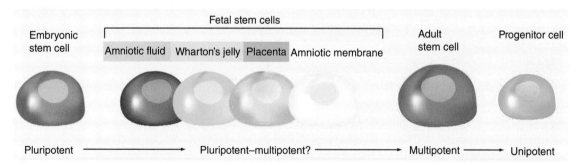

Figure 3.18 Diagrammatic illustration of fetal stem cell sources. See text for details. (Diagram courtesy Kalliopi I. Pappa, Nicholas P. Anagnou, *Regenerative Medicine* (Pappa and Anagnou, 2009); reprinted with permission).

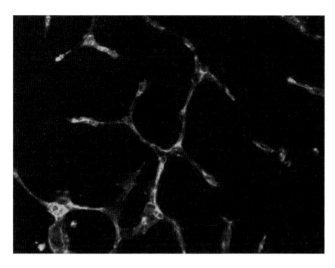

Figure 3.19 Example of AF stem cells. These cells were isolated from human amniotic fluid and have been treated with specific growth factors to induce the formation of capillary-like networks. (*Source*: Courtesy of J. Jacot and Rice University.)

layer markers, with endoderm and mesoderm levels higher if the cells are isolated at earlier versus later stages.

Mesenchymal markers such as CD29 and CD90 are typically expressed by AF stem cells along with unique markers of the endothelium CD44 and CD58. As such, AF stem cells are often referred to as AF mesenchymal stem cells or AF-MSCs. Interestingly, these cells do not express myogenic or hematopoietic surface markers. Resembling ES cells, AF stem cells also express some pluripotency markers including Oct-4, Nanog, and SSEA-4. AF stem cells have been isolated which exhibit many of the characteristics of pluripotency, including indefinite cell culture, maintenance of a normal karyotype and directed differentiation into cell types representing the three primary germ layers, yet these cells failed to produce teratomas when transplanted into nude mice: thus their true potency is still in question.

Self-renewal capacities of AF stem cells are considerable and have been shown to double in culture as fast as every 36 hours, though average doubling times are roughly 3.6 days. Given the expansion capacities and wide range of differentiation capabilities (or at least the AF stem cells mentioned here), they may represent an important source for the generation of mature, functional cells for use in therapy or diagnostics.

Wharton's Jelly Stem Cells Wharton's jelly contains a defined matrix, perivascular, and stromal regions, and stem cells have been derived and characterized from each. Stromal stem cells derived from Wharton's jelly connective tissue are fibroblasts of myogenic origin. These cells are known as **umbilical cord stromal cells (UCSCs)** and exist as two unique populations of cells depending upon the expression of either vimentin or cytokeratin filaments (Figure 3.20) (Baba et al., 2012). In some instances they also express pluripotency markers Oct-4, Sox-2, Rex-1, and Nanog, but their potency properties have yet to be fully characterized and to date they have not been shown to form teratomas when transplanted into nude mice. They can be expanded from an initial isolation of 3.6 million cells to over 11.5×10^8 cells after ~7 months of culture.

Umbilical cord perivascular cells (UCPVCs) are, by definition, isolated from the regions surrounding the vasculature of the umbilical cord, have a fibroblastic morphology, and tend to differentiate into bone derivatives in tissue culture (Figure 3.21).

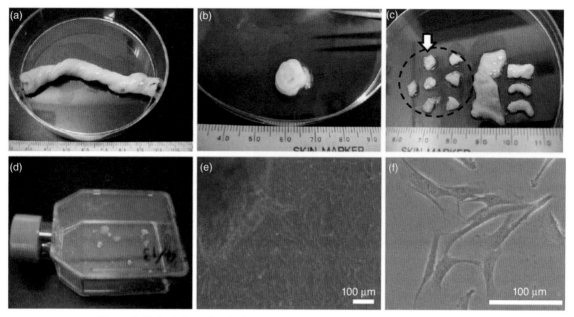

Figure 3.20 Isolation of umbilical cord stromal cells. (a) Human umbilical cord, (b) Umbilical cord cross section, (c) Wharton jelly isolates, (d) Wharton jelly isolates cultured in 10% serum, (e) outgrowths from Wharton jelly 2 weeks after initiation of culture, (f) higher magnification of outgrowth demonstrating the spindle-like shape of UCSCs (Images courtesy Kyoko Baba and the *Journal of Cranio-Maxillo-Facial Surgery* (Baba et al., 2012); reprinted with permission).

Finally, umbilical cord matrix stem cells (UCMSCs), which are derived from the matrix component of Wharton's jelly, have been shown to yield an improvement in muscle regeneration when transplanted into a murine model of severe muscle damage.

Amniotic Membrane Stem Cells The **amniotic membrane**, also known as the amnion, is defined as the innermost layer of the placenta consisting of a thick basement membrane and an avascular stromal matrix. It is a source of amniotic membrane mesenchymal stromal cells (AM-MSCs), also referred to as amnion-derived stem cells. These cells are highly proliferative and can be expanded roughly 300× after 3 weeks of cell culture. *In vitro* directed differentiation experiments have demonstrated the capacity of these cells to efficiently differentiate into cardiomyocytes, hepatocytes, and endothelial cells (Figure 3.22).

Placental Stem Cells The **placenta** is an organ that connects the developing fetus to the uterine wall to allow nutrient uptake, waste elimination, and gas exchange via the mother's blood supply. Given the developmental origins of the placenta, it is not surprising that it is a reliable source of various types of stem cells. The placenta is derived from the trophectoderm and is highly trophoblastic in composition. There are numerous stem cell types that can be derived from placental tissues, including chorionic trophoblasts as well as mesenchymal stromal cells (MSCs): either type may be referred to as placental derived stem cells (PDSCs). MSCs can be easily isolated from placental samples via the application of flow cytometry to isolate cells expressing the unique marker *frizzled 9* (CD349). Multipotency of placenta-derived trophoblastic stem cells is due to

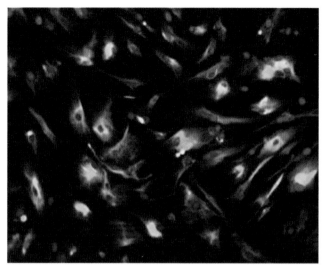

Figure 3.21 Human umbilical cord perivascular cells (HUCPVCs). Fluorescent micrograph of HUCPVCs. (*Source:* Courtesy of A. G. Aristizabal, John E. Davies lab, University of Toronto.)

the activity of the Ets2 transcription factor that regulates a number of target genes involved in both cellular self-renewal and potency. Placental stem cells, if cultured in the appropriate conditions, often express pluripotency markers such as SSEA-4, Nanog3, Oct-4, and Rex-1. In addition, they have the capacity to differentiate into many (but not all) mature lineages including endothelial cells, hepatocytes, pancreatic islet cells, neurons, and glia (Figure 3.23).

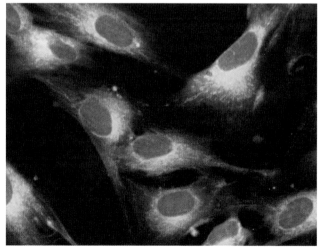

Figure 3.22 Fluorescence microscopy of human amniotic membrane mesenchymal stem cells. Cells were harvested by enzymatic digestion after the amniotic membrane had been stripped of epithelial cells. (Photograph courtesy Case Western Reserve; reprinted with permission.)

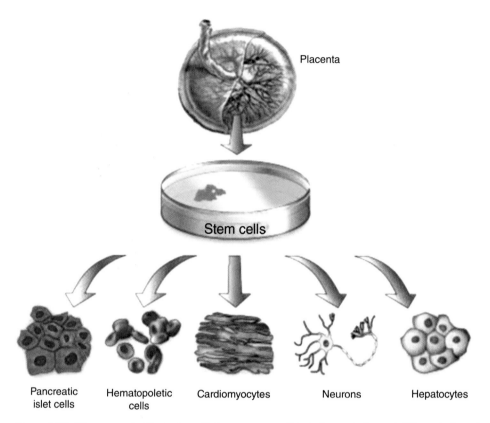

Figure 3.23 Diagrammatic illustration of placenta stem cell isolation and directed differentiation of these cells into mature lineages. (Diagram courtesy Center for Reproductive Sciences; reprinted with permission.)

CHAPTER SUMMARY

ES Cells

1. ES cells, if cultured properly, are pluripotent, giving them the capacity to generate cell types representing the three primary germ layers.
2. Both Wnt and LIF signaling molecules are essential for the maintenance of pluripotency in mouse ES cells.
3. FGF negatively regulates pluripotency in mouse ES cells.
4. Human ES cells require FGF2 and not LIF for maintenance of pluripotency.
5. Alkaline phosphatase, the cell surface antigens stage-specific embryonic antigen SSEA-3, SSEA-4, and the keratin sulfate antigens TRA-1-60 and TRA-1-81 are all markers of pluripotency.
6. Indefinite replicative capacity allows for the generation of an unlimited supply of ES cells.
7. Both feeder-dependent and feeder-independent methods have now been developed for culturing ES cells.
8. LIF activates STAT3 to drive mouse ES cell self-replication.

9. Both ethical and immunorejection concerns hinder ES cells for therapeutic use.

10. The study of human ES cells may yield important insight into mechanisms of various genetic disorders.

11. ES cells have now been derived from many species.

12. The successful isolation of murine ES cells by the Martin and Evans/Kaufman research teams had a profound impact on virtually every aspect of stem cell and even genetic research.

13. Murine ES cells differ from EC cells as they maintain normal karyotypes and have the ability to colonize the germ line.

14. Genetic background plays an important role in the derivation of mouse ES cells.

15. A variety of methods have been developed for the culture of mouse ES cells.

16. Rats mimic certain aspects of human physiology more accurately than mice and thus rat ES cells act as a powerful research platform.

17. Qi-Long Ying et al., isolated the first true rat ES cells using a kinase inhibitor cocktail for cell culture.

18. Human and nonhuman primate embryonic development are extremely similar from a morphological, biochemical, and molecular perspective.

19. Rhesus macaques are widely used in biomedical research but have properties unfavorable for the study of embryonic development.

20. The common marmoset has characteristics that make it well-suited for embryological research.

21. The three primary criteria to define a primate ES cell include derivation from a pre- or peri-implantation embryo, prolonged undifferentiated proliferation, and stable developmental potential.

22. LIF promotes the differentiation of nonhuman primate ES cells.

23. Rhesus and marmoset were the first nonhuman primate ES cells to be isolated.

24. Unlike mES cells, nonhuman primate ES cells also differentiate into trophoblast cells.

25. LIF is not required for maintaining human ES cells in an undifferentiated state.

26. BMPs promote the differentiation of human ES cells.

27. FGF and TGFβ pathways promote both the proliferation and pluripotent properties of human ES cells.

28. Over 300 hES cells lines were isolated by January, 2006.

29. The study of human ES cells will continue to play a major role in elucidating the key molecular and biochemical events associated with growth and differentiation properties.

EC Cells

1. Teratocarcinomas have provided numerous cell lines for the study of developmental biology.

2. In 1964, Kleinsmith and Pierce demonstrated that a single EC cell has the capacity to differentiate into representative cells of the three primary germ layers and also self-renew indefinitely.

3. EC cells have striking resemblance to ES cells in both morphology and differentiation potential, yet, in general, most lines characterized to date have limited or restricted developmental potential.

4. Brigid Hogan et al., isolated the first human EC cells in 1977.

5. The P19 clonal isolated is a widely studied mouse EC line that can differentiate into a variety of cell types.

EG Cells

1. EG cells are similar to EC cells but have a more restricted potentiality as they have been isolated from a more narrow developmental window.

2. A combination of BMP and Wnt signaling induces the serial expression of key genes involved in PGC specification.

Comparing Embryonically Derived Stem Cells

1. Various embryonically derived stem cells exhibit considerable variations in morphology and differentiation capabilities.

2. Human and mouse ES cells both have a high nuclear to cytoplasmic ratio.

Fetal Stem Cells

1. Fetal stem cells encompass a broad range of cell types isolated from a variety of tissues either pre- or postnatally.

2. Fetal stem cells exhibit many of the same properties regarded to ES cells including specific markers and the ability to self-renew.

3. AF stem cells are influenced by the developmental timepoint from which they are isolated.

4. AF cells often express both mesenchymal and endothelial markers as well as some pluripotency markers.

5. UCSCs are derived from Wharton's jelly and typically express either vimentin or cytokeratin.

6. UCPVCs have the capacity to differentiate into bone derivatives.

7. UCMSCs have been demonstrated to regenerate muscle in murine animal models.

8. AM-MSCs are highly proliferative and capable of differentiating into cardiomyocytes, hepatocytes, and endothelial cells.

9. PDSCs have the capacity to differentiate into endothelial cells, hepatocytes, pancreatic islet cells, neurons, and glia.

KEY TERMS

(Key terms are listed by order of appearance in the text.)

- **Wnt proteins**—proteins originally isolated in fruit flies and later shown to act through the β-catenin pathway to promote pluripotency.
- **Leukemia inhibitory factor (LIF)**—an interleukin 6 cytokine that results in stem cell differentiation if levels drop below a certain threshold in tissue culture.
- **Allogeneic**—transplant from one human to another.

- **Conditioned medium**—medium that has been prepared in the presence of co-culture with various cell types.
- **Bone morphogenetic proteins (BMPs)**—a family of growth factors that play key roles in signaling the body axis and tissue architecture of developing embryos.
- **Epiblast stem cells (EpiSCs)**—stem cells isolated from E5.5 to E6.5 postimplantation mouse embryos that have the capacity to self-renew and differentiate into a variety of lineages but are distinct from ES cells.
- **3i**—small molecule inhibitors of specific kinase-dependent signaling pathways that have allowed for the derivation of germ line-competent rat ES cells.
- **Teratocarcinomas**—malignant germ cell tumors that consist of both an undifferentiated embryonal carcinoma (EC) component and a differentiated component that may or may not include representative cell types from all three primary germ layers.
- **Aneuploid**—an abnormal number of chromosomes.
- **P19 cells**—multipotent embryonic carcinoma cell lines derived from murine embryo-derived teratocarcinomas.
- **Embryonal (embryonic) germ (EG) cells**—cells that have the capacity to give rise to gametes that also take on properties similar to embryonic stem cells.
- **Wharton's jelly**—a gelatinous substance within the umbilical cord also present in vitreous humor of the eyeball, largely made up of mucopolysaccharides (hyaluronic acid and chondroitin sulfate).
- **Amniotic fluid (AF) stem cells**—a heterogeneous population of cells isolated from amniotic fluid that are derived from the three primary germ layers endoderm, mesoderm, and ectoderm.
- **Umbilical cord stromal cells (UCSCs)**—stromal stem cells derived from Wharton's jelly connective tissue that are fibroblasts of myogenic origin.
- **Umbilical cord perivascular cells (UCPVCs)**—cells isolated from the regions surrounding the vasculature of the umbilical cord that have a fibroblastic morphology and tend to differentiate into bone derivatives in tissue culture.
- **Amniotic membrane**—also known as the amnion, is the innermost layer of the placenta consisting of a thick basement membrane and an avascular stromal matrix.
- **Placenta**—an organ that connects the developing fetus to the uterine wall to allow nutrient uptake, waste elimination, and gas exchange via the mother's blood supply.

REVIEW QUESTIONS

(Answers to select review questions can be found at www.stemcelltextbook.com.)

1. Describe the difference between pluripotency and totipotency.
2. What are the key signaling factors and pathways involved in maintaining mouse ES cells in a pluripotent state?
3. What are the two primary disadvantages of using human ES cells for therapeutic applications?
4. What two research groups independently isolated mouse ES cells?
5. What are the cell culture conditions that allow for successful mouse ES cell derivation?
6. How do epiblast stem cells differ from embryonic stem cells?

7. How did Qi-Long Ying's group successfully derive rat ES cells?

8. Why is there a need for a nonhuman primate embryonic stem cell line?

9. What properties of rhesus monkeys make them a challenge to study with respect to embryological research?

10. What properties of marmosets make them suitable for embryological research?

11. What are the basic criteria that define a primate embryonic stem cell?

12. What are the two primary differences between mouse and nonhuman primate ES cells?

13. What signaling molecules and pathways act to maintain pluripotency in human ES cells?

14. Describe the first definitive demonstration of a cancer stem cell.

15. Why is the mouse P19 embryonal carcinoma line so popular with developmental biologists?

16. What signaling molecules play a role in specifying primordial germ cell specification *in vivo* and EG cells *in vitro*?

17. What are the basic properties of fetal stem cells?

18. Compare and contrast at least three types of fetal stem cells.

THOUGHT QUESTION

If you could isolate embryonic stem cells from any species which would you choose and how would you coordinate the experiment, that is, outline the experimental plan in detail from acquisition of embryos to cell culture conditions, components, and techniques needed to optimize chances for success.

SUGGESTED READING

Allegrucci, C., L. E. Young (2007). "Differences between human embryonic stem cell lines." *Hum Reprod Update* **13**(2): 103–120.

Baba, K., Y. Yamazaki, et al. (2012). "Osteogenic potential of human umbilical cord-derived mesenchymal stromal cells cultured with umbilical cord blood-derived autoserum." *J Craniomaxillofac Surg* **40**(8): 768–772.

Ennis, J., R. Sarugaser, et al. (2008). "Isolation, characterization, and differentiation of human umbilical cord perivascular cells (HUCPVCs)." *Methods Cell Biol* **86**: 121–136.

Giritharan, G., D. Ilic, et al. (2011). "Human embryonic stem cells derived from embryos at different stages of development share similar transcription profiles." *PLoS One* **6**(10): e26570.

Guhr, A., A. Kurtz, et al. (2006). "Current state of human embryonic stem cell research: an overview of cell lines and their use in experimental work." *Stem Cells* **24**(10): 2187–2191.

Li, P., C. Tong, et al. (2008). "Germline competent embryonic stem cells derived from rat blastocysts." *Cell* **135**(7):1299–1310.

Najm, F. J., A. Zaremba, et al. (2011). "Rapid and robust generation of functional oligodendrocyte progenitor cells from epiblast stem cells." *Nat Methods* **8**(11): 957–962.

Ohinata, Y., H. Ohta, et al. (2009). "A signaling principle for the specification of the germ cell lineage in mice." *Cell* **137**(3): 571–584.

Pappa, K. I. and N. P. Anagnou (2009). "Novel sources of fetal stem cells: where do they fit on the developmental continuum?" *Regen Med* **4**(3): 423–433.

Sokol, S.Y. (2011). "Maintaining embryonic stem cell pluripotency with Wnt signaling." *Development* **138**(20):4341–4350.

Tam, P. P. and D. A. Loebel (2009). "Specifying mouse embryonic germ cells." *Cell* **137**(3): 398–400.

Thomson, J. A., J. Itskovitz-Eldor, et al. (1998). "Embryonic stem cell lines derived from human blastocysts." *Science* **282**(5391): 1145–1147.

Thomson, J. A. and V. S. Marshall (1998). "Primate embryonic stem cells." *Curr Top Dev Biol* **38**: 133–165.

Chapter 4

ADULT STEM CELLS

In Chapter 2, the various properties of stem cells were defined along with an overview of the optimal cell culture conditions for a few representative examples of embryonic and adult stem cells (ASCs). Chapter 3 dealt with descriptions, classifications, and examples of embryonic, fetal, and amniotic stem cells. This chapter describes the field of ASC research. Particular emphasis is placed on defining the basic properties of these cells and widely studied examples of different types of ASCs is introduced.

DISCOVERY AND ORIGIN OF ASCs

As mentioned in Chapter 1, researchers James Till and Ernest McCulloch inadvertently discovered the existence of ASCs exhibiting indefinite replicative capacity in a population of murine bone marrow cells. This finding set the stage for what would soon become a near revolution in stem cell research, prompting the discovery of stem cells from a variety of other adult origins. It should be noted here that many researchers utilize the term "adult" loosely to describe any derived stem cell line that is not truly embryonic in origin: that is, that are not considered embryonic stem or germ cells. Thus, the term "adult stem cells" may be used, correctly or incorrectly, to refer to stem cells derived from aborted fetuses or embryonic origins other than the blastocyst, whereby the cells are not pluripotent in nature. ASCs may also be referred to as "somatic stem cells" ("**soma**" is a Greek term meaning "body"). These cells are present throughout the body in various undifferentiated states depending on the origin and play a primary and crucial role in acting as a source for the replenishment of dead and dying differentiated cells.

BASIC PROPERTIES OF ASCs

ASCs have restricted ability to differentiate into terminal, mature lineages in comparison to embryonic stem cells. ASCs are often multi- or unipotent in nature, and an ASC line has yet to be discovered that is pluripotent. Restrictions in potency exhibited by ASCs are most

Stem Cells: A Short Course, First Edition. Rob Burgess.
© 2016 John Wiley & Sons, Inc. Published 2016 by John Wiley & Sons, Inc.

likely due to the fact that they function as a resource for only one or a few differentiated cell types. Like the original mouse bone marrow-derived stem cells discovered by Till and McCulloch, many ASCs also exhibit an unlimited self-replicative capacity, although some lines have limits on replicative capacity and tend to **senesce** (biological age to the point whereby changes in molecular and cellular structure prevent further cell division) in tissue culture after a certain number of passages. It is the virtually unlimited replicative capacity of ASCs combined with their ability to differentiate into mature lineages that may make certain types valuable for either therapeutic or diagnostic endeavors. Since they are not derived from embryos, the controversial issue of embryo destruction or damage in cellular harvesting does not come into play.

There are two basic properties of ASCs:

- Self-renewal—indefinite or near-indefinite replicative capacity
- Multipotency—the ability to generate multiple progeny of unique identities

Note, however, that some researchers believe unipotency, the ability of a stem cell to differentiate into one lineage, qualifies a cell to be classified as an ASC.

Self-Renewal

It is the self-renewal capacity of ASCs which provides a virtually unlimited supply of progenitor cells from which to generate differentiated cell types; this property warrants further consideration here. ASCs undergo two distinct types of cell division. In **symmetric division,** an ASC gives rise to two identical stem cells. In **asymmetric division**, an ASC yields one identical daughter cell and one progenitor cell that will terminally differentiate into a mature lineage (Figure 4.1).

Both ASCs and progenitor cells may divide numerous times in a symmetric fashion prior to committing to either an asymmetric division or a terminal differentiation (progenitors only).

Multipotency

True ASCs tend to exhibit the property of multipotency, or giving rise to more than one differentiated lineage. An example of this is a neural stem cell's ability to differentiate into

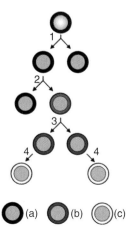

Figure 4.1 Diagrammatic illustration of adult stem cell division. 1. Symmetric cell division yielding two identical daughter stem cells. 2. Asymmetric cell division yielding one identical daughter cell and one progenitor cell. 3. Symmetric cell division yielding two identical daughter progenitor cells. 4. Terminal differentiation of progenitor cells into mature lineages. (Diagram courtesy Wikimedia Commons; reprinted with permission.) (a) daughter stem cell; (b) progenitor cells; (c) mature lineage cell

either glial cells or neurons. Numerous other examples of multipotency exist, some of which are illustrated below.

EXAMPLES OF ASCs

Many examples of ASCs exist which have been derived from an abundance of different tissues and organs. These include the blood, skeletal and cardiac muscle, gut, testis, and liver tissue, to name a few. Table 4.1 summarizes some of the ASC types isolated, along with their tissues of origin and the mature lineages they are capable of producing.

TABLE 4.1 Examples of adult stem cells.

Mature lineage produced	Adult stem cell type	Tissue of origin
Adipocyte Osteocyte	Circulatory Skeletal	Blood
Mature endothelia and newly formed blood vessel	Angioblast (endothelial precursor)	Bone marrow
Hepatocyte Cholangiocyte	Hematopoietic stem cell (HSC)	Liver
Stromal-derived cell engrafted in rat brain	Human marrow stromal	Various
Adipocyte Chondrocyte Osteocyte	Mesenchymal stem cell (MSC)	Various
Neuron	MSC	Various
Neuron	MSC	Various
Adipocyte Bone marrow stromal cell Cardiomyocyte Chondrocyte Myocyte Thymic stromal cell	MSC	Various
HSC Red blood cell lineages White blood cell lineages	HSC	Bone Marrow (Fetal)
Muscle cell	Neural stem cell (NSC)	Brain
Astrocyte Neuron Oligodendrocyte	Neural progenitor cell (NPC)	Brain (Adult and Neonatal)
Astrocyte Neuron Oligodendrocyte	Human central nervous system stem cell (hCNS-SC)	Brain (Fetal)
Adipocyte precursor Osteocyte precursor Chondrocyte precursor Myocyte precursor	Stromal vascular cell fraction of processed lipoaspirate	Fat

continued

Mature lineage produced	Adult stem cell type	Tissue of origin
Hematopoietic progenitor cell (HPC) Red blood cell lineages White blood cell lineages	HSC	Liver (Fetal)
Pancreatic Hepatic	Nestin-positive islet-derived progenitor cell (NIP)	Pancreas
Most red and white blood cell lineages	HPC	Umbilical Cord
Most red and white blood cell lineages Osteoblasts Adipocytes	HSC Mesenchymal progenitor cell (MPC)	Blood, Various

Adapted from *Stem Cells Handbook*, 2nd Edition; Humana Press.

Each of the cell types listed in Table 4.1 are unique in origin and properties. This is due to a phenomenon known as the **stem cell niche**, which is defined as the microenvironment wherein a stem cell resides. This niche provides the necessary growth and differentiation signaling events which are conducive to both cellular division and differentiation. Figure 4.2 illustrates the concept of a niche microenvironment for hematopoietic stem cells (HSCs).

The remaining sections of this chapter will outline and describe several examples of ASCs, comparing and contrasting their various properties and potential for differentiation.

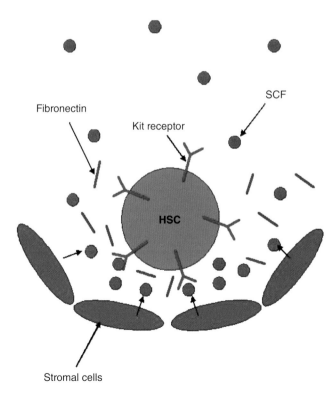

Figure 4.2 Diagrammatic illustration of a stem cell niche. Hematopoietic stem cells in the bone marrow are surrounded by stromal cells, which secrete a variety of factors, including stem cell factor (SCF). SCF, also known as kit ligand (KL), or steel factor has been suggested to both act as a cue for guiding HSCs to the proper niche as well as for HSC number maintenance. (Diagram courtesy Wikimedia Commons; reprinted with permission.)

Hematopoietic Stem Cells

HSCs are defined as a heterogeneous population of multipotent stem cells that can differentiate into myeloid or lymphoid cell types of the adult blood system. It was in 1963 that pioneering researchers James Till and Ernest McCulloch made the striking discovery of a stem cell population present within the bone marrow (see Chapter 1). While they are referred to as ASCs, HSCs initially arise as a population of **hemangioblasts**, multipotent precursors that have the ability to differentiate into either endothelial cells or hematopoietic cells. True "adult" HSCs actually arise later, although, ironically, still during embryonic development. These are unrelated to hemangioblasts. HSCs are multipotent in nature and can differentiate into

- Lymphocytes
- Monocytes
- Neutrophils
- Eosinophils
- Basophils
- Erythrocytes
- Thrombocytes (platelets)

In adults, HSCs exist as three distinct populations of cells including

- Myeloid-based HSCs
- Lymphoid-based HSCs
- Balanced HSCs

The existence of each population or class of cells is dependent upon the **L/M ratio**, the ratio of lymphoid to myeloid progenitor cells in the blood. Myeloid-based HSCs exist as a result of a low L/M ratio of less than 3, lymphoid-based HSCs are present due to a large L/M ratio, typically greater than 10, and balanced HSCs have an L/M ratio between 3 and 10. Lymphoid-based HSCs typically do not have the self-renewal capacities inherent in myeloid and balanced HSCs.

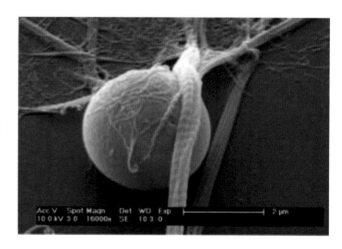

Figure 4.3 Scanning electron micrograph of a human hematopoietic stem cell present within a fibrillar/collagen I heparin sulfate matrix. (Image courtesy Biomechanism.com; reprinted with permission.)

Morphology and Marker Expression Human HSCs are smaller in size, spheroid in appearance with a spheroid nucleus, and have a strikingly smooth surface in comparison to other cell types (Figure 4.3). They have a relatively low cytoplasm-to-nucleus ratio, resemble lymphocytes, and typically do not adhere to most cell culture surfaces. It should be noted that HSCs exist as a heterogeneous population *in vivo*, and thus, this morphological description is a general consensus of all HSC types.

Sources HSCs are found mostly in the bone marrow of adults. Specific bone marrow locations include that within the sternum, ribs, femurs, and various other bones. In the bone marrow, HSCs have often already undergone significant lineage commitment and differentiation, producing a variety of mature cell types (Figure 4.4).

For scientific and therapeutic research, human HSCs are most often isolated directly from the bone marrow of the pelvic iliac crest, umbilical cord blood, and peripheral blood. Efficiency of HSC extraction from the bone marrow is improved upon treatment of patients with certain cytokines including, but not limited to, granulocyte colony stimulating factor (GCSF). GCSF was originally identified as a therapeutic for the purpose of boosting granulocyte counts in immunosuppressed patients, but it was later discovered to result in the mobilization of HSCs out of the bone marrow and into the blood.

Signaling and Multipotency **Hematopoiesis** is defined as the formation of blood cellular components. This is different from the Theory of Hematopoiesis defined in Chapter 1, which states that all blood cells are derived from a common precursor cell. Differentiation of HSCs into various mature lineages requires signaling cues from a variety of growth factors and cytokines which act to induce directed differentiation. Figure 4.5 illustrates this point and outlines the specific inductive signals required at each stage of HSC differentiation. Early events in HSC commitment result in a determination event whereby

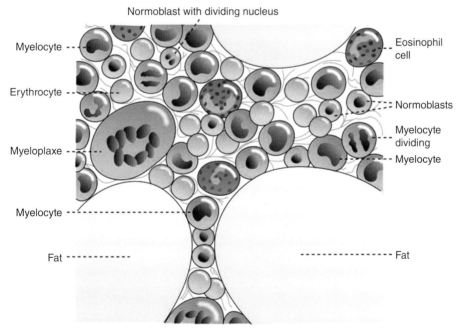

Figure 4.4 Diagrammatic illustration of human bone marrow and internal cell populations. (Courtesy Wikimedia Commons; reprinted with permission.)

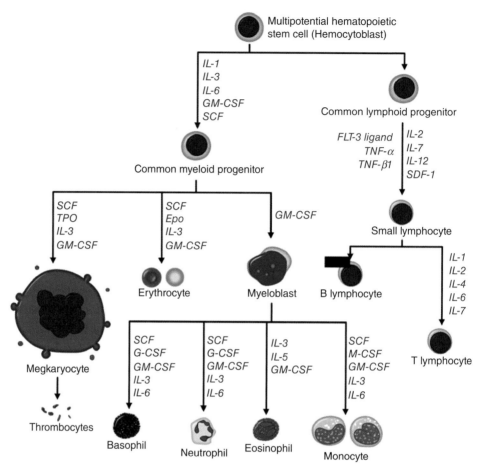

Figure 4.5 Diagrammatic illustration of hematopoietic stem cell differentiation and the signaling molecules involved. (Diagram courtesy Wikimedia Commons; reprinted with permission.)

cells become either myeloid or lymphoid progenitors. Myeloid progenitors then progress toward thrombocytic, erythrocytic, or myeloblastic fates, the latter of which ultimately becomes basophils, neutrophils, eosinophils, or monocytes. Lymphoid progenitors have a narrower range of terminal lineage options, either differentiating into B or T lymphocytes. Each cellular fate is the result of signaling from multiple molecules such as growth factors and cytokines.

THE HSC "NICHE"

The concept of a unique niche for HSC propagation was first contemplated by Ray Schofield in 1978. He hypothesized the existence of a unique microenvironment present in the bone marrow that provides the molecular signaling needed to promote such processes and HSC expansion, migration, and differentiation. HSCs arise from a unique and complex interplay of cross-talk and signaling between both bone marrow cells of the **endosteal niche** and the vascular system of the **vascular niche**. The signaling molecules emanating from each niche work to create an ideal environment for HSCs, providing a valuable source for the mature cell types listed above. This microenvironment is required throughout life and is necessary for the maintenance of proper blood cell content.

SIGNALS CONTROLLING HSC FATE

What are the major signaling events/molecules that drive HSC multipotency? The region that acts as an interface between bone marrow and bone, known as the **endosteum**, encompasses a large population of osteoblasts. The endosteum and corresponding osteoblasts are components of the endosteal niche, with osteoblastic cells secreting a variety of signaling molecules. These include thrombopoietin and angiopoietin, which act to drive HSC quiescence, defined as cellular dormancy, and other factors such as stromal cell derived factor, SDF-1. SDF-1 has been demonstrated to promote HSC self-renewal and differentiation into myeloid cells, a process known as **myelopoiesis**. SDF-1 is also expressed by other cell types such as those of endothelial origin and acts as a homing cue to drive HSC migration within the bone marrow. In addition, the transforming growth factor beta, TGF-β superfamily of growth factors, including bone morphogenetic proteins (BMPs), is released in direct correlation to osteoclast bone breakdown and act to drive HSC quiescence. Increased levels of various ions including calcium have also been implicated in affecting HSC migratory patterns. Other signals emanate from the vascular niche, a microenvironment of vasculature with a major endothelial component. The extraembryonic vascular niche, as well as the vasculature of the liver in adults, have both been shown to drive hematopoiesis (see Chapter 4 and the Theory of Hematopoiesis in Chapter 1), although the exact signaling molecules are not as clearly defined as for those of the endosteal niche. CXCL12-abundant reticular cells, **CAR cells**, are stromal in nature, found in the bone marrow, and affect both the migration and localization of HSCs through the secretion of SDF-1. Finally, both bioactive signaling molecules such as eicosanoids and oxygen levels affect HSC self-renewal, differentiation, and migration. For example, prostaglandins have been shown to promote HSC migratory capabilities while a **hypoxic** environment (less oxygen than under normal physiological conditions) drives hematopoiesis (Figure 4.6).

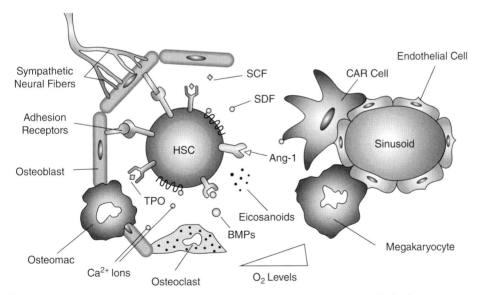

Figure 4.6 Diagrammatic illustration of endosteal and vascular niche cross-talk in the maintenance of HSCs. Communication between localized cells of the endosteal and vascular niches combined with cues from local oxygen levels regulate HSC behavior. (Artwork courtesy Connie Zhao.)

Case Study 4.1: Deterministic regulation of hematopoietic stem cell self-renewal and differentiation

Christa E. Muller-Sieburg, Rebecca H. Cho, Marilyn Thoman, Becky Adkins, and Hans B. Sieburg

Hans Seiburg and colleagues at the Sidney Kimmel Cancer Center, University of California, San Diego, were the first to demonstrate that daughter progeny from HSCs are remarkably similar in self-renewal and differentiation capacities. The researchers discovered a new subset of HSCs using a stromal cell line supportive matrix and novel serial transplant isolation technique which allowed for efficient isolation of HSCs from bone marrow (Figure 4.7). Analysis of the isolated cells and resulting progeny revealed that clonal derivatives had very similar repopulation kinetics. In addition, self-renewal, primitiveness (potency capability of a cell at an early time point), and lineage contribution were all similar. These data suggest that HSC properties are not randomly programmed but rather a coordinated signaling effect promotes a common predetermined state for HSCs *in vivo* (Muller-Sieburg et al., 2002).

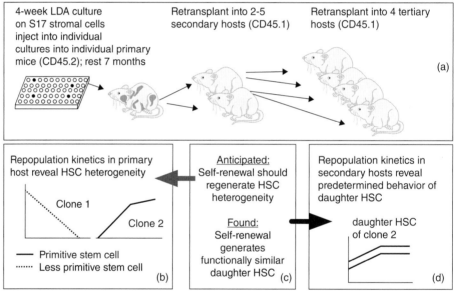

Figure 4.7 Serial transplantation method for the isolation of hematopoietic stem cells. (A) Experimental approach; (B) initial findings in primary host show a great deal of heterogeneity; (C) hypothesis tested and actual findings; (D) results of serial transplantation revealing similar predetermined behavior of progeny. (Figure courtesy Christa E. Müller-Sieburg and Blood (Muller-Sieburg et al., 2002); reprinted with permission.)

Given the complex interplay between the endosteal and vascular niches impacting HSC behavior, the generally accepted hypothesis was that self-renewal and differentiation of HSCs is a randomly regulated process, influenced by both intrinsic and environmental factors emanating from each niche. That changed in 2002 with a seminal publication by researchers at the University of California, San Diego, who closely followed the behavior of individual, clonally derived HSCs in long-term, serial repopulation experiments. In these studies it was revealed that clonally derived HSCs have remarkable similarity with respect to **repopulation** (the ability to recolonize the bone marrow) characteristics as well

---

Focus Box 4.1: Alexander Mauro and the discovery of myosatellite cells

 In 1961, a series of elegant electron microscopic analyses of frog periph-
eral skeletal muscle fibers performed by Rockefeller Institute researcher
Alexander Mauro resulted in the pivotal discovery of the presence of cells
embedded within the muscle tissue that could allow for the repair and
replacement of muscle. Mauro referred to these as **satellite cells**, and the
discovery laid the groundwork for what would become a worldwide scientific interest in the study of cellular differentiation.

Source: Courtesy of the Rockefeller University Archive Centre.

---

as contributions to both lymphoid and myeloid differentiated lineages (Jankowski et al.,
2002). Case Study 4.1 outlines these findings.

Muscle-Derived Stem Cells

The largest overall percentage of body mass is skeletal muscle. It is considerably dynamic
in nature, as it must be, for a great deal of tissue repair and replacement is needed following
injury or aging. Like other organ systems such as bone marrow, skeletal muscle is infiltrated with stem and progenitor cells that may be useful in a therapeutic setting. These cells
are the basis behind regenerative repair and replacement and can be easily isolated via
biopsy. Discovered in 1961 by Alexander Mauro and referred to as **myosatellite cells**, they
are defined as progenitor cells present in muscle capable of terminal differentiation and
fusion to enhance or augment existing muscle fibers and to form new fibers (see Focus Box
4.1). Myosatellite cells produce progeny known as **myoblasts** and represent the oldest
known ASC niche, yet there is only limited knowledge with respect to the signaling and
transcriptional control mechanisms that drive their survival and differentiation into terminal lineages *in vivo*. One of the most widely studied myoblast cell lines is the C2C12 clonal
line derived by researchers David Yaffe and Ora Saxel in the Department of Cell Biology
at The Weizmann Institute of Science in Rehovot, Israel. Drs. Yaffe and Saxel successfully
isolated clonal C2C12 cells by serially passaging cultured cells derived from adult
dystrophic mouse muscle. They demonstrated that these cells could differentiate into
mature skeletal muscle and it was later revealed that C2C12 cells possess osteogenic properties as well. Figure 4.8 is an example of C2C12 myoblasts grown in tissue culture under
differentiation-inducing conditions, differentiating into myocytes and muscle fibers.

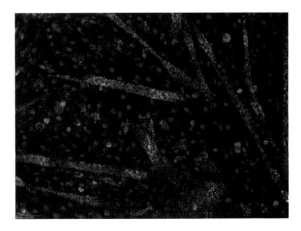

Figure 4.8 Detection of miR-133
microRNA (green) and myogenin
mRNA (red) in differentiating C2C12
cells. (Image courtesy Ryan Jeffs and
Wikimedia Commons; reprinted with
permission.)

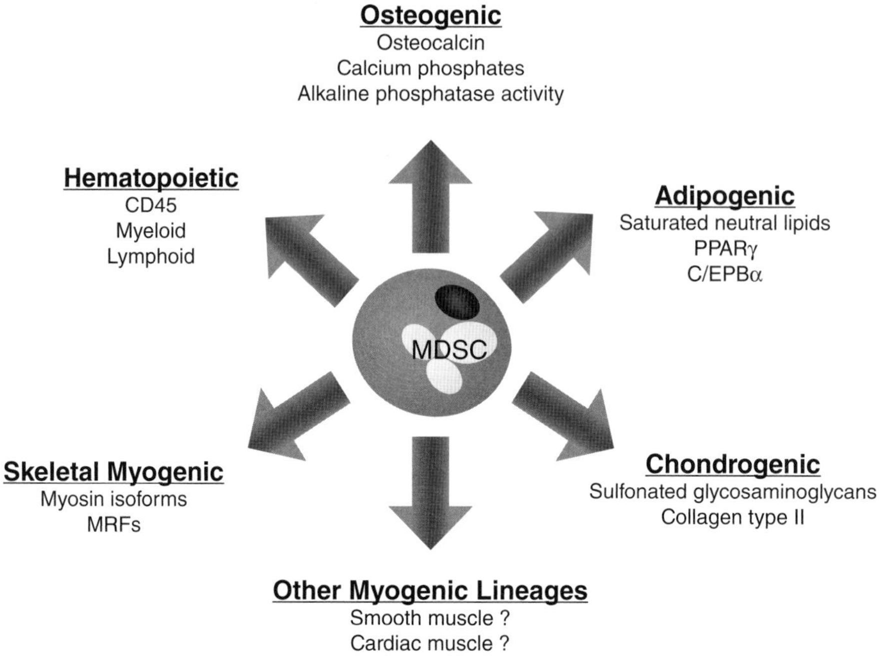

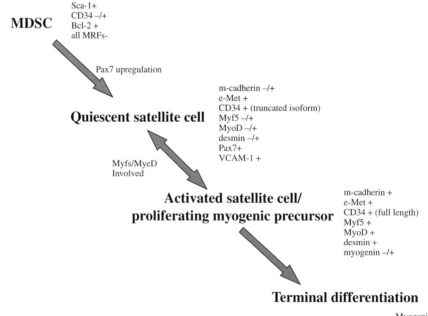

Osteogenic
Osteocalcin
Calcium phosphates
Alkaline phosphatase activity

Hematopoietic
CD45
Myeloid
Lymphoid

Adipogenic
Saturated neutral lipids
PPARγ
C/EPBα

MDSC

Skeletal Myogenic
Myosin isoforms
MRFs

Chondrogenic
Sulfonated glycosaminoglycans
Collagen type II

Other Myogenic Lineages
Smooth muscle ?
Cardiac muscle ?

Figure 4.9
Diagrammatic
illustration of the
multipotent
capabilities of
muscle-derived stem
cells. (Figure courtesy
R.J. Jankowski and
Gene Therapy
(Jankowski et al.,
2002); reprinted with
permission.)

Recently, researchers have also suggested the presence of an earlier population of stem cells—referred to as **muscle-derived stem cells (MDSCs)**—that confer the ability to differentiate into a number of mesodermal lineages (Figure 4.9) (Jankowski et al., 2002).

MDSC
Sca-1+
CD34 –/+
Bcl-2 +
all MRFs-

Pax7 upregulation

Quiescent satellite cell

m-cadherin –/+
e-Met +
CD34 + (truncated isoform)
Myf5 –/+
MyoD –/+
desmin –/+
Pax7+
VCAM-1 +

Myfs/MyeD
Involved

**Activated satellite cell/
proliferating myogenic precursor**

m-cadherin +
e-Met +
CD34 + (full length)
Myf5 +
MyoD +
desmin +
myogenin –/+

Terminal differentiation

Myogenin +
MRF4 +
CD34 –
Pax7 –
other MRFs +

Figure 4.10
Proposed pathway
for myogenesis.
See text for details.
(Figure courtesy
RJ Jankowski and
Nature (Jankowski
et al., 2002);
reprinted with
permission.)

References: Beauchamp et al. (2000)[11], Yoshida et al. (1998)[12]
Miller et al. (1999)[12], Seale and Rudnicki (2000)[14],
Seale et al. (2000)[18], Cornelison et al. (1997)[10]

MDSCs also act as the precursors to myosatellite cells. During the process of **myogenesis**, defined as the differentiation of cells into muscle, MDSCs transition from an undifferentiated state into quiescent myosatellite cells, later reentering the cell cycle and becoming committed myoblasts. Inductive signaling cues present within the muscle microenvironment cause proliferating myoblasts to exit the cell cycle, elongate, and fuse to form mature muscle (Figure 4.10) (Jankowski et al., 2002). For the purposes of this text, the remainder of this section will focus on the properties of myosatellite cells.

Myosatellite Cell Morphology and Marker Expression Myosatellite cells are small, with a single nucleus and almost no cytoplasm. Upon initial isolation, the cells are rounded, but rapidly elongate into a spindle-like morphology within 24 hours. As cellular density within the tissue culture environment increases, cells begin to line up with one another in preparation for fusion. Within 3 weeks, if cultured under high-density differentiation-inducing conditions (typically high density and low serum), myosatellite cells will fuse to form primitive fibers. At this stage the cells exhibit an ellipsoidal, cobblestone appearance (Ytteborg et al., 2010) (Figure 4.11).

Sources Typically, myosatellite cells can be found embedded between the **basement membrane**—a thin sheet of fibers underlying the epithelium that lines the cavities of various organs, including muscle, and the **sarcolemma**—which is the cellular membrane of mature individual muscle fibers.

Signaling, Transcriptional Control, and Multipotency In an undifferentiated state, it has been established that myosatellite cells do not express skeletal muscle-specific markers, and it is known that the paired box transcription factors Pax3 and Pax7 are co-expressed in myosatellite cells. In 2005, researchers in the Department of Developmental Biology at the Pasteur Institute in Paris lead by Margaret Buckingham demonstrated that, in Pax3/Pax7 double knockout mice, myosatellite cells fail to differentiate and either die or assume a non-myogenic phenotype (McCroskery et al., 2003). These data suggest a crucial regulatory role for Pax3 and Pax7 in promoting muscle formation through myosatellite cell differentiation. In addition, there is some evidence suggesting that **myostatin**, which promotes

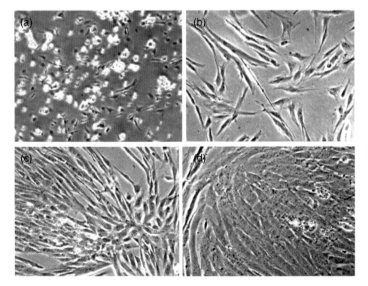

Figure 4.11 Brightfield microscopy of myosatellite cells isolated from salmon muscle at various stages of differentiation. Post-plating time: (A) One day; (B) One week; (C) Two weeks; (D) Three weeks. (Images courtesy Elisabeth Ytteborg and Biochimica et Biophysica Acta (BBA)— *Molecular and Cell Biology of Lipids* (Ytteborg et al., 2010); reprinted with permission.)

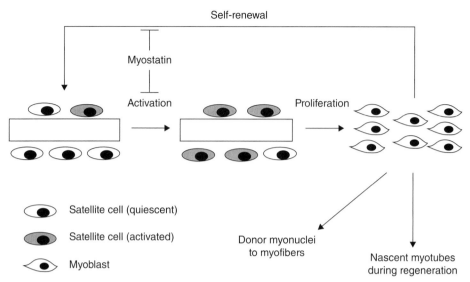

Figure 4.12 Diagrammatic illustration of the role myostatin plays in myosatellite cell differentiation. High levels of myostatin act to block the self-renewal of either proliferating myoblasts or quiescent satellite cells entering the cell cycle.

muscle differentiation, inhibits myosatellite cell self-renewal through the induction of the cyclin-dependent kinase p21 thereby driving their differentiation. Researchers at the Animal Genomics AgResearch Center in Hamilton, New Zealand, demonstrated this through the analysis of myostatin-null adult myoblasts, revealing a pathway that acts to block both the self-renewal and activation capacities of myosatellite cells, thus promoting their differentiation (Figure 4.12) (Villageois et al., 2012).

It is clear that both muscle-derived stem cells and related myosatellite cells have enormous potential with respect to cell-based therapeutics applications. The therapeutic potential of both muscle-derived stem cells and myosatellite cells will be discussed in Chapter 8.

Adipose-Derived Stem Cells

As introduced in Chapter 2, adipose-derived stem cells (also referred to as ASCs or ADSCs) have a great deal of potential with respect to acting as a resource for both cell-based therapeutics and regenerative medicine initiatives. They can be cultured through multiple passages and long periods and the differentiation potential of ADSCs is considerable. They are similar in nature to mesenchymal stem cells (MSCs) in multipotency. The following sections will focus on human multipotent adipose-derived stem (hMADS) cells.

Morphology and Marker Expression In a tissue culture setting, from a morphological perspective, hMADS and other ADSCs are strikingly similar to fibroblasts, exhibiting an elongated, spindle shape (Figure 4.13).

Undifferentiated, highly proliferative ADSCs secrete high levels of FGF2 and activin A, yet there appears to be an inverse relationship between FGF2 and activin A expression during initial *ex vivo* expansion. In addition, cellular morphology also changes during tissue culture propagation, with cells transitioning from a spindle shape with a low cytoplasmic to nuclear ratio to a flat morphology with a much higher cytoplasmic content

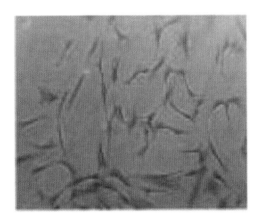

Figure 4.13 Undifferentiated adipose-derived stem cells grown in tissue culture. (Photograph courtesy Human Mobility Research Centre; reprinted with permission.)

(Figure 4.14) (Villageois et al., 2011). Activin A is expressed at higher levels in the adipose tissue of obese versus non-obese individuals. In addition, activin A levels are further increased in adipose-derived progenitor cells upon stimulation by factors secreted by macrophages in the adipose tissue of obese patients, although the reason for this regulation remains unclear.

ADSCs have a tendency to line up with one another upon reaching confluence in tissue culture and exhibit **contact inhibition**, which is when a cell exits the cell cycle upon contact with neighboring cells. Upon differentiation into fat cells they become globular in shape (Figure 4.15).

Sources ADSCs are typically isolated from adipose tissue, specifically from white adipose tissue (WAT). In humans, this is often accomplished through routine liposuction surgery, which produces a **lipoaspirate** containing hMADS. Following the collection of the lipoaspirate, the tissue is washed and blood cells are removed. A **collagenase**, which is defined as an enzyme that breaks down collagen, is applied to break up the tissue in preparation for individual cell harvesting. Centrifugation of the sample follows with the collection of the cell pellet, referred to as the **stromal vascular fraction (SVF)**. It is the SVF that contains both hMADS and endothelial cells, and after plating and serial passage, pure populations of ADSCs can be obtained which have the potential to differentiate into a variety of lineages (Figure 4.16 and see Signaling and Multipotency section below) (Locke et al., 2011).

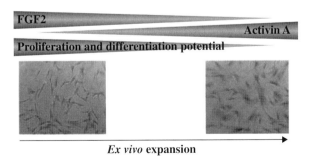

Figure 4.14 *Ex vivo* expansion characteristics of adipose-derived stem cells. (Figure courtesy Phil Villageois and the *American Journal of Stem Cells* (Villageois et al., 2011); reprinted with permission.)

ASC Grown on Tissue-Culture Polystyrene

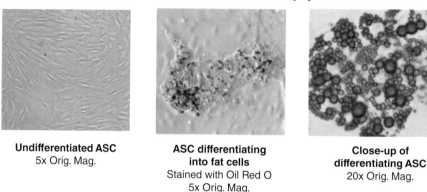

Undifferentiated ASC	**ASC differentiating**	**Close-up of**
5x Orig. Mag.	**into fat cells**	**differentiating ASC**
	Stained with Oil Red O	20x Orig. Mag.
	5x Orig. Mag.	

Figure 4.15 Undifferentiated and differentiated adipose-derived stem cells. (Photographs courtesy Human Mobility Research Centre; reprinted with permission.)

Signaling and Multipotency hMADS and other ADSCs have the potential to differentiate into a variety of mature cell types including those which form bone, cartilage, muscle, and, of course, fat (Figure 4.17) (Nauta and Fibbe, 2007).

Unique signaling and transcriptional cascades underlie the differentiation of hMADS into the different lineages. For example, medium containing steroids, cyclic AMP inducers, and fatty acids drives hMADS to become fat cells. Induction of the osteogenic pathway in hMADS has been accomplished in the presence of medium containing low serum

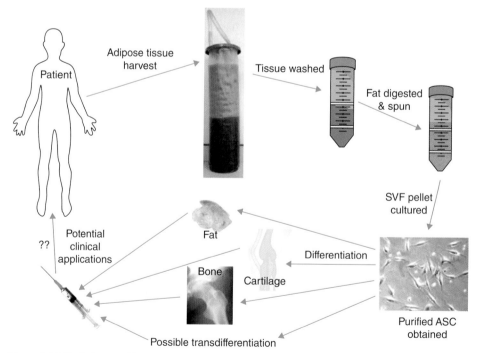

Figure 4.16 Isolation, purification, and clinical application of hMADS. (Figure courtesy Michelle Locke and *Stem Cells* (Locke et al., 2011); reprinted with permission.)

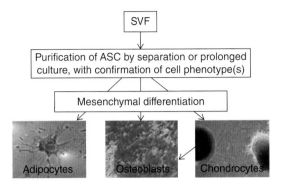

Figure 4.17 Differentiation potential of adipose-derived stem cells. (Figure courtesy Michelle Locke and *Stem Cells* (Locke et al., 2011); reprinted with permission.)

and epidermal growth factor (EGF). In addition and as mentioned above, proliferative ADSCs secrete FGF2: this is downregulated in cells committed to undergo differentiation, with a corresponding activation of activin A evident. Given the fact that ADSCs express FGF type 1 receptor, which has a high affinity for FGF2, it has been suggested that the FGF signaling pathway plays a critical role in regulating ADSCs self-renewal. This could be through **autocrine** mechanisms, defined as same-cell signaling, or **paracrine** mechanisms, defined as near-cell signaling. FGF1 has also been implicated in promoting ADSCs self-renewal and the FGF pathway is thought to act, at least partially, through the extracellular signal-regulated kinases (ERK1/2). Activin A secretion rapidly decreases as ADSCs differentiate into adipocytes, and sustained activin A expression promotes self-renewal while simultaneously inhibiting adipocyte differentiation. Conversely, inhibiting the activity or function of activin A promotes ADSC exit from the cell cycle and adipocytic differentiation. These data have resulted in the proposed model, whereby FGF2 and activin A, secreted by adipocytes, act in both autocrine and paracrine pathways to promote ADSC expansion and self-renewal while inhibiting differentiation (Figure 4.18) (Villageois et al., 2011).

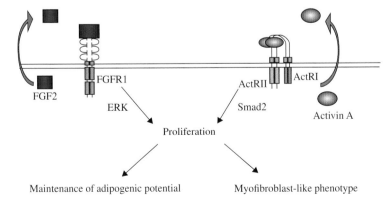

Figure 4.18 A model of autocrine/paracrine signaling regulating adipocyte stem cell proliferation and differentiation. Both FGF2 and activin A are secreted by ADSC and promote proliferation through separate ERK and Smad2 pathways, respectively. (Figure courtesy Phil Villageois and the *American Journal of Stem Cells* (Villageois et al., 2011) reprinted with permission.)

Mesenchymal Stem Cells

Mesenchyme is defined as mesodermally derived embryonic connective tissue that has the capacity to differentiate into hematopoietic lineages and mature, adult connective tissue. **Mesenchymal stem cells (MSCs)** are defined as multipotent stromal cells arising from the mesenchyme and they can be isolated from a variety of sources (Figure 4.19 and Table 4.2).

The terms "mesenchymal stem cells" and "stromal stem cells" are often used interchangeably to describe the same cells. This is inaccurate, as mesenchyme tends to be embryonic in origin while stroma is typically referred to as post-natal connective tissue. MSCs were originally discovered in 1924 by the Russian-American scientist Alexander A. Maximow (see Focus Box 1.1). In his studies, he performed an extensive analysis of tissue morphology and concluded that there must be a common precursor cell which gives rise to both hematopoietic lineages and lineages separate and unique but located within the hematopoietic microenvironment (MSCs). As discussed above and in Chapter 1 this is referred to as the Theory of Hematopoiesis.

Morphology and Marker Expression   Mesenchymal stem cells are small with a large round nucleus and low cytoplasm-to-nucleus ratio. They have thin processes and in general resemble fibroblasts. The nucleus tends to have a large, visible nucleolus (Figure 4.20).

MSCs may be identified and classified by several marker and non-marker based criteria. These include

- Plastic adherence under normal cell culture conditions
- Fibroblast-like morphology

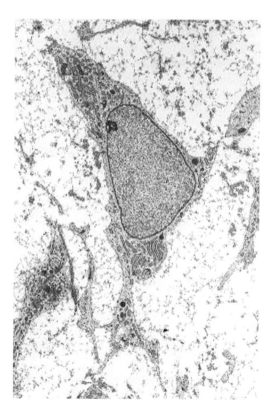

Figure 4.19 Scanning electron micrograph of mesenchymal stem cell present in tissue. (Micrograph courtesy Robert M. Hunt and Wikimedia Commons; reprinted with permission.)

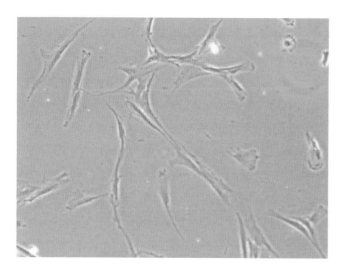

Figure 4.20 Brightfield phase contrast micrograph of human bone marrow-derived mesenchymal stem cells. Micrograph was taken after 3 weeks of culture. (Photograph courtesy C. Mahapatra and Wikimedia Commons; reprinted with permission.)

- Ability to undergo osteogenesis, adipogenesis, and chondrogenesis in cell culture
- Express the cell surface markers CD73, CD90, and CD105
- Do not express the cell surface markers CD11b, CD14, CD19, CD34, CD79a, and HLA-DR

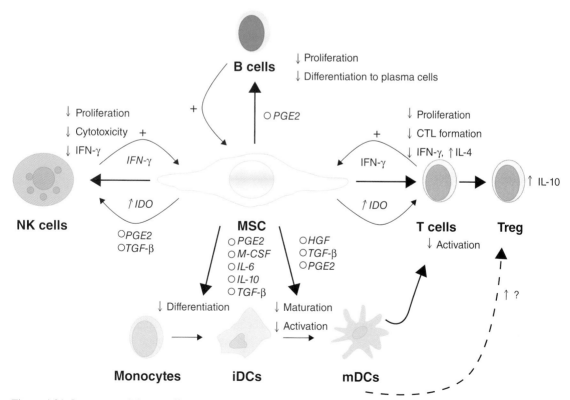

Figure 4.21 Immunomodulatory effects of mesenchymal stem cells. (Illustration courtesy Alma J. Nauta and *Blood* (Nauta and Fibbe, 2007); reprinted with permission.)

TABLE 4.2 Sources of mesenchymal stem cells.

Tissue of origin	Stage of ontogeny	Species
Adipose tissue	Post-natal	Human
Adipose tissue	Post-natal	Mouse
Pancreas	Fetal	Human
Bone marrow	Fetal	Human
Liver	Fetal	Human
Blood	Fetal	Human
Tendon	Post-natal	Mouse
Synovial membrane	Post-natal	Mouse
Amniotic fluid	Fetal	Human
Peripheral blood	Post-natal	Human
Umbilical cord blood	Fetal and post-natal	Human
Wharton's jelly	Fetal and post-natal	Human
Dental pulp	Post-natal	Human
Tooth bud	Fetal and post-natal	Human and mouse
Corneal stroma	Post-natal	Human

Mesenchymal stem cells have several unique features not exhibited by other types of stem cells. Specifically, MSCs have been demonstrated to be **immunomodulatory**, which means they have the capacity to regulate immune response in a localized microenvironment by secreting cytokines. The effect is mostly immunosuppressive, downregulating immune response by affecting the functions of B cells, T cells, monocytes, and NK cells (Figure 4.21) (Nauta and Fibbe, 2007). In addition, MSCs are referred to as **immunoprivileged**, defined as being resistant to bystander effects of local immune responses. MSCs have also been demonstrated to be highly migratory and motile in nature, infiltrating and localizing many tissues upon transplant introduction *in vivo*.

Sources Mesenchymal stem cells may be isolated from numerous embryonic and post-natal sources. Considerably rich sources of MSCs include the developing tooth bud and adipose tissue. Table 4.2 lists some examples of sources for various species studied.

It should be noted that MSCs isolated from different sources are not all alike. Factors such as the source and isolation methodology may affect the morphological and molecular makeup of isolated cells. A comparison of mesenchymal stem cells isolated from either bone marrow or adipose tissue is outlined in Table 4.3.

Signaling and Multipotency As mentioned in the Morphology and Marker section above, in order for a cell to be classified as an MSC it must at least be able to differentiate into mesodermal bone or fat derivatives. Other lineages for which MSCs may have the capacity to transdifferentiate into include those of myogenic and neuronal origins (Figure 4.22). **Transdifferentiation** is defined as a cell's ability to differentiate into lineages of different embryonic germ layers. It should be noted that transdifferentiation is also defined as the differentiation of one adult cell type into another, although this concept will not be discussed here. The following sections provide an overview of MSC properties and potency. For a more thorough description, Edda Tobiasch and colleagues in the Department of Natural Sciences at Bonn-Rhine-Sieg University of Applied Sciences in Rheinbach, Germany, have written an excellent comprehensive review on the mechanisms underlying mesenchymal stem cell differentiation (Zhang et al., 2012).

TABLE 4.3 Comparison of the differences between bone marrow- and adipose tissue-derived mesenchymal stem cells.

Characteristic	Bone marrow-derived MSCs	Adipose tissue-derived MSCs
Marker Expression	CD106, MEST Higher than in Adipose-Derived MSCs	CD49b, CD54, CD34, Ki-67, CDCA8, CCNB2 higher than in bone marrow-derived MSCs
Proliferation Capacity	Low	High
Cell Size	Large	Normal
Apoptotic Threshold	Low	High
Isolation Method	Invasive via bone marrow extraction	Limited invasiveness via lipoaspiration

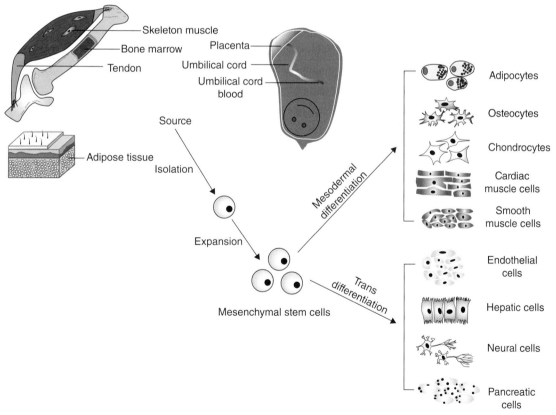

Figure 4.22 Diagrammatic illustration of human mesenchymal stem cell multipotency. Sources of MSCs include, but are not limited to, skeletal muscle, bone marrow, tendon, adipose tissue, umbilical cord blood, umbilical cord, and placenta. These cells retain the ability to differentiate into various mesodermal lineages such as osteocytes and chondrocytes but also may transdifferentiate into non-mesodermal lineages such as neurons. (Illustration courtesy Yu Zhang and the *Scientific Word Journal* (Zhang et al., 2012); reprinted with permission.)

What drives MSCs to differentiate into certain mature lineages? Given the therapeutic relevance, much research has been conducted on the osteogenic potential of MSCs and several inducing molecules and platforms have been outlined which efficiently drive MSC osteoblast commitment and maturation. These include

- 1,25-Dihydroxyvitamin D3 + ascorbate-2-phosphate + β-glycerophosphate
- β-Glycerophosphate + ascorbic acid + dexamethasone + pulsed electromagnetic fields (PEMFs)
- Ascorbate-2-phosphate + dexamethasone + β-glycerophosphate cultured on Matrigel
- Three-dimensional type I collagen matrices combined with tensile strain
- Hydrostatic pressure combined with culture in the presence of collagen-containing extracellular matrices

OSTEOGENESIS

How the molecules and cell culture conditions described above specifically drive osteogenic commitment by MSCs remains a mystery, but several key factors have been noted to initiate the osteogenic cascade of events. The signaling molecule Wnt10b and the runt-related transcription factor Runx2 have been identified as key regulators of the decision between commitment to adipocytic or osteogenic lineages. **Runx2**, also known as CBF-α-1, is a member of the Runt DNA-binding domain family of transcription factors and has been shown to be essential for

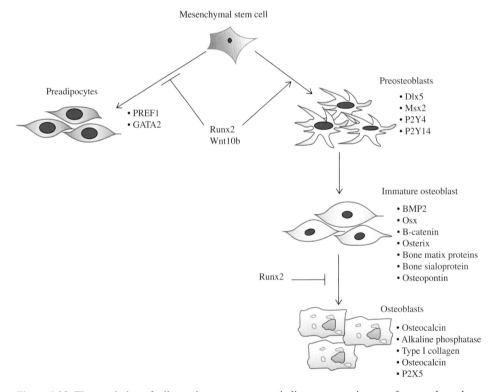

Figure 4.23 The regulation of adipocytic versus osteogenic lineage commitment of mesenchymal stem cells. Wnt10b and Runx2 drive undifferentiated MSCs toward an osteoblastic fate. (Illustration courtesy Yu Zhang and the *Scientific Word Journal* (Zhang et al., 2012); reprinted with permission.)

both the differentiation of cells into osteoblasts and later stage skeletal morphogenesis. Induced expression of Runx2 by steroids such as dexamethasone both inhibits adipogenesis and initiates the commitment of MSCs to a preosteoblastic fate (Figure 4.23) (Zhang et al., 2012).

PEMF has also been demonstrated to promote MSC osteogenesis and it is speculated that electromagnetic pulses enhance DNA synthesis, the expression of bone markers, and calcified matrix production. Extracellular matrix proteins such as vitronectin and collagen have also been demonstrated to induce MSC osteogenesis in tissue culture. Interestingly, as demonstrated by researchers in the Department of Chemical Engineering and Materials Science, at the University of California, Irvine, these proteins act through completely different signaling events, with vitronectin stimulating the activation of focal adhesion kinas (FAK) and collagen I activating the ERK and phosphatidylinositol 3-kinase (PI3K) pathways (Zhang et al., 2012).

ADIPOGENESIS

It is clear that different inducing factors and mechanisms are required for *in vivo* versus *in vitro* differentiation of mesenchymal stem cells into either bone or fat lineages. This is most likely due to differences between the natural and artificial environments of the two systems.

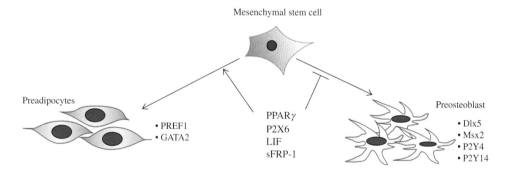

Mesenchymal stem cell

Preadipocytes
• PREF1
• GATA2

PPARγ
P2X6
LIF
sFRP-1

Preosteoblast
• Dlx5
• Msx2
• P2Y4
• P2Y14

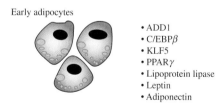

Early adipocytes
• ADD1
• C/EBPβ
• KLF5
• PPARγ
• Lipoprotein lipase
• Leptin
• Adiponectin

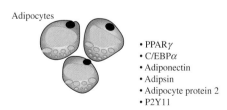

Adipocytes
• PPARγ
• C/EBPα
• Adiponectin
• Adipsin
• Adipocyte protein 2
• P2Y11

Figure 4.24 The regulation of mesenchymal stem cell adipogenesis. PPARγ is a key transcription factor regulating adipogenesis at multiple steps along the differentiation pathway. Its activity is stimulated by both insulin and indomethacin. Many other signaling and transcription factors promote adipogenesis and are listed. (Illustration courtesy Yu Zhang and the *Scientific Word Journal* (Zhang et al., 2012); reprinted with permission.)

For example, cytokines are readily available for signaling induction *in vivo*, yet are not present unless artificially added in a cell culture environment. Studies *in vitro* have therefore elucidated three primary inducers of MSC-based adipogenesis that are required including the synthetic glucocorticoid dexamethasone, the anti-inflammatory drug **indomethacin** and insulin. While dexamethasone induces late stage MSC terminal differentiation into adipocytes indomethacin drives the activation of the transcription factor **peroxisome proliferator activated receptor PPARγ**, which initiates the commitment of MSCs to preadipocytes (Figure 4.24) (Zhang et al., 2012). PPARγ is a nuclear receptor transcription factor which heterodimerizes with the retinoid X receptor to regulate the transcription of target genes involved in cellular development, differentiation, metabolism, and even tumor development. Insulin activates adipogenesis through multiple pathways via binding to IGF-1, which results in the phosphorylation of cAMP response element-binding protein (CREB) by PI3K. CREB also acts to stimulate PPARγ. In addition, insulin promotes the phosphorylation of the anti-adipogenic forkhead transcription factors, FOXO1 and FOXA2, thus resulting in their translocation from the nucleus and alleviating their inhibition of MSC differentiation into adipocytes.

Finally, a comparison of osteogenesis versus adipogenesis *in vivo* reveals other non-signaling environmental cues that are unique which may result in the activation of mesenchymal stem cell adipogenesis. For example, cell-to-cell contact promotes adipogenesis over osteogenesis as typically more cells are required locally for the former to occur efficiently. In addition, the three-dimensional structure of the extracellular matrix may influence adipogenic versus osteogenic events. In contrast to MSC osteogenesis, which is stimulated by rigid structural scaffolds, adipogenesis is promoted by a soft structural geometry. pH, ionic concentrations, and oxygen tension also play roles in the guidance of MSC differentiation. In humans, for example, MSCs reside in a **hypoxic** (less oxygen than normal) environment and which tends to inhibit osteogenesis while chondrogenesis is unaffected.

Neural Stem Cells

Neurogenesis is defined as the process by which neurons and glial cells are generated from neural stem and progenitor cells. It was widely accepted to occur prevalently during embryogenesis for many years and in 1965 was confirmed to also take place in the rat adult brain by pioneering Massachusetts Institute of Technology researchers Joseph Altman and Gopal Das. This finding was in direct contradiction to Santiago Ramon y Cajal's "no new neurons" hypothesis (see Chapter 1 and Case Study 4.2). Two years later adult brain neurogenesis was confirmed in non-human primates and later in many organisms, including humans. Thus, there exists neural stem and neural progenitor cells in the adult brain that act to repopulate the specific regions with new neural cells. Neural stem cells are perhaps the most intensely studied of all ASC types due to their potential therapeutic and drug screening applications.

Morphology and Marker Expression Adult neural stem cells have a morphology resembling astroglial cells. The morphology is bipolar in nature, resembling astrocytes more than oligodendrocytes, and the cells are typically elongated and star-shaped, sending out numerous projections from a main cell body (Figure 4.26).

A plethora of markers are expressed by undifferentiated neural stem cells. Markers for the different phases of NSC behavior and function will be discussed in the section Signaling and Multipotency. Perhaps the most widely studied marker for undifferentiated, multipotent NSCs is nestin. **Nestin** is a protein component of nerve cell type VI intermediate filaments that play a role in axonic radial growth. In adult organisms, nestin is expressed by NSCs in the subgranular zone of the brain and rapidly becomes downregulated upon

Case Study 4.2: Post-natal origin of microneurones in the rat brain

Joseph Altman and Gopal D. Das

The "no-new-neurons" hypothesis was widely accepted for over 30 years until pioneering researchers Joseph Altman and Gopal D. Das of MIT's Psychophysiological Laboratory definitively demonstrated the presence of mitotically active **microneurones** in the adult rat brain. They accomplished this by implementing autoradiographic procedures. Specifically, they injected radiolabeled thymidine, a precursor required for DNA synthesis, directly into the brains of adult rats and measured its uptake which marks proliferating, mitotically active cells. The subsequent radiolabeled nature of the cells then allowed for their tracking (Figure 4.25). Gopal and Das demonstrated the presence of newly generated neurons and glial cells in the adult rat brain, with a particularly high concentration of new cells in the dentate gyrus in newborns and infants, yet the amount of these cells declined in adults. These findings set the stage for an explosion of research in the field of adult neural stem cells and provided the basis for the pursuit of adult neural stem cells in cell-based therapeutic applications (Altman and Das, 1965).

Figure 4.25 Photomicrograph of a section of an adult rat brain demonstrating newly generated glial cells and neurons. Black dots represent labeled cells. A, Ammon's horn; C, Cortex; D, dentate gyrus; F, fimbria hippocampi; P, choroid plexus; T, thalamus; V, lateral ventricle. Magnification X68. (Photo courtesy Joseph Altman and Nature (Altman and Das, 1965); reprinted with permission.)

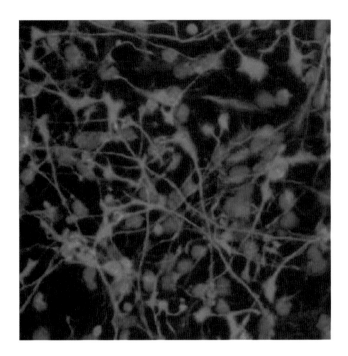

Figure 4.26 Fluorescence microscopy of human neural stem cells in culture. Nuclei are stained blue with cytoplasm and projections stained red. (Photo courtesy R. Cassiani; reprinted with permission.)

differentiation of these cells into mature lineages (Figure 4.27). CD133 is also a common marker of undifferentiated neural stem cells. It should also be noted that this marker also delineates HSCs. Table 4.4 summarizes some of the more widely studied markers expressed in undifferentiated neural stem cells.

Figure 4.27 Fluorescence microscopy of neural stem cells demonstrating nestin and GFAP expression. Cells were tripled-labeled to stain for nuclei (blue), GFAP (red), an astroglial marker and nestin (green). (Photos courtesy John Sinden; reprinted with permission.)

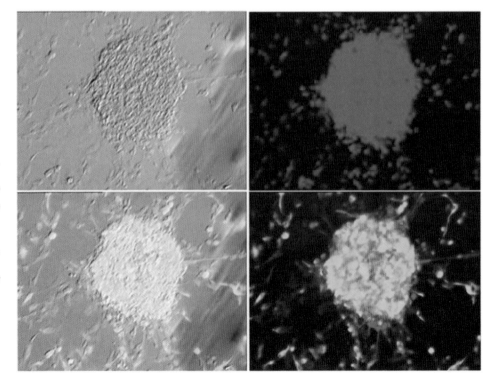

TABLE 4.4 Markers of undifferentiated neural stem cells.

Marker	Description
CD133	The CD133 (also named AC133) antigen is a 97 kDa glycoprotein, with 5 transmembrane domains.
FGFR4 (CD334)	A member of FGFR family, and preferentially binds αFGF/FGF1.
Fibronectin	Fibronectin is a glycoprotein present in a soluble dimeric form in plasma, and in a dimeric or multimeric form at the cell surface and in extracellular matrix.
GLUT	Glucose transporters are integral membrane glycoproteins involved in transporting glucose into most cells.
Islet-1	Islet 1 is encoded by ISL1. Islet 1 is a transcription factor containing two amino-terminal LIM domains and one carboxy-terminal homeodomain.
Musashi-1	The Musashi family is important for cell fate determination, playing roles in maintenance of the stem-cell state, differentiation and tumorigenesis.
Nestin	Nestin is a Class VI intermediate filament expressed in the developing central nervous system (CNS) in early embryonic neuroepithelial stem cells.
Nerve Growth Factor (NGF)	Nerve growth factor (NGF) is a polypeptide involved in the regulation of growth and differentiation of sympathetic and some sensory neurons.
Noggin	A secreted homodimeric glycoprotein whose scaffold contains a cysteine-knot topology similar to that of BMPs. It has a complex pattern of expression during embryogenesis, and is expressed in the central nervous system, as well as in several peripheral tissues. Noggin can diffuse through extracellular matrices more efficiently than members of the TGF-β superfamily, and thus may have a role in creating morphogenic gradients.
Pax3	This protein is a member of the paired box (PAX) family of transcription factors. Members of the PAX family typically contain a paired box domain and a paired-type homeodomain. These proteins play critical roles during fetal development.
PDGFA (CD140a)	A tyrosine kinase receptor for members of the platelet-derived growth factor family, probably involved in kidney development.
SLAIN1	SLAIN motif family, member 1
Vimentin	Vimentin is the major subunit protein of the intermediate filaments of mesenchymal cells. It is believed to be involved with the intracellular transport of proteins between the nucleus and plasma membrane.

Marker descriptions courtesy Molecular Diagnostics Services.

Sources and Origins *In vivo*, neurogenesis during embryonic development occurs, for the most part, in the germinal ventricular zone (VZ), with progenitor populations rapidly expanding in concert with neural tube thickening (Figure 4.28) (Kirsch et al., 2008). Signaling cues emanating from the neural tube result in a conversion of NSC morphology from columnar cells to more of a star-shaped radial appearance. These cells will ultimately become **radial glia**, which are defined as neurogenic precursor cells present in the VZ that divide asymmetrically to give rise to more radial glia, intermediate progenitor cells, and post-mitotic neurons. It should be noted that embryonically derived neural stem cells are still considered to be adult, as they have limited capacity to differentiate. Many researchers use the term "adult" interchangeably to describe neural stem cells isolated from either embryonic or adult brain.

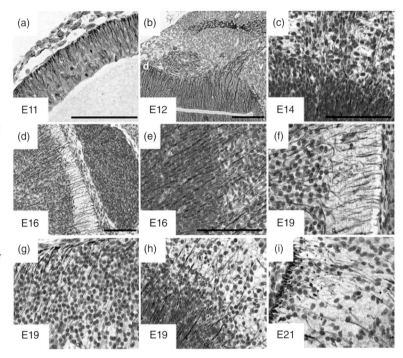

Figure 4.28 Presence of radial glial cells in the developing embryonic nervous system as marked by staining for granulocyte colony stimulating factor (G-CSF). (a) E11 forebrain; (b) E12 spinal cord with dorsal root ganglion, axon root and muscle; (c) E14 hindbrain; (d) E16 spinal cord with dorsal root ganglion; (e) E19 spinal cord; (f) E19 spinal cord; (g) E19 hindbrain; H, E21 olfactory bulb; (i) E21 diencephalon (immunohistochemical staining of 10 μm paraffin sections, scale bar = 50 μm, d: dorsal, E: embryonic day, v: ventral). (*Source:* Kirsch et al., 2008. Licensed under the Creative Commons Attribution License CC-BY-2.0.)

In the adult brain, neurogenesis occurs primarily in the subventricular zone (SVZ) of the lateral ventricles (Figure 4.29). Neurogenesis is also evident in the subgranular zone of the dentate gyrus in the adult brain hippocampus (Figure 4.30). This is discussed in more detail in the section Signaling and Multipotency.

The sections above outline the embryonic and adult anatomical origins of neural stem cells, yet they may be isolated from a variety of sources for study *in vitro* or possible expansion and use in cell-based therapies. One of the first successful examples of neural stem cell

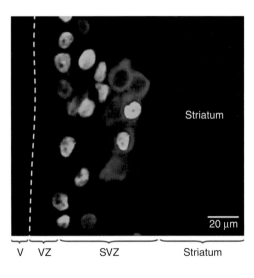

Figure 4.29 Fluorescence microscopy demonstrating the presence of adult neural stem cells in the rat brain subventricular zone. NSC nuclei were labeled with BrdU (green). V, Ventricle; VZ, Ventricular zone; SVZ, Subventricular zone. (Photo courtesy Oscar Arias-Carrión and Wikimedia Commons; reprinted with permission.)

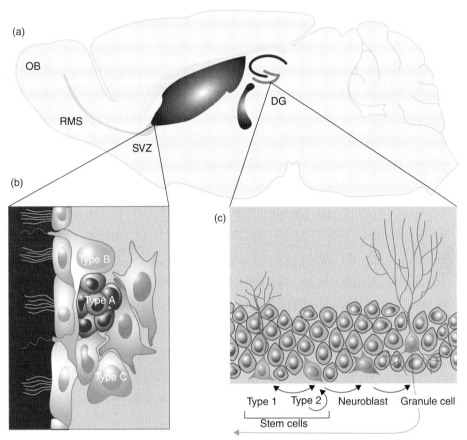

Figure 4.30 Neurogenesis in the adult brain. (a) Adult NSCs are primarily located in two germinal zones of the brain: the SVZ of the lateral ventricles and the SGZ of hippocampal DG. (b) A subset of relatively quiescent GFAP+ radial cells (type B cells) in the SVZ has the potential to serve as adult NSCs and generate rapidly dividing, transit-amplifying non-radial NSCs (type C cells), which in turn give rise to neuroblasts (type A cells) that migrate through the RMS toward the OB. (c) In the adult SGZ, a population of GFAP+ Sox2+ radial cells corresponds to quiescent NSCs (type 1 cells). They coexist with actively proliferating, GFAP- Sox2+ non-radial NSCs (type 2 cells) that generate both astrocytes and neuroblasts. Neuroblasts then migrate into the granule cell layer and mature into neurons. (Figure courtesy Fred H. Gage and Elsevier, Ltd.; figure legend text courtesy Rob Burgess and *Stem Cells Handbook 2nd Edition*, reprinted with permission.)

isolation and culture occurred in 1989 when Sally Temple (see Focus Box 4.2), a researcher in the Department of Physiology and Biophysics at the University of Miami Medical School, isolated what she referred to as "blast" cells (cells capable of dividing) from E13.5-14.5 rat forebrain. These cells could not only divide in tissue culture but could be manipulated to differentiate into a variety of clonal cell types including neurons, glia, and non-neuronal cells. Dr. Temple demonstrated that for this particular system live conditioning cells must also be used in co-culture, suggesting that environmental signaling events induce proliferation. In addition, she determined that heterogeneous clones were capable of developing in homogeneous culture conditions which revealed that cell autonomous signaling events are also key to determining clonal progeny numbers and cell types.

Focus Box 4.2: Sally Temple and the isolation of "blast" cells

In 1989 University of Miami pioneering neural stem cell researcher Sally Temple developed a cell culture system that allowed for the efficient isolation and characterization of what she termed "blast" cells from the rat embryonic forebrain. This was the first successful isolation of what are now known as neural stem cells and opened the door for many researchers worldwide to more efficiently study neurogenesis. Dr. Temple is currently the co-founder and Scientific Director of the Neural Stem Cell Institute and Professor of Neuroscience and Neuropharmacology at Albany Medical College. (Photo courtesy NIH; reprinted with permission.)

Three primary sources for the large-scale production of neural stem cells have been exploited, including

- Immortalization using oncogenes
- Generation using neurospheres (Figure 4.31)
- Derivation from embryonic stem cells (not discussed here)

These sources provide a virtually unlimited supply of neural stem cells due to the fact that replication and self-renewal is virtually indefinite in each case. Immortalization of neural progenitors may be accomplished via the introduction of nucleic acid sequences coding for **SV40 large T antigen (Simian Vacuolating Virus 40 TAg)**. SV40 large T antigen is a proto-oncogene originating in the polyomavirus SV40 that acts through an alteration in the naturally occurring tumor suppressor proteins p53 and retinoblastoma (RB), thereby disrupting the regulation of cell cycle and apoptosis. **Neurospheres** are defined as free-floating cultures of neural stem cells. The three-dimensional nature of the culture system is conducive to large-scale expansion of NSCs. Derivation of neural stem cells from pluripotent ES cells will not be discussed here.

Signaling and Multipotency Although this section focuses on ASC characteristics, the vast majority of research directed at deciphering not only the multipotent capabilities of neural stem cells but also the signaling mechanisms regulating their behavior has been undertaken using embryogenesis as a model system. Many of the signaling cascades involved in guiding such NSC properties as migration and commitment are conserved in embryos and post-natally. During embryonic development NSC behavior can be broken down chronologically into three phases as discussed below.

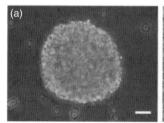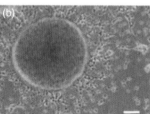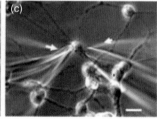

Figure 4.31 Neurospheres consisting of rat SVZ cells in culture. A, 2 days post-plating; B, 4 days post-plating demonstrating cell migration out of and away from the neurosphere. C. Cells at the periphery of neurospheres were chosen for electrophysiological recording. Most of the recorded cells extended processes. (Photographs courtesy Wikimedia Commons; reprinted with permission.)

EXPANSION

Expansion is defined as the growth of a population of cells through symmetric cell division. During the expansion phase, which occurs at very early stages of embryogenesis, **neuroepithelial cells**, which are early-stage columnar neural stem cells, divide symmetrically for the purpose of expanding the population of progenitors needed to populate the developing embryonic CNS with much needed neurons and glial cells. In vertebrates, neuroepithelial cell expansion occurs near the embryonic ventricle and pial surfaces, and even at this early stage NSCs begin to differentiate into mature neurons.

NEUROGENESIS

Neurogenesis occurs after cell populations have expanded to the point when numbers are sufficient to begin commitment to terminal lineages. At this point, neuroepithelial cells are referred to as neural stem cells and they begin to divide asymmetrically, with one cell remaining mitotically active and the other exiting the cell cycle for differentiation into a mature cell type. It is through asymmetric cell division that populations of NSCs continue to increase while mature neuronal lineages are produced simultaneously. It is the germinal VZ of the embryo that is rapidly populated with undifferentiated NSCs needed to produce mature neuronal cell types. A thickening of the **neural tube**, which is defined as the embryonic precursor to the central nervous system comprising the future brain and spinal cord, coincides with inductive signaling cues provided by it and surrounding tissues to drive a transition in NSC morphology from a columnar to a radial appearance. This morphological transition and the related signaling events driving it result in the development of radial glial cells, defined above, which will differentiate into astrocytes and cortical neurons. This process occurs not only during embryonic development but has been confirmed to occur in the adult vertebrate brain throughout life, specifically in the SVZ of the lateral ventricles and the subgranular zone of the dentate gyrus (see Figure 4.30).

Neurogenic signaling cues that induce changes in NSC behavior described above are complex and often involve multiple factors acting in concert. For example, the Wnt/β-catenin signaling pathway has been confirmed as a major driving force for neurogenesis in both embryos and adults through the activation of the basic helix-loop-helix transcription factor NeuroD1. **Basic helix-loop-helix proteins** are transcription factors containing a unique protein structural motif characterized by two α helices separated by a loop. This structure allows for efficient homo- and hetero-dimerization with other bHLH family members and provides a potent mechanism for tight regulation of neurogenesis (and other pathways) (Figure 4.32).

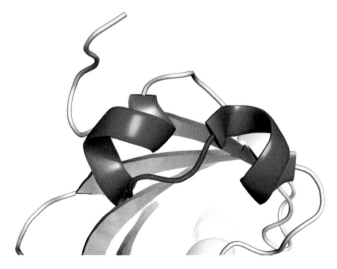

Figure 4.32 Diagrammatic illustration of the basic helix-loop-helix motif. (Diagram courtesy Thomas Splettstoesser and Wikimedia Commons; reprinted with permission.)

Other bHLH transcription factors that influence neurogenesis include neurogenin1 and 2, specifically regulating waves of neurogenesis which occur in the developing **dorsal root ganglia (DRG)**, which is defined as a nodule on the afferent sensory root of the central spinal nerve containing bodies of nerve cells. Mash1 (Asc1), another bHLH transcription factor, has been implicated in promoting the development of olfactory bulb GABAergic **interneurons**, which are defined as neurons that form connections with other neurons. In the juxtaglomerular region, both neurogenin2 and Tbr2 drive glutamatergic neuronal development. Finally, with respect to bHLH transcription factor regulation of neurogenesis, the Hes family of bHLH proteins is negative regulators that antagonize the differentiation promoting effects for other positive bHLH regulators in order to regulate the numbers of neural stem cells needed for the production of different progeny. In addition, both the sonic hedgehog (shh) pathway and the transcription factor Sox2 have been demonstrated to control NSC numbers in the adult brain, with shh acting as a potent mitogen with the negative regulator Sox2 inhibiting the expression of **glial fibrillary acidic protein (GFAP)**. GFAP is an intermediate protein filament expressed in glial cells and represents early gliogenic events in NSCs. It should be noted that shh also promotes NSC mitogenic activity in the developing embryo.

GLIOGENESIS

Gliogenesis is defined as the generation of glial cells such as astrocytes and radial glia from neural stem cells. In the adult vertebrate brain, it occurs after most neurogenesis has completed within the SVZ and involves intermediate glial progenitor cells. An elegant *in vitro* model system has been developed to study gliogenesis based upon embryoid bodies. **Embryoid bodies** are three-dimensional structures composed of pluripotent stem cells (Figure 4.33).

Researchers in the Department of Physiology, Graduate School of Medicine, Keio University, Japan, lead by Hideyuki Okano have shown through the application of *in vitro* knockdown of key genes in embryoid bodies that the transcription factor COUP-TFII acts to activate the expression of GFAP by eliminating its silencing by epigenetic effects. Gliogenesis

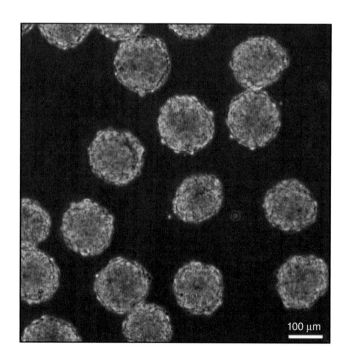

Figure 4.33 Phase contrast microscopy of mouse embryoid bodies. Embryoid bodies (EBs) were cultured in suspension and photos taken after 24 hours of formation. EBs were created from approximately 1000 mouse embryonic stem cells (D3 cell line). Scale bar = 100 μm. (Photo courtesy Wikimedia Commons; reprinted with permission.)

100 μm

is also positively regulated by the bHLH protein neurogenin2 and negatively regulated by various homeobox-containing transcription factors such as Emx2, Sox5, and Sox6. This negative regulation actually acts to promote what has been referred to as the "default" neurogenic pathway. Signaling factors that act through a number of the transcription factors mentioned to promote gliogenesis include those of the Notch, BMP, sonic hedgehog, EGF, and FGF families. Much of this signaling is concentration-dependent, and without appropriate levels of the individual molecules, gliogenesis will not occur (Figure 4.34) (Temple, 2001).

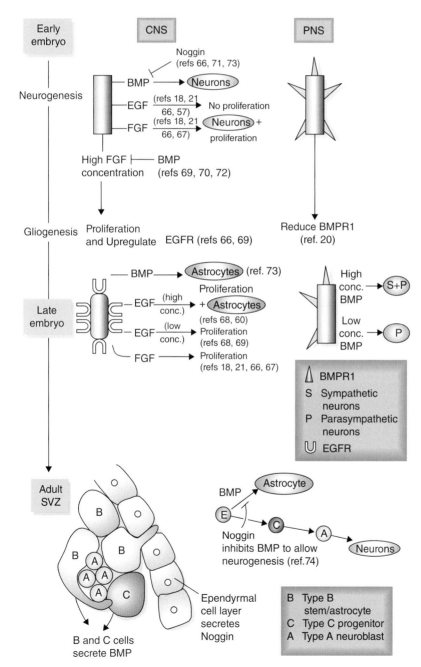

Figure 4.34 Signaling cascades affecting neural stem cells activity *in vivo*. Multiple signaling molecules act at different points in the neurogenic and gliogenic phases of nervous system development in the embryo and its maintenance in the adult SVZ. Many of these act in a concentration-dependent manner to either promote or inhibit the differentiation of neuronal or glial cells. (Diagram courtesy Sally Temple and *Nature* (Temple, 2001); reprinted with permission.)

Endothelial Stem Cells HSCs are not the only multipotent cell type present within the bone marrow. **Endothelial stem cells**, also referred to as endothelial progenitor cells (EPCs), reside in the bone marrow as well, and are comprised of a multitude of progenitors that can differentiate into numerous endothelial adult cell lineages. EPCs are thought to be a later-stage version of endothelial stem cells, but for the purposes of this description the terms will be used interchangeably. It is thought that these cells are derived from hemangioblasts. EPCs are known to migrate to the peripheral regions of bone marrow, proliferate, and terminally differentiate into adult endothelial cells that have the primary purpose of creating a thin **endothelium** lining the walls of the vasculature.

Morphology and Marker Expression Numerous markers provide a valuable resource for the isolation of endothelial stem cells. EPCs express CD34 (Figure 4.35), VEGFR-2, and CD133. It is **CD133**, a trans-membrane glycoprotein, which is a true marker of early-stage endothelial stem cells, as mature endothelial cells lack in it expression.

In addition, EPCs have a transcriptional profile that is unique in comparison to other stem cell types. Researchers at the Centre for Vision and Vascular Science, School of Medicine, Dentistry & BioMedical Science, Queen's University, Belfast, performed a comprehensive analysis of two distinct populations of EPCs: early EPCs (eEPCs) and outgrowth endothelial cells (OECs). From a morphological perspective, endothelial stem cells have either a spindle- (eEPCs) or cobblestone-shaped (OECs) appearance depending upon the type. As Case Study 4.3 illustrates, these two populations are quite unique in molecular, biochemical, and morphological makeup. Specifically, the researchers identified numerous transcripts differentially expressed between eEPCs and OECs including some representing unexpected genes such as EGF-containing fibulin-like extracellular matrix protein 1 (see Table 4.5). Interestingly, eEPCs expressed more endothelial cell-related transcripts at much higher levels than OECs but took on a hematopoietic phenotype when software-based transcriptome analysis was performed (see Table 4.5 and text below) (Medina et al., 2010).

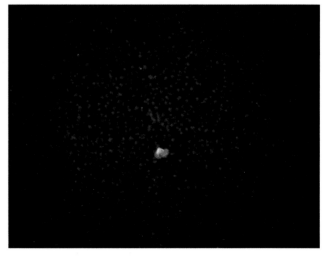

Figure 4.35 Fluorescence microscopy of endothelial stem cells expressing the cell surface marker CD34. (Photo courtesy Louis Gagné and Wikimedia Commons; reprinted with permission.)

TABLE 4.5 Differential expression of key transcripts between early endothelial progenitor cells and outgrowth endothelial cells.

Gene name; symbol	Fold increase in OECs vs. eEPCs	Fold increase in eEPCs vs. OECs
Caveolin 1; CAV1	216.27	
EGF-containing fibulin-like extracellular matrix protein 1; EFEMP1	199.8	
Connective Tissue Growth Factor; CTGF	106.24	
VE-Cadherin; CDH5	100.55	
Cysteine-rich, angiogenic inducer, 61; CYR61	88.06	
Claudin 5; CLDN5	79.78	
Endothelial cell-specific molecule 1; ESM1	69.92	
Laminin, alpha5; LAMA5	61.16	
Complement component 1, q subcomponent, C chain; C1QC		257.10
TYRO protein kinase binding protein; TYROBP		250.38
CD163; CD163		244.85
Secreted phosphotein 1; SPP1		239.94
Fc fragment of IgE; FCER1G		222.58
Lymphocyte cytosolic protein 1; LCP1		220.90
ADAM-like, decysin 1; ADAMDEC1		206.59
CD14; CD14		176.01
Matrix metallopeptidase 9; MMP9		173.02
Colony stimulating factor 1 receptor; CSF1R		164.51

Source: Adapted from Medina et al. (2010).

Upon further analysis of the **transcriptome**, defined as the set of all RNA molecules produced by one or a population of cells, of the endothelial stem cell subtypes it was revealed that each exhibits similarities to entirely different cell types. Specifically, the **interactome network**—the network of interactions between genes—of eEPCs resembled that of hematopoietic cells, whereas that of OECs was much more similar to endothelial cells (Beltrami et al., 2003).

Sources and Origins As mentioned above, endothelial stem cells can be isolated from the bone marrow, peripheral blood (isolated from peripheral blood mononuclear cells), or umbilical cord blood. The latter are often referred to as **cord blood-derived embryonic-like stem cells (CBEs)**. Endothelial stem cells are also present in the developing embryo and play a crucial role in **vasculogenesis,** the development and maturation of the circulatory system. It is generally accepted that EPCs are specialized cell types directly derived from HSCs, yet growing evidence is pointing to EPCs originating from other non-hematopoietic sources such as the mesenchyme and even adult tissues (Figure 4.37) (Shantsila et al., 2007).

Case Study 4.3: Molecular analysis of EPC subtypes reveals two distinct cell populations with different identities

Reinhold J. Medina, Christina L. O'Neill, Mark Sweeney, Jasenka Guduric-Fuchs, Tom A. Gardiner, David A. Simpson, and Alan W. Stitt

A team of researchers in Belfast, UK, led by Alan W. Stitt performed a comprehensive characterization of two distinct subtypes of EPCs. This study is one of the most thorough of its kind with respect to the analysis of EPC identity and revealed striking differences in the two subpopulations. From a morphological perspective, eEPCs exhibited a spindle-shaped appearance whereas OECs were more cobblestone-like. Organelle analysis via transmission electron microscopy revealed lysosome-like structures in eEPCs not found in OECs, while OECs exhibited abundant caveolae not found in eEPCs. In addition, extensive transcriptional profiling of both cell populations revealed striking differences in gene expression patterns, with the transcriptome of eEPCs resembling that of hematopoietic cells while OECs exhibited a transcriptional profile much more similar to endothelial cells (see main text and Figure 4.36) .

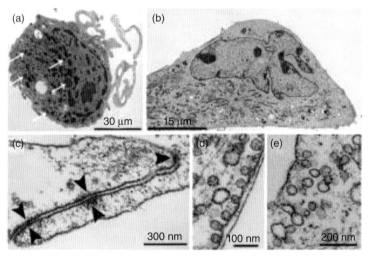

Figure 4.36 Ultrastructure characterization of EPCs via transmission electron microscopy. a, eEPCs; b, OECs; c, adherin junctions; d, caveolae in single pits; e, caveolae in clusters. (Photos courtesy Alan W. Stitt and *BMC Medical Genomics* (Medina et al., 2010); reprinted with permission.)

For example, stem cells capable of producing endothelial lineages, referred to as **cardiac stem cells**, have been isolated from the adult heart. These cells express the stem cell-related surface antigen c-kit and the crucial myogenic homeodomain-containing myogenic transcription factor Nkx2.5 (Figure 4.38) (Beltrami et al., 2003. Given the utility of EPCs in providing differentiated, mature cell types for the vasculature and other organ systems, it is anticipated that yet other sources of these cells will be discovered in the near future.

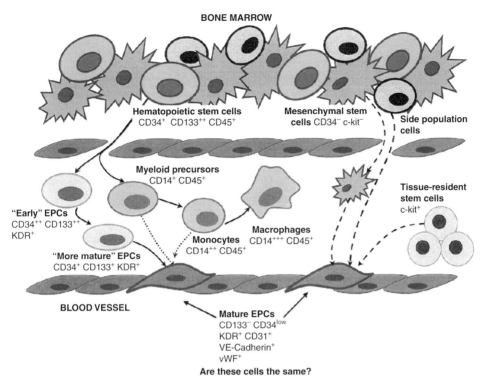

BONE MARROW

Hematopoietic stem cells
CD34⁺ CD133⁺⁺ CD45⁺

Mesenchymal stem
cells CD34⁻ c-kit⁻

Side population
cells

Myeloid precursors
CD14⁺ CD45⁺

Tissue-resident
stem cells
c-kit⁺

"Early" EPCs
CD34⁺⁺ CD133⁺⁺
KDR⁺

Macrophages
CD14⁺⁺⁺ CD45⁺

"More mature" EPCs
CD34⁺ CD133⁺ KDR⁺

Monocytes
CD14⁺⁺ CD45⁺

BLOOD VESSEL

Mature EPCs
CD133⁻ CD34^low
KDR⁺ CD31⁺
VE-Cadherin⁺
vWF⁺

Are these cells the same?

Figure 4.37   Schematic illustration of endothelial progenitor cell sources and differentiation. (Schematic courtesy Eduard Shantsila and the *Journal of the American College of Cardiology* (Shantsila et al., 2007); reprinted with permission.)

Signaling and Multipotency

COMBINATORIAL REGULATION OF EPC BEHAVIOR

Endothelial stem cells play crucial roles in both **angiogenesis**, the formation of new blood vessels from pre-existing blood vessels, and **neovasculogenesis**, the development of new blood vessels through *de novo* production of endothelial cells. The signaling mechanisms that result in the mobilization of EPCs from the bone marrow for the purpose of driving

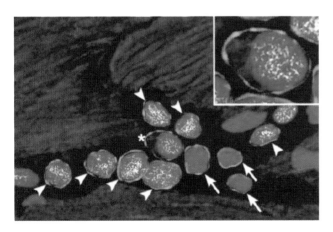

Figure 4.38   Fluorescence microscopy of cardiac stem cells in the heart. Green = c-kit positive cells; white dots = Nkx2.5; blue = nuclei; red = α-sarcomeric actin. (Photo courtesy Antonio P. Beltrami and *Cell* (Beltrami et al., 2003); reprinted with permission.)

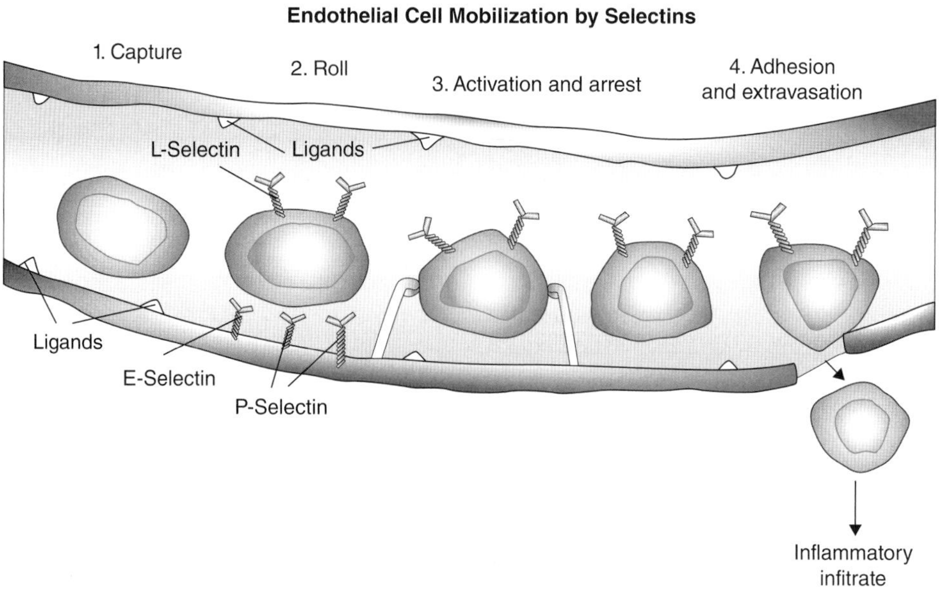

Endothelial Cell Mobilization by Selectins

Figure 4.39 E- and P-selectin mobilization of endothelial cells. (Artwork courtesy Connie Zhao.)

neovasculogenesis are poorly understood, with the exception of selectin interactions. **Selectins** are a family of single transmembrane domain glycoprotein cell surface receptors that bind to sugar moieties present on glycoproteins. It has been demonstrated that the cell surface receptor P-selectin glycoprotein ligand (PSGL-1) interacts specifically with P-selectin and E-selectin, which are expressed on existing mature endothelial cells allowing for mobilization in response to numerous physiological conditions including, but not limited to, inflammation (Figure 4.39).

This is closely followed by the activation of EPCs via both autocrine and paracrine signaling, which ultimately result in the control of EPC proliferation and differentiation. Mobilization and homing of EPCs to the tissue periphery is the result of β integrin expression. **Integrins** are heterodimeric proteins consisting of a and b subunits. They play roles in cell migration through both recognition and adhesion mechanisms. Specifically, the β2 integrins LFA-1 and Mac-1 as well as β1 integrin drive peripheral migration of EPCs. EPCs are mobilized from the bone marrow in response to the paracrine signals emanating from ischemic tissue and tumors including GM-CSF and VEGF. In addition, a hypoxic (low oxygen) environment, which results in the activation of the transcription factor **Hypoxia Inducible Factor (HIF-1)** and hence activation of the target gene VEGF also promotes EPC recruitment locally. **Vascular endothelial growth factor (VEGF)** is a key modulator of EPC homing. VEGF may act in a paracrine fashion locally or through autocrine mechanisms as a hormone via secretion into the circulatory system. VEGF is known to tightly interact with receptors VEGFR1 and VEGFR2 on both endothelial and HSCs, resulting in the activation of the matrix metalloproteinase MMP-9. Activated MMP-9 cleaves Kit ligand, which further induces EPC proliferation and migration.

ANGIOGENIN AND ANGIOPOIETIN

Numerous other pro-angiogenic factors also influence EPC behavior. These include **angiogenin**, a 14-kDa ribonuclease that cleaves certain tRNAs in cells, resulting in decreased protein synthesis and an induction of angiogenesis. It has been speculated that angiogenin may directly interact with follistatin, another pro-angiogenic protein. **Angiopoietins** are a family of angiogenic factors involved in the regulation of neovasculogenesis, both during embryonic development and post-natally. The angiopoietins Ang-1 and Ang-2 are thought to act through direct binding to, and activation of, the tyrosine kinase receptors Tie-1 and Tie-2, which subsequently activates the PI3 kinase/Akt pathway in EPCs, resulting in a mobilization of the cells from bone marrow reserves into the periphery for such processes as synovial vasculature repair (Figure 4.40).

Cytokines and Chemokines

Cytokines have been widely demonstrated to be potent regulators of EPC behavior and function. For example, granulocyte colony-stimulating factor (G-CSF) and granulocyte/monocyte colony-stimulating factor (GM-CSF) not only stimulate the generation of granulocytes in the bone marrow, but also promote the proliferation, migration, and differentiation of EPCs. Interleukin 8 (IL-8) has also been demonstrated to promote EPC proliferation concomitant with the formation of capillary tubes. It has been shown that EPC proliferation and migration are heavily influenced by the local presence of chemokines (see Figure 4.40). The chemokines RANTES/CCL5 and stromal cell derived factor-1α (SDF-1α) both affect EPC behavior. RANTES binds to CCR5 receptors expressed on the surface of EPCs resulting in a modulation of the cells for atherogenesis. SDF-1α interacts with EPC CXCR4 receptors to promote the recruitment of cells to regions exhibiting gradients of low oxygen concentration (hypoxia).

Tumor Effects on EPC Behavior

Tumors also recruit EPC to enter the circulatory system through the secretion of various factors such as FGF, osteopontin, CCL2, and CCL5. Circulating EPCs are known to play a

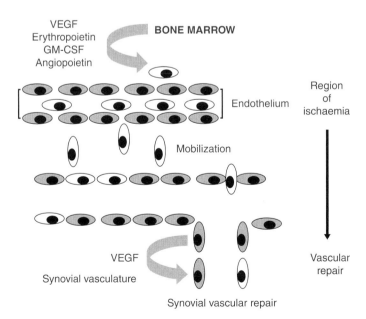

Figure 4.40 Diagrammatic illustration of endothelial cell mobilization by angiopoietin and other signaling molecules. See text for details.

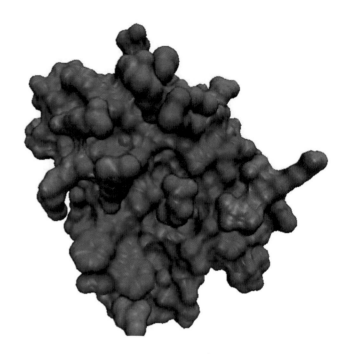

Figure 4.41   Molecular structure of endostatin. (Image courtesy Wikimedia Commons; reprinted with permission.)

role in the maintenance of vascular integrity at tumorigenic sites, thus ensuring a supply of nutrients to growing tumors. Anti-angiogenic therapeutics have been studied intensely as avenues to block this process and choke off nutrient supply to tumors. An example of this is **endostatin**, which is a 20-kDa monomer protein derived from type XVIII collagen and has been shown to have anti-angiogenic properties (Figure 4.41). Endostatin is classified as an angiogenesis inhibitor and is speculated to interfere with the pro-angiogenic properties of basic FGF and vascular endothelial growth factor (VEGF).

CHAPTER SUMMARY

Discovery and Origin of ASCs

1. James Till and Ernest McCulloch inadvertently discovered the existence of ASCs present in a population of murine bone marrow cells exhibiting indefinite replicative capacity.
2. ASCs are present throughout the body in various undifferentiated states depending upon origin and play a primary and crucial role in acting as a source for the replenishment of dead and dying differentiated cells.

Basic Properties of ASCs

1. The two basic properties of ASCs are self-renewal and multipotency.
2. Self-renewal can occur in a symmetric or asymmetric fashion.
3. ASCs can give rise to more than one lineage and are hence multipotent.

Examples of ASCs

1. There are many sources of ASCs.

2. ASCs often exist in specific "niches."

3. HSCs have the capacity to differentiate into all blood cell lineages and exist as three distinct populations.

4. Human HSCs are smaller in size, spheroid in appearance, with a spheroid nucleus, and have a strikingly smooth surface in comparison to other cell types.

5. HSCs are found mostly in the bone marrow of adults.

6. The Theory of Hematopoiesis defined in Chapter 1 states that all blood cells are derived from a common precursor cell.

7. HSCs arise from a unique and complex interplay of crosstalk and signaling between both bone marrow cells of the endosteal niche, and the vascular system of the vascular niche.

8. Myosatellite cells produce progeny known as **myoblasts** and represent the oldest known ASC niche.

9. During the process of myogenesis muscle-derived stem cells transition from an undifferentiated state into quiescent myosatellite cells and later become activated to reenter the cell cycle and become committed myogenic precursors known as myoblasts.

10. Myosatellite cells can be found in embedded between the basement membrane.

11. Human multipotent ASCs and other ASCs are strikingly similar to fibroblasts, exhibiting an elongated, spindle shape.

12. ASCs are typically isolated from adipose tissue, specifically from WAT.

13. hMADS and other ASCs have the potential to differentiate into a variety of mature cell types including those which form bone, cartilage, muscle, and fat.

14. In 1924 the Russian-American scientist Alexander A. Maximow discovered mesenchymal stem cells.

15. Mesenchymal stem cells are small with a large round nucleus and low cytoplasm-to-nucleus ratio.

16. Mesenchymal stem cells have several unique features not exhibited by other types of stem cells including immunomodulatory and immunoprivileged capacities.

17. Mesenchymal stem cells may be isolated from numerous embryonic and post-natal sources.

18. Mesenchymal stem cells isolated from different sources are not all alike.

19. Several inducing molecules and platforms have been outlined which efficiently drive MSC osteoblast commitment and maturation.

20. *In vitro* studies have elucidated three primary inducers of MSC-based adipogenesis including dexamethasone, indomethacin, and insulin.

21. Cell-to-cell contact promotes MSC adipogenesis over osteogenesis.

22. Joseph Altman and Gopal Das were the first to demonstrate neurogenesis in the adult brain.

23. *In vivo*, neurogenesis during embryonic development occurs, for the most part, in the germinal ventricular zone (VZ).

24. In the adult brain neurogenesis occurs primarily in the subventricular zone (SVZ) of the lateral ventricles.

25. Sally Temple developed a cell culture system that allowed for the efficient isolation and characterization of what she termed "blast" cells, now known as neural stem cells, from the rat embryonic forebrain.

26. There are three primary sources for the large-scale production of neural stem cells including immortalization using oncogenes, generation of neurospheres, and derivation from embryonic stem cells.

27. The three phases of neural stem cell maturation during embryonic development include expansion, neurogenesis, and gliogenesis.

28. Neurogenesis and gliogenesis are regulated by basic helix-loop-helix transcription factors.

29. Endothelial stem cells reside in the bone marrow as well and are composed of a multitude of progenitors that can differentiate into numerous endothelial adult cell lineages.

30. EPCs are known to migrate to the peripheral regions of bone marrow, proliferate, and terminally differentiate into adult endothelial cells that have the primary purpose of creating a thin endothelium lining the walls of the vasculature.

31. Two distinct populations of EPCs exist—early EPCs (eEPCs) and outgrowth endothelial cells (OECs)—and they each exhibit unique transcriptional profiles.

32. EPCs are specialized cell types directly derived from HSCs and other non-hematopoietic sources such as the mesenchyme and even adult tissues.

33. Endothelial stem cells play crucial roles in both angiogenesis and neovasculogenesis.

34. EPCs are activated via both autocrine and paracrine signaling which ultimately result in the control of EPC proliferation and differentiation.

35. Integrins, hypoxia, growth factors, and cytokines all regulate EPC behavior.

36. Tumors are known to recruit EPCs to enter the circulatory system for the maintenance of vascular integrity at tumorigenic sites.

KEY TERMS

(Key terms are listed by order of appearance in the text.)

- **Soma**—Greek for "body."
- **Senescence**—changes in molecular and cellular structure resulting in the prevention of further cell division.
- **Symmetric division**—an adult stem cell gives rise to two identical daughter cells.
- **Asymmetric division**—an adult stem cell gives rise to one identical daughter cell and one progenitor cell that will terminally differentiate into a mature lineage.
- **"Stem cell niche"**—the microenvironment wherein a stem cell resides which provides the necessary growth and differentiation signaling events which are conducive to both cellular division and differentiation.
- **Hematopoietic stem cells (HSCs)**—a heterogeneous population of multipotent stem cells that can differentiate into the myeloid or lymphoid cell types of the adult blood system.
- **Hemangioblasts**—multipotent precursors that have the ability to differentiate into either endothelial cells or hematopoietic cells.

- **L/M ratio**—the ratio of lymphoid to myeloid progenitor cells in the blood.
- **Hematopoiesis**—the formation of blood cellular components.
- **Endosteal niche**—the interface between marrow and bone that is lined by osteoblasts.
- **Vascular niche**—sinusoidal endothelium encompassing a microenvironment of vasculature.
- **Endosteum**—a thin layer of connective tissue lining the surface of long bone medullary cavities.
- **Myelopoiesis**—the differentiation of hematopoietic stem cells into myeloid cells.
- **CAR cells**—CXCL12-abundant reticular cells found in the bone marrow that affect both the migration and localization of HSCs through the secretion of SDF-1.
- **Hypoxia**—less oxygen than under normal physiological conditions.
- **Repopulation**—the ability to recolonize the bone marrow.
- **Myoblasts**—mesodermally derived progenitor cell that differentiates to give rise to muscle cells known as myocytes.
- **Myosatellite cells**—progenitor cells present in muscle capable of terminal differentiation and fusion to enhance or augment existing muscle fibers and to form new fibers.
- **Myoblasts**—the progeny of myosatellite cells.
- **Muscle-derived stem cells (MDSCs)** —progenitor cells derived from muscle which have the ability to differentiate into muscle cells as well as other mesodermally-derived lineages such as bone and fat cells.
- **Basement membrane**—a thin sheet of fibers underlying the epithelium that lines the cavities of various organs, including muscle.
- **Sarcolemma**—the cellular membrane of mature individual muscle fibers.
- **Myostatin**—a secreted molecule, also known as growth differentiation factor 8, which promotes muscle differentiation and inhibits myosatellite cell self-renewal through the induction of the cyclin-dependent kinase p21.
- **Contact inhibition**—an exiting from the cell cycle upon contact with neighboring cells.
- **Lipoaspirate**—material acquired during the surgical removal of subcutaneous fat.
- **Stromal vascular fraction (SVF)** —the cell pellet collected after centrifugation of enzyme digested lipoaspirate.
- **Autocrine**—same-cell signaling.
- **Paracrine**—near-cell signaling.
- **Mesenchyme**—mesodermally-derived embryonic connective tissue that has the capacity to differentiate into hematopoietic lineages and mature, adult connective tissue.
- **Mesenchymal stem cells (MSCs)**—multipotent stromal cells arising from the mesenchyme and they can be isolated from a variety of sources.
- **Immunomodulatory**—the capacity to regulate immune response in a localized microenvironment by secreting cytokines.
- **Immunoprivileged**—resistance to bystander effects of local immune responses.
- **Transdifferentiation**—a cell's ability to differentiate into lineages of different embryonic germ layers.
- **Runx2**—also known as CBF-α-1, a member of the Runt DNA binding domain family of transcription factors and has been shown to be essential for both the differentiation of cells into osteoblasts as well as later stage skeletal morphogenesis.

- **Indomethacin**—a non-steroidal anti-inflammatory drug.
- **Peroxisome proliferator activated receptor PPARγ**—a type II nuclear receptor involved in glucose metabolism and fatty acid storage.
- **Hypoxia**—less oxygen content than under normal physiological conditions.
- **Neurogenesis**—the process by which neurons and glial cells are generated from neural stem and progenitor cells.
- **Microneurones**—mitotically active cells present in the adult rat brain initially discovered by Joseph Altman and Gopal D. Das now known as neural stem cells.
- **Nestin**—a protein component of nerve cell type VI intermediate filaments which play a role in axonic radial growth.
- **Radial glia**, which are defined as neurogenic precursor cells present in the ventricular zone that divide asymmetrically to give rise to more radial glia, intermediate progenitor cells and post-mitotic neurons.
- **SV40 Large T antigen**—a proto-oncogene originating in the polyomavirus SV40 which acts through an alteration in the naturally occurring tumor suppressor proteins p53 and retinoblastoma, RB, thereby disrupting the regulation of cell cycle and apoptosis.
- **Neurospheres**—free-floating cultures of neural stem cells.
- **Expansion**—the growth of a population of cells through symmetric cell division.
- **Neuroepithelial cells**—early-stage columnar neural stem cells, divide symmetrically for the purpose of expanding the population of progenitors needed to populate the developing embryonic CNS with much needed neurons and glial cells.
- **Neural tube**—the embryonic precursor to the central nervous system.
- **Basic helix-loop-helix proteins**—transcription factors containing a unique protein structural motif characterized by two α helices separated by a loop that regulate a variety of processes involved in differentiation.
- **Dorsal root ganglia (DRG)** —a nodule on the afferent sensory root of the central spinal nerve that contains bodies of nerve cells.
- **Interneurons**—neurons that form connections with other neurons.
- **Glial fibrillary acidic protein (GFAP)** —an intermediate protein filament expressed in glial cells and represents early gliogenic events in NSCs.
- **Gliogenesis**—the generation of glial cells such as astrocytes and radial glia from neural stem cells.
- **Embryoid bodies**—three-dimensional structures comprised of pluripotent stem cells.
- **Endothelial stem cells**—also referred to as endothelial progenitor cells (EPCs), these cells reside in the bone marrow and are comprised of a multitude of progenitors that can differentiate into numerous endothelial adult cell lineages.
- **Endothelium**—a thin layer of cells lining the interior surface of blood vessels and lymphatic vessels.
- **Transcriptome**—the set of all RNA molecules produced by one or a population of cells.
- **Interactome network**—the network of interactions between genes.
- **Cord blood-derived embryonic-like stem cells (CBEs)**—endothelial stem cells isolated from umbilical cord blood.
- **Vasculogenesis**—the development and maturation of the circulatory system.
- **Cardiac stem cells**—endogenous tissue-specific progenitor cells found in the heart that can differentiate into cardiac lineages.

- **Angiogenesis**—the formation of new blood vessels from pre-existing blood vessels.
- **Neovasculogenesis**—the development of new blood vessels through *de novo* production of endothelial cells.
- **Selectins**—a family of single trans-membrane domain glycoprotein cell surface receptors that bind to sugar moieties present on glycoproteins.
- **Integrins**—heterodimeric proteins consisting of a and b subunits which play roles in cell migration through both recognition and adhesion mechanisms.
- **Hypoxia inducible factor (HIF-1)** —a transcription factor activated by low local oxygen content.
- **Vascular endothelial growth factor (VEGF)**—a signaling molecule that promotes vasculogenesis and angiogenesis, often through the recruitment of endothelial progenitor cells.
- **Angiogenin**—a 14kDa ribonuclease which acts to cleave certain tRNAs in cells resulting in decreased protein synthesis and an induction of angiogenesis.
- **Angiopoietins**—a family of angiogenic factors involved in the regulation of neovasculogenesis, both during embryonic development and post-natally.
- **Endostatin**—a 20 kDa monomer protein derived from type XVIII collagen and has been shown to have anti-angiogenic properties.

REVIEW QUESTIONS

(Answers to select review questions can be found at www.stemcelltextbook.com.)

1. List and describe the two basic properties of adult stem cells.
2. List at least four different examples of adult stem cells, outlining their tissues of origin and mature lineages produced.
3. What is a "stem cell niche" and why is it important for regulating stem cell behavior?
4. List the seven different lineages produced by hematopoietic stem cells and describe the L/M ratio necessary for the existence of each population of HSCs.
5. What are some sources of hematopoietic stem cells?
6. Describe the interplay and crosstalk of molecular signaling between the two niches which regulate HSC behavior.
7. How did Seiburg and colleagues demonstrate the similarity in self-renewal and differentiation properties of HSC daughter progeny?
8. What seminal discovery was made in 1961 regarding muscle development and who made it?
9. Outline the proposed pathway for myogenesis.
10. What homeodomain-containing transcription factors are crucial for myosatellite cell differentiation?
11. Describe the process of adipose-derived stem cell isolation and purification.
12. What lineages are ASCs competent to differentiate into and what are some of the inducing factors identified for each pathway?
13. List at least three criteria for the identification and classification of mesenchymal stem cells.

14. What key transcription factor and signaling protein drives MSC osteogenic versus adipogenic differentiation?

15. How does PPARγ act to drive adipogenesis?

16. Describe how Altman and Das identified microneurons in the adult rat brain.

17. What is the most widely studied/used marker for neural stem cells?

18. Where does neurogenesis occur in the developing mammalian embryo and in the adult?

19. How can immortalization of neural stem cells be accomplished?

20. Describe the three phases of neural stem cell behavior during embryonic development.

21. What is the primary function of endothelial stem cells?

22. Compare and contrast early endothelial progenitor cells and outgrowth endothelial cells from a morphological and molecular perspective.

23. What are three sources of endothelial stem cells?

24. How do endothelial progenitor cells mobilize and home in on peripheral tissue?

25. How do tumors affect endothelial progenitor cell behavior?

THOUGHT QUESTION

If you were going to establish a new adult stem cell population, which lineage would you choose and how would you go about the isolation? Include methods to isolate the primary cells as well as culture, expansion, and confirmation of differentiation capacities.

SUGGESTED READINGS

Altman, J. and G. D. Das (1965). "Post-natal origin of microneurones in the rat brain." *Nature* **207**(5000): 953–956.

Beltrami, A. P., L. Barlucchi, et al. (2003). "Adult cardiac stem cells are multipotent and support myocardial regeneration." *Cell* **114**(6): 763–776.

Jankowski, R. J., B. M. Deasy, et al. (2002). "Muscle-derived stem cells." *Gene Ther* **9**(10): 642–647.

Kirsch, F., C. Kruger, et al. (2008). "The receptor for granulocyte-colony stimulating factor (G-CSF) is expressed in radial glia during development of the nervous system." *BMC Dev Biol* **8**: 32.

Kundu, A. K. and A. J. Putnam (2006). "Vitronectin and collagen I differentially regulate osteogenesis in mesenchymal stem cells." *Biochem Biophys Res Commun* **347**(1): 347–357.

Locke, M., V. Feisst, et al. (2011). "Concise review: human adipose-derived stem cells: separating promise from clinical need." *Stem Cells* **29**(3): 404–411.

McCroskery, S., M. Thomas, et al. (2003). "Myostatin negatively regulates satellite cell activation and self-renewal." *J Cell Biol* **162**(6): 1135–1147.

Medina, R. J., C. L. O'Neill, et al. (2010). "Molecular analysis of endothelial progenitor cell (EPC) subtypes reveals two distinct cell populations with different identities." *BMC Med Genomics* **3**: 18.

Muller-Sieburg, C. E., R. H. Cho, et al. (2002). "Deterministic regulation of hematopoietic stem cell self-renewal and differentiation." *Blood* **100**(4): 1302–1309.

Nauta, A. J. and W. E. Fibbe (2007). "Immunomodulatory properties of mesenchymal stromal cells." *Blood* **110**(10): 3499–3506.

Paleolog, E. (2005). "It's all in the blood: circulating endothelial progenitor cells link synovial vascularity with cardiovascular mortality in rheumatoid arthritis?" *Arthritis Res Ther* **7**(6): 270–272.

Relaix, F., D. Rocancourt, et al. (2005). "A Pax3/Pax7-dependent population of skeletal muscle progenitor cells." *Nature* **435**(7044): 948–953.

Shantsila, E., T. Watson, et al. (2007). "Endothelial progenitor cells in cardiovascular disorders." *J Am Coll Cardiol* **49**(7): 741–752.

Temple, S. (2001). "The development of neural stem cells." *Nature* **414**(6859): 112–117.

Villageois, P., B. Wdziekonski, et al. (2012). "Regulators of human adipose-derived stem cell self-renewal." *Am J Stem Cells* **1**(1): 42–47.

Ytteborg, E., A. Vegusdal, et al. (2010). "Atlantic salmon (Salmo salar) muscle precursor cells differentiate into osteoblasts in vitro: polyunsaturated fatty acids and hyperthermia influence gene expression and differentiation." *Biochim Biophys Acta* **1801**(2): 127–137.

Zhang, Y., D. Khan, et al. (2012). "Mechanisms underlying the osteo- and adipo-differentiation of human mesenchymal stem cells." *ScientificWorldJournal* **2012**: 7938–7923.

Chapter 5

NUCLEAR REPROGRAMMING

The ability to alter a cell's fate for either research purposes or the generation of a more desirable cell line to be used in a therapeutic setting has been a goal of researchers almost since the inception of the scientific discipline of cell biology. Technologies that could allow for the transition of a cell from one phenotype to another could enable not only new avenues for the treatment of human disease but also usher in a new era of biological research driven by advancements in science that could provide a limitless supply of virtually any cell type. In addition, whole organisms such as mice, rats, or porcine animal models created from single cell isolates could yield a wealth of information related to gene function and even epigenetic effects on lifespan and mortality. Most importantly, technologies that would allow for the reprogramming of genetic material from adult cells could allow for the creation of pluri- or multipotent stem cell populations that are patient-specific, that is genetically identical to the cell type donated by the patient for reprogramming. Patient-specific stem cell populations provide the unique and critical advantage of immunocompatibility and eliminate risks of immunorejection by the patient's own immune system of the cells, tissues, or organs used in the treatment regimen. As an example, the derivation of stem cells from a skin cell biopsy of a patient in need of liver repair might provide a valuable cell-based platform for the unlimited generation of patient-specific hepatocytes for use in cell-based therapy. This is an example of **autologous stem cell-based therapy** and is considered to be the holy grail of stem cell research. Many medical disorders may be addressed with autologous cell replacement therapy including Parkinson's, Alzheimer's, and diabetes, but the genetic material of donor cells must be properly and effectively reprogrammed to release the potency needed for the generation of new cell types. **Nuclear reprogramming** is defined as the induction of changes in gene activity that result in a comprehensive transformation of cellular phenotype. This chapter deals with three groundbreaking technologies developed to reprogram nuclei and alter cell fate: cell fusion, somatic cell nuclear transfer, and induced pluripotency (Figure 5.1). These technologies have had profound impacts on biological scientific research and the quest for disease intervention. The basics behind technological design, studied and developed model systems, and therapeutic implications will be discussed.

Stem Cells: A Short Course, First Edition. Rob Burgess.
© 2016 John Wiley & Sons, Inc. Published 2016 by John Wiley & Sons, Inc.

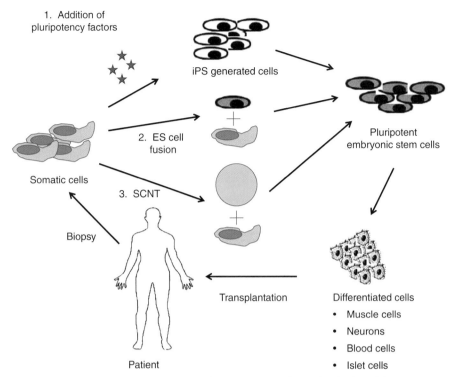

1. Addition of pluripotency factors

iPS generated cells

2. ES cell fusion

Somatic cells

3. SCNT

Biopsy

Pluripotent embryonic stem cells

Transplantation

Patient

Differentiated cells
- Muscle cells
- Neurons
- Blood cells
- Islet cells

Figure 5.1 Diagrammatic illustration of nuclear reprogramming technologies. (Diagram courtesy Jinnuo Han, Kuldip S. Sidhu and InTech; redrawn with permission.)

Case Study 5.1: Reprogramming adult Schwann cells to stem cell-like cells by leprosy bacilli promotes dissemination of infection

Toshihiro Masaki, Jinrong Qu, Justyna Cholewa-Waclaw, Karen Burr, Ryan Raaum, and Anura Rambukkana

Differentiated cells may be manipulated through reprogramming mechanisms toward new cell fates. Researchers studying the leprosy bacterium demonstrated that it acts to disseminate infectivity through the reprogramming of Schwann cells, which are defined as the principle glial cells of the peripheral nervous system that function to support neurons in a myelinating and non-myelinating fashion. Schwann cells are reprogrammed by leprosy bacterium into a progenitor-/stem cell-like phenotype. Reprogrammed Schwann cells, referred to as progenitor/stem-like cells (pSLC) are highly plastic. In addition, they migrate and secrete immunomodulatory factors that further drive and exacerbate leprosy spread through the promotion of bacteria-laden macrophage release. Finally, the researchers demonstrated that pSLCs subsequently redifferentiate into meso-dermal lineages including skeletal and smooth muscle. How does leprosy bacteria accomplish the dedifferentiation and subsequent mesodermal redifferentiation of Schwann cells? Rambukkana's team deciphered an epigenetic mechanism promoting changes in Schwann cell identity. Specifi-cally, genes involved in driving mesodermal fate and function such as Twist and Tbx18 were significantly less methylated in reprogrammed Schwann cells when compared to uninfected con-trols. Less methylation tends toward less transcriptional silencing, and indeed these genes became active after infection in Schwann cells. These findings reveal a striking epigenetic mechanism driven by bacterial infection for enhancement of infectivity (Masaki et al., 2013).

EXAMPLES OF NUCLEAR REPROGRAMMING IN NATURE

In nature, nuclear reprogramming is not a new concept and numerous instances have been described in which the genomic global genomic organization and the discrete genetic activity of cells have been altered or reprogrammed to induce changes such as dedifferentiation or **transdifferentiation**, which is defined as the change of a differentiated cell into another unique differentiated cell (also referred to as **metaplasia**). An example of reprogramming-based metaplasia includes the transdifferentiation of cells in the iris to that of a lens phenotype. Often transdifferentiation in nature occurs as a response to an external stimulus. For example, lung exposure to cigarette smoke is known to drive the reprogramming and transdifferentiation of columnar epithelial cells into stratified squamous epithelia. One of the most shocking examples of nuclear reprogramming occurs naturally upon infection by leprosy bacterium. As Case Study 5.1 highlights, Anura Rambukkana and her team at the Medical Research Council in London and the Rockefeller University in New York studied leprosy mechanism of action at the cellular level and demonstrated that the bacterium has the unique ability to hijack cell fate.

CELL FUSION

Cell fusion is the joining of two unique cell types to form a single entity. The resulting cell can be classified as either a **hybrid** (if dividing) or a **heterokaryon** (if non-dividing). The hybrid or heterokaryon typically becomes reprogrammed due to influence of both intranuclear changes such as the activity of transcription factors and intracytoplasmic changes such as organelle makeup. The multinuclear cell may also be referred to as a **syncytium**. Cell fusion not only has been devised as a method for the cellular reprogramming but also occurs naturally—for example, during embryonic development and the differentiation of bone and muscle. Mature muscle cells are syncytia of fused mononuclear myocytes resulting from successive cell fusion (Figure 5.2).

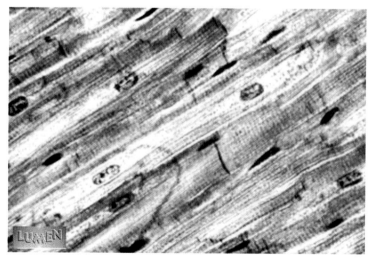

Figure 5.2 Photograph of multinucleated mature cardiac muscle cells formed by cell fusion. (Photo courtesy Michael J. Farabee and Estrella Mountain Community College; reprinted with permission.)

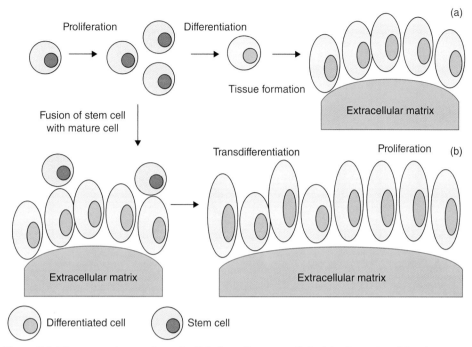

Figure 5.3 Diagrammatic overview of cell fusion effects on cellular identity and proliferation. Differentiated, mature cells may fuse to aid in tissue formation (a) while stem cells may fuse with mature cells to drive a reprogramming of cellular identity and result in transdifferentiation or an alteration of cell cycle (b).

In embryonic development and in the maintenance of mature adult tissues and organs, cell fusion is thought to affect both cellular identity and proliferative capacities (Figure 5.3).

Like vertebrates, a muscle in the fruit fly *Drosophila melanogaster* is actually a syncytium of fused myoblast cells, the result of a successive fusion of mesodermally derived individual cells arising during embryonic development. Vertebrate bone differentiation proceeds through successive fusion of what are referred to as osteoclast precursor cells, which ultimately results in multinucleated mature bone. Placental trophoblasts are also known to fuse, resulting in the formation of **syncytiotrophoblasts** which is the epithelial covering of the embryonic placental villi (Figure 5.4).

Many non-vertebrate species depend upon the formation of syncytium for the proper development of organs and tissues. For example, the hypodermis, excretory glands, uterus, and other tissues in the nematode *Caenorhabditis elegans* require extensive syncytia formation. Fusion of epithelial cells in certain instances may cause rearrangements of **adherens junctions**, which are protein complexes that occur at the contact points of cells. In these junctions, the cytoplasmic face of the cell membrane is linked to the actin cytoskeleton and its conformational state determines the spatial order of the cells involved in the fusion process. Cell fusion is also known to occur naturally in bacteria, yeast, and fungi.

From an experimental perspective, it should be noted that the application of cell fusion for the generation of a new and unique cell types is not a new or recent concept, and has

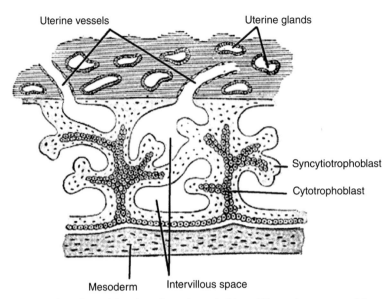

Figure 5.4 Illustration of a multinucleated syncitiotrophoblast. (Illustration courtesy Magnus Manske and Wikimedia Commons; reprinted with permission.)

been successfully accomplished for a number of different types of cells. Cell fusion studies can be traced back to 1960 when Georges Barski, Serge Sorieul, and Francine Cornefert of the Institut Gustave Roussy, Laboratorie de Culture de Tissus et de Virologie in Villejuif, France, first observed the *in vitro* fusion tumor cells derived from two different strains of mice (Figure 5.5).

It was Sorieul who observed the occurrence of a new type of cell in the mixed cultures that exhibited a karyotype containing nearly twice the normal number of chromosomes with chromosomal markers unique to each strain. These initial findings were later expanded upon by the cell biologist Leo Sachs at the Weizmann Institute of Science in Israel, where he and his colleague David Gershon successfully fused murine tumor cells and confirmed fusion by demonstrating the expression of histocompatibility antigens originating from both parental strains. This was further confirmed *in vivo* as the resulting fused cells were immunorejected by both parental strains but could be grown in F1 mice. As discussed below, perhaps the biggest impact resulting from experimental cell fusion has been for the generation of research-grade and therapeutic monoclonal antibodies.

Cell Fusion for the Generation of Hybridomas Reports of successful *in vitro*-driven cell fusion resulted in an explosion of research in this area given the obvious impact it would have on genetics research. In 1975, a full 15 years after the initial report on cell fusion in murine cells and following the accumulation of a wealth of technological know-how in the area, Cesar Milstein in the Laboratory of Molecular Biology, Cambridge University, and Georges Kohler of the Max Planck Institute for Immunobiology in Freiburg, Germany, successfully produced the world's first **hybridoma** cell line, a cell line resulting from the fusion of an antibody-producing B cell with a myeloma cancer cell (see Focus Box 5.1 and Figure 5.6) (Kohler and Milstein, 1975).

Figure 5.5 Institut Gustave Roussy, Laboratorie de Culture de Tissus et de Virologie in Villejuif, France, where cell fusion was first reported in 1960. (Image courtesy the Institut Gustave; reprinted with permission.)

Milstein and Kohler demonstrated that the resulting immortalized (cancerous) cell line could produce an unlimited supply of high-affinity monoclonal antibodies—antibodies of an identical class (Sapir et al., 2008). They won the 1984 Nobel Prize in Medicine and Physiology for this research. Milstein refused to patent hybridoma technology, strongly believing that it was the intellectual property of mankind and should be freely pursued by all involved in science and medicine.

Focus Box 5.1: Milstein, Kohler, and the invention of hybridomas

 In 1975, Argentine cell biologist Cesar Milstein and German biologist Georges Kohler produced the world's first hybridoma cell line which produced monoclonal antibodies of unlimited supply. This finding revolutionized the study, use and application of antibodies in research and therapeutics and resulted in an award of the 1984 Nobel Prize in Medicine and Physiology. Prior to his death in 2002, Milstein was named Fellow of the Royal Society and Fellow of Darwin College and received both the Copley Medal and Louisa Gross Horwitz Prize. In 1984, Kohler became Director of the Max Planck Institute for Immunobiology where he worked until his death in 1995. (Photo courtesy Life Sciences Foundation; reprinted with permission.)

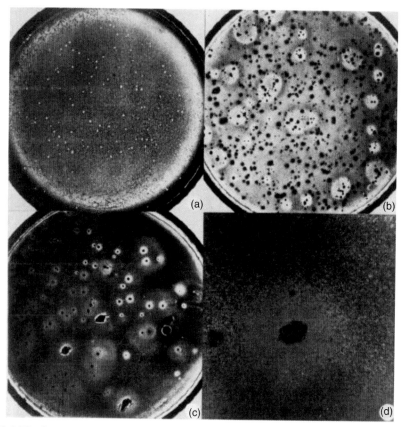

Figure 5.6 The first hybridoma cell lines. (a) 6,000 hybrid cells grown in tissue culture. (b) Soft agar growth of 2,000 hybrid lines. (c) The first recloning of a positive line. (d) Higher magnification of a positive hybridoma. (Figure courtesy G. Kohler, C. Milstein and *Nature* (Kohler and Milstein 1975); reprinted with permission.)

The procedures and methodologies employed for the production of hybridoma cell lines involve a combination of *in vivo* and *in vitro* applications. After a suitable antigen has been chosen, laboratory animals, most often mice, are immunized via a series of antigen injections over a period of several weeks (Figure 5.7 (1)). This allows for the development of an immune response, specifically activating **splenocytes**, which are defined as white blood cells consisting of populations of T and B lymphocytes, dendritic cells, and macrophages. It is the B lymphocytic population that is of primary interest in hybridoma production. After effective immunization of the animal, these cells are isolated from the spleen, cultivated (Figure 5.7 (2)), and fused with immortalized myeloma cells, which have been preselected for an absence of antibody secretion (Figure 5.7 (3)). Fusion of B and myeloma cells may be accomplished either via the use of polyethylene glycol (PEG), electroporation or by infection with the **Sendai virus**. PEG treatment results in an increase in membrane permeability. This is thought to occur through PEG interaction with water molecules, resulting in dehydration of lipid headgroups, thereby making them more susceptible to fusion. Sendai virus (SeV) is an RNA virus well known to induce the formation of eukaryotic syncytia *in vitro*.

Figure 5.7 Diagrammatic illustration of the procedure for generating a hybridoma. See text for details. (Diagram courtesy Wikimedia Commons: reprinted with permission.)

The mechanism is thought to occur as a result of cellular expression of viral F protein post infection coupled with the expression of the protein HN that allows for the fusion of adjacent cells (a detailed explanation of fusion mechanisms for the different techniques is described below). Following fusion, cells are cultured in the presence of hypoxanthine-aminopterin-thymidine (HAT) medium. HAT contains the drug aminopterin which blocks the synthesis of nucleotides, thus allowing for the selection of **hypoxanthine-guanine-phosphoribosyl transferase (HGPRT)**, an enzyme that drives *de novo* nucleotide synthesis and expressed by B cells. The presence of HGPRT activity allows for nucleotide synthesis, and, thus, cell survival in the presence of HAT. Following fused cell selection (Figure 5.7 (5)), selection of clonal cell lines which

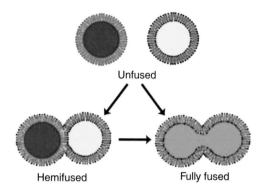

Figure 5.8 Diagrammatic illustration of the different states of cell fusion. (Diagram courtesy Wikimedia Commons; reprinted with permission.)

express antibodies of the appropriate titer and specificity is undertaken and desired clones are expanded for large-scale antibody production (Figure 5.7 (6)). Expansion of desired cell lines can occur either *in vitro* or *in vivo* (Figure 5.7 (7)). The last step involves the harvesting of biologically active monoclonal antibodies for research, diagnostic, or therapeutic use (Figure 5.7 (8)).

Mechanisms of Cell Fusion

The formation of eukaryotic syncytia requires the fusion of cellular membranes between two cells. With lipid bilayers the fusion process is less than straightforward, and can proceed to various extents, including **hemifusion**, in which only the outer cell membrane lipid layers are fused, or full fusion in which both layers are fused (Figure 5.8).

Various proteins have now been identified which are involved in the cell fusion process. Often referred to as **cell–cell fusogens**, these proteins actively mediate the fusion process. The EFF-AFF family of type I membrane glycoproteins have been demonstrated to be essential in the formation of syncytia in the nematode *C. elegans*, with mutants effectively blocking the fusion process. These proteins possess an N-terminal signal sequence juxtaposed to an extracellular transmembrane domain and an intracellular tail. Given the conservation between different FF family members, especially with respect to the extracellular domain, it has been speculated that all family members fold in a similar manner and that the three-dimensional nature of the glycoproteins is essential for proper cell–cell fusion (Figure 5.9) (for a comprehensive review see (Sapir et al., 2008)). Upon completion of full fusion, both the outer and inner lipid bilayers are continuous between both cells, resulting in an aqueous bridge between the merged cells.

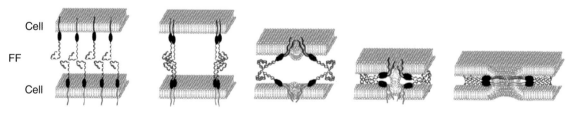

Figure 5.9 Diagrammatic illustration of the hypothetical cell fusion process mediated by FF proteins in *Caenorhabditis elegans*. Brown ellipses represent protein extracellular domains proximal to the cysteine-rich domain thought to form an "S-S" based structure (blue). It is speculated that protein homodimers are formed via a zippering mechanism that results in membrane tightening and hemifusion (far right diagram). (Diagram courtesy Amir Sapir and *Cell* (Sapir et al., 2008); reprinted with permission.)

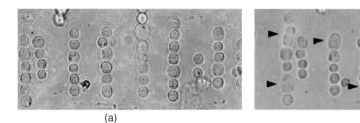

Figure 5.10 Brightfield microscopy of cell–cell electrofusion. (a) Aligned cells after application of electropulses. (b) Arrows denote fusion products. (Images courtesy Wikimedia Commons; reprinted with permission.)

Comparison of Cell Fusion Techniques

Electrofusion A comparison of the three methods for generating cell fusions reveals different mechanisms for generating resulting homo-, hetero- or synkaryons. In **electrofusion**—which in this context is the use of an electric field or pulse to drive the fusion of cells—an electric pulse is generated in a low-conductivity medium containing cells. The resulting dielectrophoresis aligns cells via a high-frequency alternating current (Figure 5.10). Changes in ionic charge and conductivity result in an increase in membrane permeability and breakdown, thus promoting new pore formation between membranes of adjacent cells, and, ultimately, membrane fusion.

Pegylation **Polyethylene glycol (PEG)** is a polyether compound with many applications from industrial manufacturing to medicine. It was first utilized to fuse eukaryotic cells in 1975 when Q.F. Ahkong and colleagues in the Department of Biochemistry and Chemistry at the University of London successfully joined hen erythrocytes with yeast protoplasts. Since then it has been widely used for the fusion of a vast array of both plant and animal cells. PEG may promote cellular fusion via several mechanisms. PEGs unique molecular structure, consisting of linear ethylene oxides combined with hydroxyl terminals allows for efficient binding to water molecules. This is thought to result in dehydration of regions proximal to PEG, thus bringing juxtaposed membranes together in a tight complex. Second, PEG provides a positive state of osmotic pressure that is thought to stabilize intermediate membrane fusion complexes. It has been demonstrated that PEG promotes the fusion of a variety of different cell types, including tumor with non-tumor based lines, and is generally accepted as the most effective of the three primary fusion technologies (Figure 5.11) (Kang et al., 2006).

Viral Induction The most popular method of virally induced cell fusion is based on the application of the Sendai virus (Figure 5.12). Also known as the Hemagglutinating Virus of Japan (HVJ), Sendai binds specifically to target membrane cell surface receptors through viral envelope glycoproteins. The binding process is most effective at a neutral pH. Cell–cell fusion induced by the Sendai virus was first reported in tumor cells in 1962 by Yoshio Okada and Jun Tadokoro, researchers in the Department of Preventative Medicine, Osaka University, Osaka, Japan. These studies revealed that successful fusion required the adsorption of many viral particles, with fusion percent efficiencies increasing with increasing numbers of adsorbed viral particles above 1,300 (Okada, 1962). Three years later, in 1965, researchers at the University of Oxford demonstrated the successful fusion of murine and human cells, and characterized their mitotic activity. This opened the door to the use of Sendai for interspecies cell fusion studies. The mechanism of Sendai-mediated cell fusion is well understood. Inactivated viral particles bind to cells through the acetyl type sialic acid receptor via recognition by viral neuraminidase (HN) protein. Binding results in cellular

Figure 5.11 Fluorescence microscopy of PEG-based cell fusion. (Red) F344 rat bone marrow tumor cells. (Green) NuTU-19 cells. (Orange) Resulting cell fusions. (Figure courtesy Yu Kang and *Vaccine* (Kang et al., 2006); reprinted with permission.)

membrane distortion due to viral F protein membrane penetration. An influx of ions into bound cells promotes cell–cell fusion. The application of HJV for mediating cell–cell fusion is considered to be moderately successful at best and is dependent upon pH levels.

Focus Box 5.2: Helen Blau and cell fusion-based nuclear reprogramming

In 1983, a seminal publication by Stanford researcher Helen Blau revealed that mature, differentiated cells exhibited a plastic phenotype which could be altered via the introduction of new genetic material by cell fusion. These studies laid the groundwork for many future endeavors related to cell fusion-mediated somatic cell nuclear reprogramming. Dr. Blau has received numerous awards and honors including the Senior Career Recognition Award of WICB of the American Society of Cell Biology, the FASEB Excellence in Science Award and an Honorary Doctorate from the University of Nijmegen, Holland. She is currently the Director of the Baxter Laboratory for Stem Cell Biology at Stanford University School of Medicine. (Photo courtesy Stanford University; reprinted with permission.)

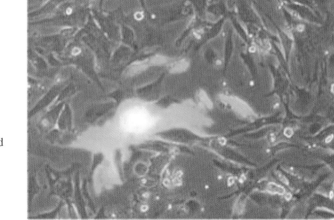

Figure 5.12 Fluorescence microscopy of HJV-induced fused neonatal rat primary cardiac myocytes and P19 cells. (Photo courtesy CosmoBio, Tokyo, Japan; reprinted with permission.)

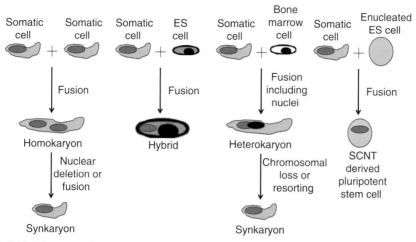

Figure 5.13 Diagrammatic illustration of cell fusion between similar or different cell types and the resulting phenotypes. (Figure courtesy Jinnuo Han, Kuldip S. Sidhu and InTech; redrawn with permission.)

Mechanism of Nuclear Reprogramming in Cell Fusion

How does cell fusion result in genetic reprogramming and ultimately a change in cellular identity? Two primary mechanisms are thought to be involved in the reprogramming process. At the macro-level, changes in chromosomal makeup result in a global change in gene expression simply due to the fact that some chromosomes are lost and others are retained in the re-sorting process. Alterations in the epigenetic makeup of cells also may play a role. **Epigenetics** is defined as gene expression or cellular phenotype patterns resulting from properties other than underlying DNA sequences. Examples of these properties include the DNA methylation or histone modification status. Cell fusion often has a profound impact on the epigenetic status of the resulting fused cell, thereby promoting changes in gene expression and ultimately cellular phenotypes. For example, fusion between identical cell types results in the formation of a **homokaryon**, defined as two genetically identical nuclei. The result is often a fusion between the two nuclei and extensive genomic rearrangement, resulting in the formation of a **synkaryon**. Synkaryons may also be formed by the fusion of two unique cell types such as two non-identical somatic cells or a somatic cell and an embryonic stem cell. An intermediate stage in the formation of a synkaryon is the development of a **heterokaryon**, which contains multiple nuclei. Synkaryons often undergo extensive genomic rearrangement, resulting in the activation of inactive or silenced genes (Figure 5.13). Synkaryons usually exhibit significant chromosomal loss, rearrangement or resorting, thus extensively changing the genetic identity of the resulting cell type and thus its overall morphological and molecular phenotype.

The reactivation of silenced genes in cell fusions was first reported in 1983 by Helen Blau and her colleagues in the Department of Pharmacology at Stanford University when they successfully fused muscle and amniotic cells. This groundbreaking research revealed that differentiated cell types possessed an inherent plasticity in phenotype that could be manipulated via the global introduction of new genetic material in the formation of heterokaryons. Silencing genes in differentiated cells can be reactivated resulting in new phenotypes (see Focus Box 5.2).

SOMATIC CELL NUCLEAR TRANSFER

Somatic cell nuclear transfer (SCNT) is a laboratory technique for the generation of a cloned cell utilizing a donor nucleus. **Cloning** is the process of generating similar populations of genetically identical entities, whether individual cells or whole organisms. In SCNT, the nucleus of a somatic cell is removed and isolated intact from the cellular cytoplasm and membrane and subsequently transplanted into an enucleated host oocyte. The host egg, due to loss of the original nucleus, is considered to be "deprogrammed," and hence receptive to reprogramming via the introduction of a nucleus. Nuclear transfer is accomplished via direct injection of the nucleus into the host egg or by fusion. Environmental cues emanating from the enucleated oocyte drive a reprogramming of the epigenetic status of the cell. Cellular division is subsequently stimulated via a shock or series of shocks to produce genetically identical daughter cells. These cells can either be allowed to continue growth into an early-stage embryo for transplantation into a surrogate mother, defined as **reproductive cloning**, or dispersed and expanded in tissue culture for purposes of generating therapeutically advantageous cells, defined as **therapeutic cloning** (Figure 5.14).

The successful cloning of whole organisms is not a new concept and was first accomplished in amphibians in the early 1950s. It has been well established that amphibious adult keratinocyte nuclei can be transferred into enucleated oocytes and support embryonic development to the tadpole stage. However, as discussed below, the definitive, transformational breakthrough in this field occurred in 1996 when Sir Ian Wilmut, Keith Campbell, and colleagues at the Roslin Institute, in collaboration with Scottish company PPL therapeutics, optimized somatic cell nuclear transfer technology for the cloning of an adult sheep (Figure 5.15). Wilmut and Campbell went on to win numerous awards for their seminal studies including the prestigious Shaw Prize for Stem Cell Discoveries which they split with stem cell researcher Shinya Yamanaka (see below) (Byrne et al., 2007).

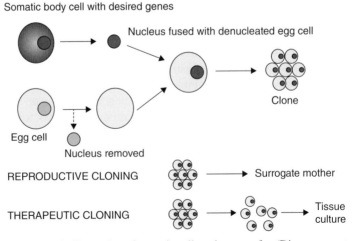

Figure 5.14 Diagrammatic illustration of somatic cell nuclear transfer. (Diagram courtesy Jurgen Groth and Wikimedia Commons; reprinted with permission.)

Figure 5.15 Photograph of Dolly the cloned sheep and her lamb, Bonnie. *Source:* Courtesy of The Roslin Institute, The University of Edinburgh.

The application of somatic cell nuclear transfer became increasingly popular in the early 2000s as it remained difficult to identify reliable, renewable multipotent stem cell populations in neonatal or adult organisms that had therapeutic potential—or that at least produced a multitude of desired, differentiated cell types for study on a large scale. SCNT was optimized as a way to reprogram the nucleus of an adult, differentiated cell via the presence of cytoplasmic environmental cues. SCNT is most relevant with respect to the development of pluripotent or multipotent stem cells for therapeutic use derived from the original donor source, which negates the need to suppress the patient's immune system when performing cell-based therapy (implants or grafts). The following sections outline the methodologies employed and the properties exhibited by SCNT cells as they apply to both research and therapy.

Method for the Production of SCNT-Derived Cells

Methods for the production of SCNT-derived cells and corresponding organisms are varied and depend not only on the organism of choice for study but also the desired outcome, that is, new cell types or entire organism development. Thus, for the purposes of this text, the strategy developed by Wilmut and colleagues for cloning Dolly the sheep will be described. In these studies, the researchers transferred mammary cell nuclei from Finn-Dorset donors to enucleated eggs of Scottish Blackface sheep, thus providing a coat color identification scheme for the determination of successful cloning. The researchers chose to utilize mammary cells from a 6-year-old Finn-Dorset ewe in the last trimester of pregnancy as nuclear donor cells. Cells were induced to enter the G0 phase of the cell cycle (**quiescence**), thus priming them and it was speculated that this transition results in a chromatin structure change that may promote nuclear reprogramming. Nuclear transfer was accomplished utilizing protocols optimized by the same group in a previous study for the transfer of nuclei from an established cell line. A direct current pulse was applied to enhance the efficiency of the system (Figure 5.16). See also Case Study 5.2 for further details.

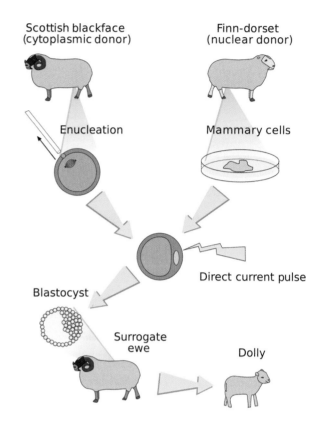

Figure 5.16 Diagrammatic illustration for the SCNT-based cloning of Dolly the sheep. (Diagram courtesy Wikimedia Commons; reprinted with permission.)

Somatic Cell Nuclear Transfer for the Creation of Stem Cells

The studies on the generation of Dolly the sheep described above focused on somatic cell nuclear transfer as a means for the production of genetically identical whole animals from adult cells. The remainder of this section will describe the use and application of SCNT as a means to produce populations of multi- or pluripotent stem cells that are genetically identical to the donor (patient), referred to as therapeutic cloning (also discussed above). As these cells would not elicit an immune response, their relevance from a cell-based therapeutics clinical perspective is significant. In a manner similar to that applied for reproductive cloning in animals, therapeutic cloning involves the transfer of an intact somatic cell nucleus into an enucleated oocyte. The resulting embryo is cultured, typically to the blastocyst stage, followed by the isolation and characterization of embryonic stem cells. Given the controversial nature of the system and the inherent destruction of viable embryos in the generation of SCNT-derived embryonic stem cells, the vast majority of efforts in this area have been focused on the use of non-human primates as a model system. SCNT-derived non-human primate embryonic stem cells were first successfully isolated by Shoukhrat Mitalipov and his team at the Oregon Stem Cell Center and Oregon Primate Research Center, Oregon Health & Science University in Beaverton. In these studies, a modified SCNT approach was designed and implemented to produce ES cells from rhesus macaque blastocyst-stage embryos that were generated from the transfer of adult fibroblast nuclei into enucleated oocytes. The modified protocols were designed to prevent a decline in **maturation promoting factor (MPF)** activity. MPF is a heterodimer of cyclin-dependent

Case Study 5.2: Viable offspring derived from fetal and adult mammalian cells

I. Wilmut, A. E. Schnieke, J. McWhir, A. J. Kind, and K. H. S. Campbell

In 1997, Ian Wilmut and colleagues successfully cloned viable offspring from adult mammalian cells. In this study, the authors demonstrated the need for induction of cellular quiescence as a prerequisite to "prime" the epigenetic and heterochromatic nature of the donor cells genome. It was speculated that induction of quiescence might provide increased genomic access to reprogramming factors present in the cytoplasm of the oocyte. In addition to deriving offspring from adult mammary cells, the authors derived offspring from fetal and embryonically derived cells, suggesting an optimized procedure had been developed for cloning from cells of multiple origins and that nuclei from these cells have the capacity to be totipotent (Figure 5.17). These studies are consistent with the generally accepted hypothesis that mammalian differentiation is the result of changes in gene expression which are heavily influenced by the mitotic state of the cells as well as the cytoplasmic environment. They also confirmed that cellular terminal differentiation does not result in an irreversible modification of the genetic material necessary for proper embryonic development to full term and post-natal viability (Wilmut et al.,1997).

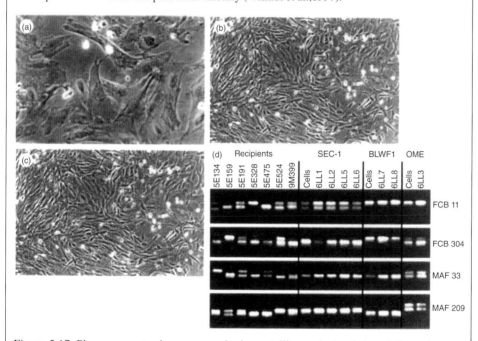

Figure 5.17 Phase contrast microscopy and microsatellite analysis of cloned sheep donor cells. (a) Embryo-derived cells. (b) Fetal fibroblasts. (c) Adult mammary cells. (d) Microsatellite marker analysis of donor cells, recipient ewes, and resulting lambs confirming resulting clones. (Figures courtesy Ian Wilmut and *Nature* (Wilmut et al., 1997); reprinted with permission.)

kinase, CDK1, and cyclin B. It promotes mitosis through the phosphorylation of key proteins involved in mitotic activity. Removal of fluorochrome bisbenzimide (a.k.a. Hoechst 33342) and UV light from the procedure played a key role in the promotion of blastocyst formation. This simple modification allowed for a significant increase in the success rate of

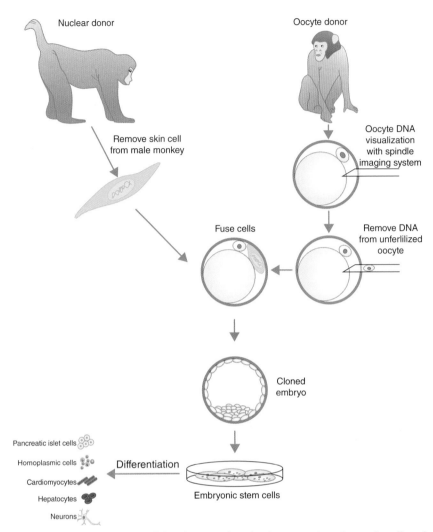

Figure 5.18 Schematic diagram outlining the procedure for the generation of somatic cell nuclear transfer-derived non-human primate embryonic stem cells. See text for details. (Diagram courtesy J.A. Byrne, S.M. Mitalipov and *Nature* (Byrne et al., 2007); reprinted with permission.)

blastocyst development and ultimately the generation of cloned primates. The general procedure, however, remained the same as with other species, including mice, and is outlined in Figure 5.18 (Byrne et al., 2007).

Basic Properties of SCNT-Derived Stem Cells Somatic cell nuclear transfer is performed primarily for the production of embryonic stem cells that are genetically identical to adult cell populations derived from the donor organism. Having been isolated from blastocyst-stage embryos in most cases, these cells are considered to be embryonic stem in origin and thus pluripotent (Figure 5.19) (Byrne et al., 2007). They possess all of the hallmarks of true embryonic stem cells while having been derived from a somatic cell. Table 5.1 outlines some of the properties exhibited by non-human primate SCNT-derived embryonic stem cells.

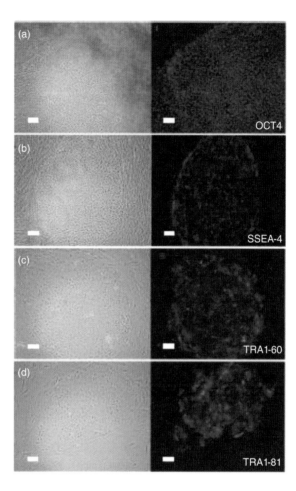

Figure 5.19 Phase contrast and fluorescence microscopy of SCNT-derived non-human primate embryonic stem cells. Morphological and molecular similarities to murine and embryonic stem cells derived from other species are apparent. (Figure courtesy J.A. Byrne, S.M. Mitalipov and *Nature* (Byrne et al., 2007); reprinted with permission.)

Examples of SCNT-Derived Stem Cells

Somatic cell nuclear transfer has now been used to generate stem cells from a variety of species. As discussed above, SCNT is not a new concept or technological advancement, and has been successfully implemented for the generation of clones from numerous species for more than 60 years (Table 5.2).

TABLE 5.1 Properties of somatic cell nuclear transfer-derived non-human primate embryonic stem cells.

Characteristic	Properties
Morphology	Compact, three-dimensional, highly refractory, tend to cluster, high nucleus-to-cytoplasm ratio; similar to that for embryonic stem cells derived from other species (see Figure 5.2).
Marker expression	Positive for Oct4 (POU5F1), SSEA-4, TRA1-60, TRA1-81, Nanog, Sox2, LeftyA, TDGF, TERT (see Figure 5.2).
Karyotype	Normal euploid
Differentiation potential	Capable of differentiation into cardiomyocytes and neuronal lineages in vitro; Teratoma formation when injected into SCID mice corresponding to lineages representing all three primary germ layers.

TABLE 5.2 History of cloning.

Cloned species	Year of successful cloning
Frog	1952
Carp	1963
Sheep	1986
Mouse	1987
Sheep	1995
Mouse, Cattle, Sheep	1997
Pig, Rhesus Monkey	2000
Cat, Guar, Mouflon, Goat	2001
Rat, Rabbit, Horse, Mule, Deer	2003
Ferret	2004
Dog, Water Buffalo, Fruit Fly	2005
Ferret	2006
Rhesus Monkey, Wolf	2007
Camel, Pyrenean ibex, Zebrafish	2009
Pashmina Goat	2012

Adapted from Stem Cells Handbook, 2nd Edition, Chapter 1. (Courtesy: Rob Burgess, Stewart Sell and Springer; reprinted with permission.)

The concept, however, of applying SCNT for the derivation of stable stem cell lines for research or therapeutic use is relatively new. Due to its controversial nature, the application of SCNT for the derivation of human ES cells will not be discussed in this text. More recently, and as discussed in the following section, new technologies have been developed that eliminate the need for the manipulation and destruction of human embryos for the generation of patient-specific embryonic stem cells.

A Note Regarding Genomic Abnormalities in SCNT-Derived Clones

Telomeric Length All clones derived via somatic cell nuclear transfer—whether they be entire organisms as in the case of Dolly the sheep or embryonic stem cell lines—have the same genetic age as the somatic cell providing the donor nucleus. As such, considerable differences in gross chromosomal makeup have been noted when compared to naturally derived organisms or cell lines of a similar age. It has been postulated that incomplete epigenetic reprogramming of the donor somatic cell genetic material is perhaps the primary cause behind postnatal abnormalities observed in domestic animal clones. This includes, for example, the composition of **telomeres**, which are repetitive nucleotide sequences of noncoding G-rich DNA $(TTAGGG)^n$ combined with specific binding proteins present on the ends of chromosomes that protect the chromosome from degradation or fusion with neighboring chromosomes. Telomeres buffer the inherent error-prone nature of DNA end terminal replication. Telomeric length is maintained by the ribonucleoprotein enzyme **telomerase**, which is an RNA-dependent DNA polymerase. Shortened telomeres have been demonstrated to result in a decrease in cellular proliferation by activating cell cycle arrest through either length or structural alterations that inhibit effective chromosomal

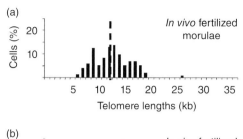

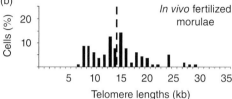

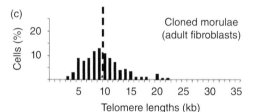

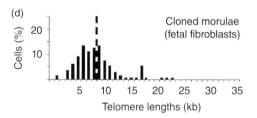

Figure 5.20 Analysis of telomere length in natural versus cloned bovine embryos by Q-FISH. (a) *In vivo* or (b) *in vitro* fertilized embryonic telomere lengths were compared to (c) cloned fetal or (d) cloned adult fibroblast-derived embryos. Telomeric lengths were observed to be considerably shorter in cloned compared to non-cloned fertilizations. (Data courtesy Heiner Neimann and the Society for Reproduction and Fertility (Niemann et al., 2008); reprinted with permission.)

end-capping. In the case of Dolly, telomeric length in SCNT-derived clones was considerably shorter than normal, and it has been speculated that the genetic age of Dolly (equivalent of 6 years at birth) could have contributed to her death. Shortened telomeres have also been found in a number of other bovine embryonic (morula-stage) clones compared to natural embryos (Figure 5.20) (Niemann et al., 2008). Roslin Institute researchers who led the study behind the cloning of Dolly dispute the hypothesis of telomeric length as a direct or indirect cause of her death and note that comprehensive health screens did not reveal age-related abnormalities or associated disease.

In addition, it should be noted that many SCNT-derived ruminants post-Dolly exhibited telomere lengths comparable to that for age-matched naturally bred control animals. These data held even in cases where nuclear donor cells were senescent and suggest that telomeric length is reset during embryogenesis as well as donor cell type dependence.

DNA Methylation and Epigenetics **Epigenetics** is the study of heritable changes in gene expression resulting from environmental factors other than the underlying DNA sequence of said genes (Figure 5.21). One of these factors (perhaps the primary factor)

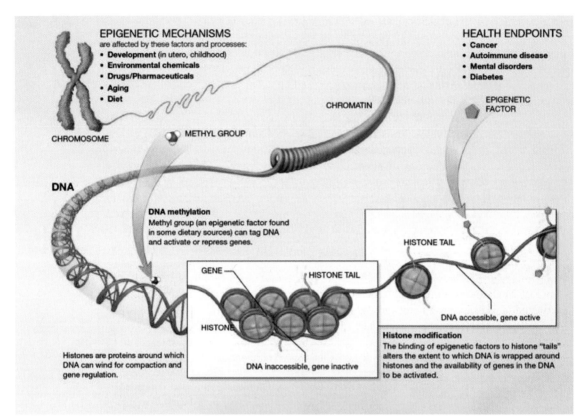

Figure 5.21 Illustration of epigenetic mechanisms involved in gene regulation. Various environmental factors influence the methylation status of DNA thereby altering DNA structure and gene expression characteristics. Epigenetic patterns are heritable but do not depend upon underlying DNA sequence and are often vastly different in somatic versus embryonic cell types. (Illustration courtesy National Institutes of Health and Wikimedia Commons; reprinted with permission.)

is the methylation state of DNA. It has been well documented that widespread changes in DNA methylation patterns are observed when bovine embryos are derived through either *in vitro* fertilization or somatic cell nuclear transfer. In many cases, methylation patterns are grossly different in SCNT-derived versus normal embryos with, not surprisingly, the methylation patterns of SCNT-derived embryos similar to that of the donor somatic cells, which suggests that nuclear reprogramming during embryogenesis was incomplete. Thus, epigenetic properties of somatic cell donors are not effectively erased. As live births do result from a low percentage of SCNT derivations, it is clear that at least some epigenetic reprogramming is occurring, and it has been speculated that the erasing of epigenetic modifications spans genes critical to embryonic development, thereby allowing for full maturation and postnatal survival.

It must be noted that epigenetic identity and DNA methylation patterns are highly dynamic during embryonic development. These patterns can change in response to various environmental factors such as exposure to nutrients, altering hormonal levels, and temperature. Thus, SCNT-derived unique epigenetic codes can be the result of the SCNT process, the local developmental environment or a combination of the two.

X-Chromosome Inactivation In mammals a unique system determines biological sex, with the presence of two X chromosomes defining a female and that of an X chromosome and Y chromosome defining a male. The X chromosome is genetically active and comprises thousands of functional genes spanning a region of ~160 megabases. The Y chromosome is largely devoid of functional genes: to date, only ~100 have been identified, and play small roles in defining sex and promoting fertility. Given that females possess two copies of the X chromosome, there is an inherent lethal risk of double the expression of X-linked genes. Thus the process of **X-chromosome inactivation (XCI)** evolved, which is defined as an epigenetically regulated process of gene dosage compensation resulting in the silencing of genes on the X chromosome. XCI occurs through a complex series of events that are not completely understood, but the non-coding RNA **Xist** is known to play a key role in this process. Xist binds the X chromosome, and its presence promotes global silencing. Given the low survival rates of SCNT-derived embryos, many researchers have speculated that X chromosome inactivation events have been disrupted as a result of the SCNT procedure and even the somatic source of the donor nucleus. Marisa Bartolomei and colleagues in the Department of Cell and Developmental Biology at Howard Hughes Medical Institute and the University of Pennsylvania School of Medicine in Philadelphia demonstrated that the inactive X chromosome was reactivated in SCNT-derived embryos. Reactivation was heterogeneic, occurring in some cells of the developing embryos but not all. Their studies also revealed abnormal Xist presence in cloned versus naturally derived embryos, which may at least partially explain aberrant X chromosome silencing seen in some, but not all, blastomeres of cloned embryos (Figure 5.22) (Nolen et al., 2005).

In addition, in many cases the researchers noted the presence of more than two X chromosomes in cloned versus naturally derived embryos, suggesting a failure in chromosomal segregation and duplication (Figure 5.23) (Nolen et al., 2005).

They concluded that embryos derived via cloning methodologies fail to consistently regulate the process of X chromosome inactivation, and that this crucial lapse in gene silencing could play an integral role in the low rates of cloned embryo survival. It is apparent that the development of somatic cell nuclear transfer technology is an important and seminal breakthrough in the quest to harness the utility of adult cell genetic material from a research and agricultural perspective, yet the low embryonic survival most likely driven by epigenetic abnormalities may limit its utility in the latter area. In addition, the controversial aspect of the technology involving the destruction of embryos prevents this technology from being a viable option for the generation of cell-based therapeutics platforms.

INDUCED PLURIPOTENCY

As the issue of patient immunorejection cannot realistically be addressed for the creation of stem cells from those of somatic origin via somatic cell nuclear transfer procedures, other technologies have been pursued. Many of these do not involve the manipulation or destruction of human embryos, thus alleviating this controversial aspect of SCNT methods. One groundbreaking method for creating stem cells derived from adult somatic cells is induced pluripotency. **Induced pluripotency** is the artificial promotion of pluripotent properties in non-pluripotent cells via forced expression of specific genes. The application of induced pluripotency—for example, in the context of an adult somatic cell—results in the generation of **induced pluripotent stem (iPS)**

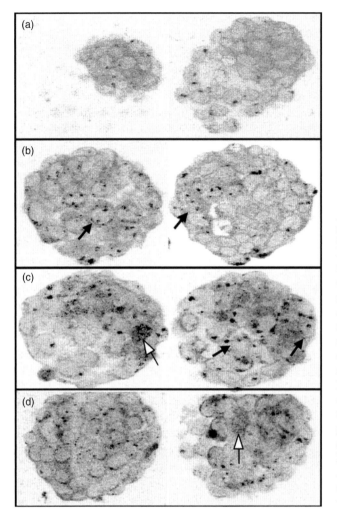

Figure 5.22 Fluorescence in situ hybridization of Xist RNA in cloned blastocysts. (a) Natural embryos exhibiting a single Xist signal in each blastomere; (b), (c), and (d) SCNT-derived embryos demonstrating abnormal multiple Xist signals in single blastomeres or blastomeres exhibiting no Xist signal. Black arrows denote two foci staining in the same blastomere, white arrow indicates diffuse staining. (Photos courtesy Marisa Bartolomei and *Developmental Biology* (Nolen et al., 2005); reprinted with permission.)

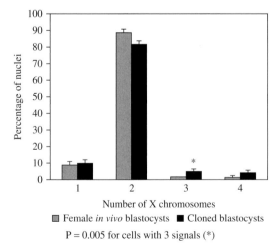

Figure 5.23 **Graphic analysis of X chromosome number in natural versus cloned blastocysts.** (Graph courtesy Marisa Bartolomei and *Developmental Biology* (Nolen et al., 2005); reprinted with permission.).

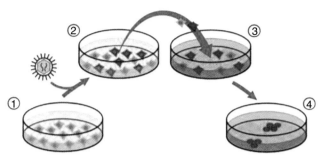

<u>Figure 5.24</u> Illustrative overview of the procedure for generating patient-specific iPS cells. (1) Isolate and culture host cells, for example, adult human dermal fibroblasts. (2) Introduce ES-specific genes (iPS factors) into the cells by using retroviral vectors. Red cells indicate the cells expressing the exogenous genes. (3) Harvest and culture the cells according to methods for ES cell culture using feeder cells (gray). (4) A subset of the cells generates ES-like colonies, that is, iPS cells. (Illustration courtesy Wikimedia Commons; reprinted with permission.)

cells. Given the fact that a patient's own somatic cells can be used as a source for iPS cell generation, allogeneic immunorejection of resulting cell or tissue transplant material is no longer an issue. Figure 5.24 illustrates the procedure for generating patient-specific embryonic stem cells.

Breakthrough in the Production of iPS Cells

The first example of the power of a single gene to alter cellular fate can be traced back to a seminal study published in 1987 by Walter (see Focus Box 5.3) and colleagues at the University of Basel in Switzerland. It was their discovery of the *Drosophila melanogaster* transcription factor **Antennapedia** and its ability to transform antennae into legs when overexpressed in developing fly embryos that shed light on the possibilities of transcription factor-based cell fate and modification. Specifically, the researchers demonstrated that loss-of-function mutations in the *Antennapedia* locus resulted in the transformation of the second leg pair into antennae. Gain-of-function studies yielded the opposite effect, with the conversion of antennae into legs (Figure 5.25) (Schneuwly and Klemenz et al., 1987).

Focus Box 5.3: Walter Gehring and the homeobox

In the early 1980s, Walter Gehring and colleagues discovered a unique DNA sequence known as a homeobox which has now been defined as a classic hallmark of homeotic genes, key regulators of body plan architecture and cellular identity. Through their studies on the homeobox-binding transcription factor Antennapedia, the researchers demonstrated the potent ability of transcription factors to alter cellular and body plan fate. Gehring's research established the importance and utility of transcription factor-based manipulation of cell fate and laid the groundwork for the advancements seen today in the use of transcription factors to induce somatic cell pluripotency. He was most recently a Professor at the Biozentrum of the University of Basel until his death in 2014.

Source: Courtesy of the University of Basel.

Figure 5.25 Photograph of a *Drosophila melanogaster* specimen exhibiting the mutation *Antennapedia*. See text for details. (Photograph courtesy Wikimedia commons; reprinted with permission.)

The studies by Gehring and colleagues were the first to demonstrate both the power and the utility of transcription factor-based induction of cell and body plan fate. The findings opened the door to the dissection of transcription factor function as it pertains to the hierarchical genetic control of embryonic development and the post-natal regulation of adult stem cell fate and function. Following the publication of Gehring's findings, an onslaught of research in transcriptional regulation has provided a wealth of information pertaining to transcription factor function. Numerous transcription factors have since been identified that have the regulatory power to guide cellular fate. These include the formation of muscle cells (MyoD), neuronal lineages (neurogenin) and even blood (C/EBP). In addition, in the field of stem cell research several key transcription factors have been identified and characterized as in essence having the ability to drive the opposite phenotype. These factors have the ability—when working in concert—to induce pluripotency in adult somatic cells. It was Shinya Yamanaka (see Focus Box 5.4) and colleagues at the Institute for Frontier Medical Sciences, Kyoto University in Japan who identified these factors through a comprehensive and systemic analysis of 24 individual genes well established as being active and expressed at high levels in pluripotent, undifferentiated embryonic stem cells. Case Study 5.3 summarizes these seminal findings.

Focus Box 5.4: Shinya Yamanaka and induced pluripotency

 The reprogramming of somatic cells to exhibit pluripotent characteristics has long been a goal for many stem cell researchers. This was successfully accomplished in 2006 for murine cells and again in 2007 for human cells by pioneering researcher Shinya Yamanaka at Kyoto University in Japan. By defining the key transcription factors driving pluripotency in embryonic stem cells and subsequently reintroducing these same factors into cells of somatic origin he was able to induce pluripotency, thus creating an avenue for the generation of patient-specific stem cells without the need for the destruction of human embryos. Dr. Yamanaka has received numerous awards for his research including the 2009 Lasker Award, the 2010 Kyoto Prize for Advanced Technology, and the 2012 Nobel Prize in Physiology or Medicine.

Source: Courtesy of J. David Gladstone Institutes. Photo by Chris Goodfellow/Gladstone Institutes.

Case Study 5.3: Induction of pluripotent stem cells from mouse embryonic and adult fibroblast cultures by defined factors

K. Takahashi and S. Yamanaka

Although it had been well established through somatic cell nuclear transfer and other meth-odologies that differentiated cells can be reprogrammed to have the pluripotent characteris-tics of embryonic stem cells, the actual genes involved remained elusive until 2006, when Shinya Yamanaka (see Focus Box 5.4) and colleagues performed a systematic functional characterization of 24 loci known to be active and highly expressed in embryonic stem cells. The researchers developed a unique assay that utilized resistance to the aminoglycoside anti-biotic G418 (neomycin) as the marker for driving pluripotency. DNA sequences coding for the neomycin resistance gene were engineered into the mouse ES cell marker gene Fbx15 by homologous recombination in a stable mouse embryonic fibroblast (MEF) cell line. After retroviral transduction of sequences coding for the 24 candidate loci in various combina-tions, clonal surviving cell lines resistant to antibiotic selection were analyzed for pluripo-tency characteristics and confirmed by morphology to resemble murine embryonic stem cells (Figure 5.26). Analysis of individual candidate genes as well as combinations of the 24 loci introduced into the cells revealed the ability of a "cocktail" of four factors, Oct3/4, Sox2, c-Myc, and Klf4, to be necessary and sufficient for cellular reprogramming. Definitive induc-tion of pluripotency by these factors was confirmed by both characterization of the presence of pluripotency markers and through *in vivo* teratoma formation (Takashi and Yamanaka 2006).

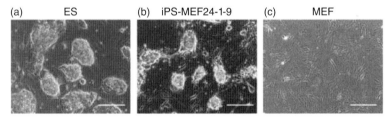

| (a) | ES | (b) | iPS-MEF24-1-9 | (c) | MEF |

Figure 5.26 Brightfield microscopy demonstrating resemblance of iPS cells to embryonic stem cells. (a) Embryonic stem cells. (b) MEF cells induced for pluripotency. (c) Uninduced MEF cells. (Photos courtesy K. Takahashi, Shinya Yamanaka and *Cell* (Takashi and Yamanaka, 2006); reprinted with permission.)

While Case Study 5.2 illustrates the first demonstration of transcription factor driven induction of pluripotency in a somatic cell, the model system used in the experiments was murine. In 2007, Yamanaka and his team expanded upon his initial findings in mice by inducing pluripotency in human dermal fibroblasts utilizing the same four factors—Oct 2/4, Sox2, c-Myc, and Klf4—which drove the induction of pluripotency in the mouse model system. An optimized retroviral transduction system was utilized in a manner simi-lar to that implemented for murine somatic cell induced pluripotency, and the resulting human iPS cells resembled human embryonic stem cells in morphology, marker expression (Figure 5.27), proliferation rates, epigenetic status, and telomerase activity. The cells were also definitively shown to differentiate into lineages representing all three primary germ layers thus confirming pluripotency characteristics (Takahashi et al., 2007).

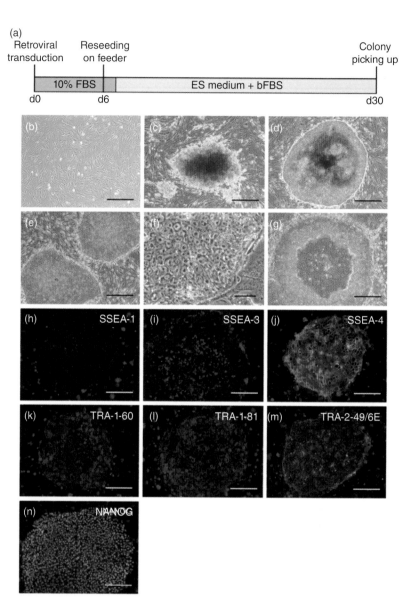

Figure 5.27 Induction of pluripotency in adult human dermal fibroblasts. (a) Diagram of timeline for induction. Brightfield microscopy revealing the morphology of (b) human dermal fibroblasts, (c) non-ES cell like colony, (d) hES cell colony, (e) established iPS cell line at passage 6, (f) higher magnification of iPS line, (g) iPS colony indicating spontaneous differentiation in the colony center. (h–n) Fluorescence microscopy of pluripotency marker expression in human dermal fibroblast iPS colonies. (Images courtesy K. Takahashi, S. Yamanaka and *Cell* (Takahashi et al., 2007); reprinted with permission.)

Since the 2006 and 2007 groundbreaking studies by Yamanaka and colleagues, a variety of somatic cell types have been successfully reprogrammed. Some examples include:

- Keratinocytes
- T cells
- Fibroblasts
- Adipose-derived stem cells (ADSCs)
- Mesenchymal stem cells (MSCs)
- Dental pulp stem cells

- Germline stem cells
- Neural stem cells
- Cord blood stem cells
- Hair follicle cells
- Retinal cells
- Skeletal muscle cells

TABLE 5.3 Overview of cellular physiological response and key factor expression during induced pluripotency reprogramming.

Physiological response to pluripotency induction	Factors upregulated	Factors downregulated
Proliferation	Cyclin D1, Klf4, c-Myc, Rem2	p16, p21, p53
Epigenetics	N/A	HDAC, histone demethylase, DNMT1
Signal transduction	N/A	TGFb, Wnt1/b catenin, PI3/AKT
Chromatin remodeling	SWI/SNF	N/A

What are the underlying alterations that occur in response to the introduction of reprogramming transcription factors that might drive an acquiescence of pluripotency? The specific mechanisms are still unknown, but there are some inherent and obvious changes that have been noted and are outlined in Table 5.3.

Methods for the Production of iPS Cells

Since the groundbreaking discoveries of Yamanaka and colleagues in 2006 in mice and 2007 in humans, a variety of methodologies have been developed for the induction of pluripotency in somatic cells (Figure 5.28). In each case the desired effect is the same: to achieve expression of the key transcription factors required for driving reprogramming and promoting pluripotent properties. For effective application of iPS cell technology in research,

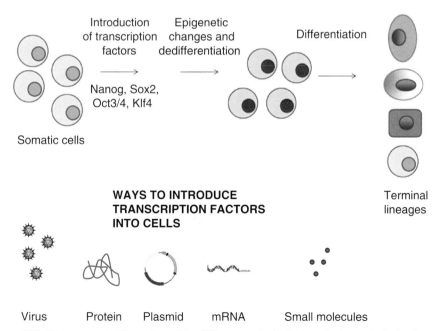

Figure 5.28 Diagrammatic illustration of the different methods for introducing or inducing key factors involved in reprogramming. See text for details.

diagnostics, and potential cell-based therapeutics endeavors, the proper reprogramming methods must be employed. The particular method of choice may depend upon the ultimate final use of the reprogrammed cells. For example, if the cells are to be used for tissue or cell transplant therapy, the procedure for introducing/activating key transcription factor activity must be both safe and effective. The following sections outline unique platforms developed since the inception of induced pluripotency for transcription factor delivery into somatic cells, and discuss the advantages and limitations of each.

Retroviral and Lentiviral Gene Delivery **Retroviruses** are RNA viruses that replicate in a host cell. Retroviruses have been used for several decades to transduce genetic material into cells for ensuring sustained and high levels of expression. Indeed, the studies described above by Yamanaka and colleagues employed retroviruses for transcription factor gene delivery. Reprogramming efficiencies utilizing retroviruses can range from 0.01% to 0.25% depending upon the cells infected and the combination of transcription factors used for reprogramming. For example, George Q. Daley's team at the Harvard Stem Cell Institute in Boston combined the original transcription factors discovered by Yamanaka with the catalytic subunit of human telomerase, hTERT, and SV-40 Large T antigen. These factors are known to aid in the establishment of human cells in culture. It was observed that human dermal fibroblasts exhibited a 10-fold increase in the level of reprogramming efficiency to that observed by Yamanaka's group. **SV40 Large T antigen** (Simian Vacuolating Virus 40 TAg) was speculated to play a crucial role in driving key transcription factor expression as it is a proto-oncogene and known to stimulate replication of viral genomes in host cells. Human induced pluripotent cells were confirmed to contribute to teratoma formation when transplanted into immune-deficient mice and could be directed *in vitro* to become blood cells (Figure 5.29) (Lengerke et al., 2009).

However, retroviruses have one major limitation for use in reprogramming: they only infect dividing cells. The inability to infect non-dividing cells has considerable implications with respect to inducing pluripotency in patient-specific stem cells, as only a source of dividing cells could be utilized in the context of the retroviral system and non-dividing, mature adult somatic cells are thus not candidates for retrovirus-mediated induction of pluripotency. **Lentiviruses** are a genus of the retroviridae family characterized by the unique ability to infect non-dividing cells and exhibit extended incubation periods. A lentiviral transduction system was first devised and implemented for reprogramming non-dividing human cells by James A. Thomson and colleagues at the Genome Center of Wisconsin and the National Primate Research Center, University of Wisconsin-Madison. Thomson's group sought to improve pluripotency induction efficiencies and eliminate the use of c-Myc, which is known to drive differentiation and death in human ES cell models.

| dH1f-iPS3-12 | dH1cf16-iPS-5 | MSC-iPS1 |

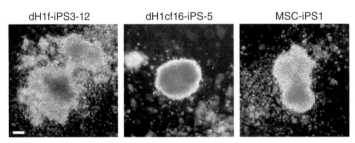

Figure 5.29 Photographs of blood cells derived from human iPS cells generated via lentiviral transduction. Hematopoietic colony-forming assays demonstrate the formation of erythroid colonies. (*Source:* Lengerke et al., 2009. Reproduced with permission from John Wiley & Sons).

IMR90   iPS(IMR90)-3

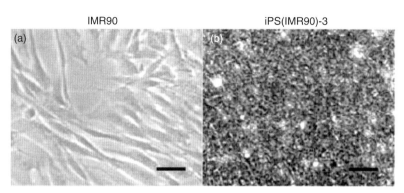

Figure 5.30 Brightfield microscopy of lentiviral driven induced pluripotency. (a) p18 IMR90 human somatic cell fibroblasts. (b) The same fibroblasts induced to become pluripotent, resembling embryonic stem cells in morphology and structure. (Images courtesy James A. Thomson and *Science* (Yu et al., 2007); reprinted with permission.)

While different factors were utilized in this study (Oct4, Sox2, Nanog, and Lin28) compared to those of the Yamanaka and Daley groups, reprogramming without the introduction of c-Myc and via the application of a lentiviral-based system was successful (Figure 5.30) (Yu et al., 2007). These studies have therefore demonstrated that different factors may be utilized to induce pluripotency and that a lentiviral system may be valuable for the introduction of genes encoding these key factors in non-dividing adult somatic cells.

The application of retroviral- or lentiviral-based systems for introducing key transcription factors to induce pluripotency in somatic cells has one major drawback: **transgenes**, which are defined as a gene or genetic material transferred from one organism to another, are stably incorporated into the genome of the host cell. The existence of these transgene integrants poses several disadvantages, the most critical of which is the possibility, however remote, that transgene integration may either activate a proto-oncogene or inactivate (repress or silence) a tumor suppressor locus. This could result in a tumorigenic phenotype and as such is not safe in regards to cell-based therapy. Various strategies have since been developed for the removal of retroviral/lentiviral integrated transgenes. One unique method is based on the application of the **P1 bacteriophage Cre recombinase**. Cre is an enzyme known as a topoisomerase which has the ability to carry out **site-specific recombination**, a DNA strand exchange occurring between nucleotide sequences possessing only a limited degree of sequence homology. In the case of Cre activity, the homologous recognition sequences are referred to as **loxP sites**. Cre catalyzes a site-specific recombination event between two loxP sites, thus allowing for the excision of sequences flanked by loxP (Figure 5.31). The short nature of the loxP sites (34 nucleotides) makes the system an excellent tool for the removal of transgenes from endogenous genomes.

The Cre-lox system was first used for removal of transgene integrants in iPS studies in 2009 in an elegant study generating iPS cells from patients with Parkinson's disease. Rudolf Jaenisch's group at the Whitehead Institute and Department of Biology, Massachusetts Institute of Technology obtained fibroblasts from five patients with idiopathic Parkinson's disease, and infected them with a lentivirus encoding the reprogramming factors Oct4, Sox2, c-myc, and Klf4 to induce reprogramming and pluripotency. Pluripotent cells could be efficiently differentiated into dopaminergic neurons, a potential cell-based therapeutic platform (Figure 5.32). In addition, the researchers accomplished the effective removal of loxP-flanked transgene integrants using an inducible Cre-loxP system (Figure 5.33) (Soldner et al., 2009).

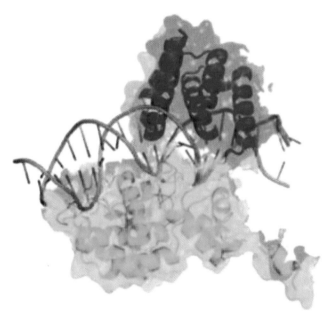

Figure 5.31 Cartoon model of Cre recombinase bound to its loxP recognition sequence. The recombinase amino-terminal is depicted in blue and the carboxyl terminal is depicted in green. (Cartoon courtesy Wikimedia Commons; reprinted with permission.)

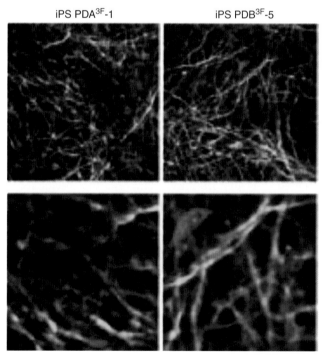

Figure 5.32 Fluorescence microscopy of dopaminergic neurons generated from Parkinson's disease patient-specific induced pluripotent cells. (Micrographs courtesy Frank Soldner, Rudolf Jaenisch and *Cell* (Soldner et al., 2009); reprinted with permission)

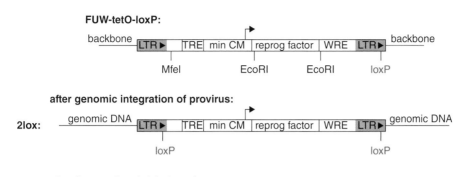

Figure 5.33 Diagrammatic illustration of inducible Cre-mediated transgene removal. See text for details. (Diagram courtesy Frank Soldner, Rudolf Jaenisch and *Cell* (Soldner et al., 2009); reprinted with permission.)

Surprisingly, the researchers observed that even post-transgene removal, residual reprogramming factor expression maintained the pluripotent phenotype. The reprogrammed, factor-free cells even more closely resembled embryo-derived stem cells than parental non-excised cell lines (Seki et al., 2010). These studies suggest that human patient-specific adult somatic cells can be reprogrammed toward a self-sustaining pluripotent state and this phenotype can be maintained, even in the absence of integrated transgenes or exogenous reprogramming factors. Therefore, it is tempting to speculate that induction of pluripotency is a transient event, and need not be actively maintained long term to retain a pluripotency phenotype.

Adenoviral Gene Delivery One method for the introduction of gene sequences coding for the key reprogramming factors that does not result in a transgene genomic integrant is the application of adenovirus. **Adenovirus** only infects non-mitotic cells and introduced nucleic acid sequences do not integrate into the host genome. Given the fact that no genomic integrant is established, there exists a narrow window of opportunity for expression of the factors that is conducive to inducing pluripotency. As such, the efficiency of generating iPS lines using adenoviral systems has been demonstrated to be low at around 0.001–0.0001% in mouse fibroblasts (Figure 5.34) (Stadtfeld et al., 2008) and 0.002% utilizing human fetal fibroblasts for reprogramming. However, neither mouse nor human adenovirus-generated

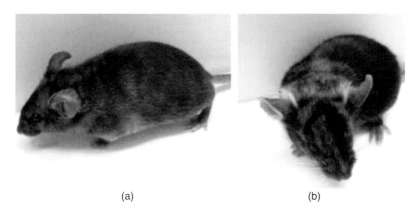

(a)   (b)

Figure 5.34 Chimeric mice produced from adenovirus-generated iPS cells. (a) Fetal liver and (b) hepatocyte iPS cells. The mosaic coats indicate a mixture of wild-type and iPS-derived cells (Images courtesy Matthias Stadtfeld, Konrad Hochedlinger and *Science* (Stadtfeld et al., 2008); reprinted with permission.)

iPS cells exhibited any transgene integration—crucial if cell-based therapies are to be developed from iPS cells.

Sendai Viral Gene Delivery An ideal technique for the induction of pluripotency in somatic cells might involve the introduction and maintenance of nucleic acid sequences coding for the key factors within the cytoplasm but not the nucleus of the cell. This is the mechanism of infection for the RNA-based **Sendai virus**. Sendai ribonucleic acid sequences remain in the cytoplasm post infection and are present through the first 5–10 cell passages. These allow for: (1) no chance of genomic integration; (2) complete loss of all viral sequences long term; and (3) enough time for key factor expression and induction of pluripotency. A variety of somatic cell types have been induced to pluripotent capacities using Sendai virus including murine neonatal fibroblasts, adult human dermal fibroblasts, and CD34+ cells derived from both umbilical cord and adult blood. The latter cell type is routinely sampled in medical settings worldwide from adults and children making it an ideal cell type for the generation of patient-specific iPS cells or derivatives. Induction of a pluripotent phenotype in these cells was accomplished by Keiichi Fukuda and colleagues in the Department of Cardiology at Keio University School of Medicine in Tokyo. The researchers used a temperature-sensitive version of the Sendai virus to reduce levels of transgene expression and eliminate viral proliferation at below standard cell culture temperatures. These studies allowed for the generation of iPS cells at high efficiencies, which were directly correlated with viral **multiplicity of infection (MOI)** (Figure 5.35) (Seki et al., 2010). The induction efficiency range of 1.0–0.1% makes Sendai virus an ideal choice for pluripotency induction if therapeutic strategies are to be considered for the resulting cell lines.

Plasmid-Based Gene Delivery In addition to his studies based on the use of retroviral or adenoviral vectors for the introduction of key factor sequences that induce pluripotency, Shinya Yamanaka also developed an iPS plasmid transfection system. In this system cDNAs encoding Oct4, Sox2, and Klf4, three of the key transcription factors required for inducing pluripotency, were cloned into a single high-level expression vector. A **self-cleaving peptide sequence** derived from the foot-and-mouth disease virus 2A was used to allow for efficient, independent polycistronic expression of the constructs. A separate plasmid designed to express c-Myc was also constructed and simultaneous or sequential transfection of the plasmids into mouse primary hepatocytes was carried out. Analysis of resulting transfectants

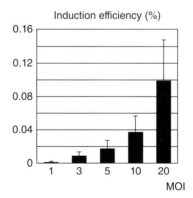

Figure 5.35 Graph of Sendai viral MOI effect on pluripotency induction efficiency (Graph courtesy Tomohisa Seki, Keiichi Fukuda and *Cell Stem Cell* (Seki et al., 2010); reprinted with permission.)

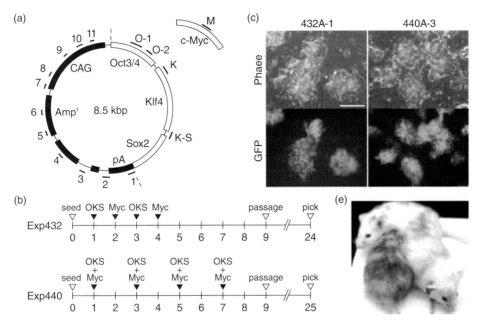

Figure 5.36 Plasmid-based induction of pluripotency. (a) Diagram of the plasmid constructs used. (b) Timeline for the induction of pluripotency via sequential or simultaneous transfection. (c) Morphology of two independent iPS lines. (e) Chimeric mouse with coat coloring indicative of iPS cell germline contribution. (Figure courtesy Keisuke Okita, Shinya Yamanaka and *Science* (Okita et al., 2008); reprinted with permission.)

revealed cellular morphology and marker expression indistinguishable from mouse ES cells (Figure 5.36) (Okita et al., 2008).

Yamanaka's group subsequently performed extensive polymerase chain reaction (PCR) studies to determine if plasmid-host genome integration had occurred and the results were negative. The group went on to perform 10 separate and independent transfection experiments using this system and achieved pluripotency induction in 7 cases. It was noted that the efficiency of iPS cell generation using plasmid transfection was much lower than when retroviral constructs were used, most likely due to timing or key factor expression level issues.

Episomal plasmids (a.k.a. episomes) are defined as closed circular DNA molecules that replicate within the nucleus of cells. Viral genomes are often packaged in an episomal format and replicate within the nucleus of a cell following infection. Episomes referred to as oriP/EBNA1 derived from the Epstein–Barr virus can be transfected into cells in the absence of viral packaging (i.e. no transduction needed). Given their stable replicative capacity, episomes behave in a manner similar to that of plasmids in prokaryotes. The lack of genomic integration by episomes provides a valuable strategy to stably express the key factors required for inducing pluripotency without the worry of harmful mutations in the host genome. James A. Thomson's group at the Genome Center of Wisconsin and Wisconsin National Primate Research Center (discussed above and throughout this book as the team responsible for discovering human embryonic stem cells) devised a system to maintain the presence of episomal vectors in cells encoding the key factors through antibiotic selection. This allowed for induction of pluripotency in human embryonic, neonatal, and adult fibroblasts. Subsequent withdrawal of selectable antibiotics resulted in an elimination of the episomes following several rounds of mitosis, yet the cells still retained an embryonic stem-like phenotype (Figure 5.38) (Yu et al., 2009). It should be noted that this method was considerably inefficient at iPS cell

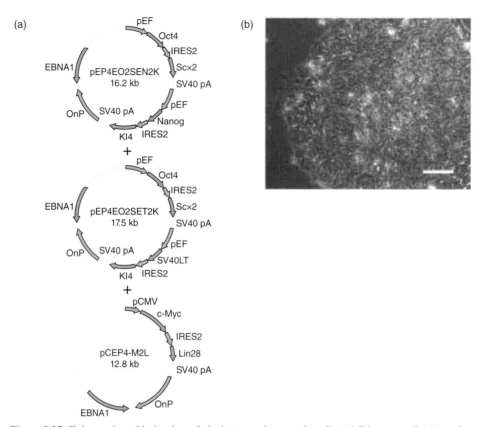

Figure 5.37 Episome-based induction of pluripotency in somatic cells. (a) Diagrams of episomal constructs expressing the key factors required for pluripotency. (b) iPS cell morphology post-episomal transfection, cellular induction, and removal of the episomal vectors through multiple rounds of cell division. (Figure courtesy Junying Yu, James A. Thomson and *Science* (Yu et al., 2009); reprinted with permission.)

generation, with success rates ranging from 0.0003% to 0.0006%. This initial episome-based strategy has since been refined and optimized by a number of groups to achieve pluripotency induction in a variety of cell types including bone and cord blood mononuclear cells.

mRNA Delivery As discussed above, the desired final phenotype of an iPS cell or downstream-differentiated progeny is one without the presence of lingering inducing molecules. It is only in this case that cell-based therapeutics such as cell or tissue transplantation can realistically be pursued. The transfection of mRNA coding for the key factors which drive pluripotency provides yet another unique avenue for leaving no footprint post induction. Synthetic mRNA often induces a severe immune reaction and subsequent cell death in transfected cells, thus limiting its utility in the generation of iPS cells. Derrik J. Rossi's team at the Immune Disease Institute and Harvard Stem Cell Institute in Boston addressed this problem by replacing the synthetic nucleic acid bases pseudouridine with uridine and cytidine with 5-methylcytidine to convert the mRNA sequences into those less resembling that of a virus. Transfection of these optimized mRNA sequences coding for the key factors into either human keratinocytes or fibroblasts revealed a striking 1.4% induction efficiency compared to that for retroviral based induction of 0.004%. In addition, mRNA-driven iPS colonies were observed 8 days prior

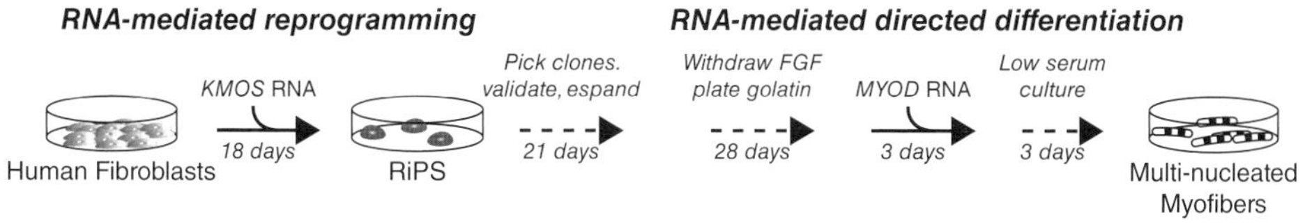

RNA-mediated reprogramming   **RNA-mediated directed differentiation**

KMOS RNA   Pick clones.   Withdraw FGF   Low serum
                validate, espand   plate golatin   MYOD RNA   culture
Human Fibroblasts   18 days   21 days   28 days   3 days   3 days   Multi-nucleated
                RiPS   Myofibers

Figure 5.38 Diagrammatic illustrations of the strategy for mRNA transfection-based generation of iPS cells and differentiation of these cells into myofibers. Human fibroblasts are cultured to a specific confluence and transfected with mRNA coding for key factors KLF4, c-Myc, Oct4, and Sox2 (KMOS) followed by clonal selection for RiPS cells. Individual clones were expanded and directed to differentiate into multi-nucleated myotubes via the transfection of mRNA coding for the potent basic helix-loop-helix factor MyoD. (Diagrams courtesy: Luigi Warren, Derrick J. Rossi and *Cell Stem Cell* (Warren et al.,2010); reprinted with permission.)

to the observance of colonies in retroviral induction experiments. In an elegant series of additional experiments the researchers also demonstrated that the same mRNA transfection strategy could be utilized to direct the differentiation of these RiPS (RNA-induced pluripotent stem) cells into mature, multi-nucleated myofibers (Figure 5.38) (Warren et al., 2010). Since the publication of this research, mRNAs coding for the key factors involved in pluripotency induction are now commercially available.

MicroRNA Induction **MicroRNAs (miRNAs)** are defined as short, non-coding RNA molecules of about 22 nucleotides in length found naturally in both prokaryotic and eukaryotic cells that function to control gene expression, both transcriptionally and post-transcriptionally. Researchers in the Department of Gastroenterological Surgery, Osaka University Graduate School of Medicine in Japan identified three miRNAs highly expressed in murine embryonic stem cells but not differentiated cells. These miRNAs were demonstrated to affect the same pathways as other microRNAs previously identified as having a positive effect on reprogramming including inhibition of embryonic epithelial to mesenchymal transition and TGFβ signaling. When the miRNAs were introduced into both mouse and human cells, pluripotency was induced. The resulting **mi-iPS cells** were subsequently demonstrated to exhibit all the hallmarks of iPS cells generated by other methods including the formation of teratomas in nude mice (Figure 5.39). This is the first example

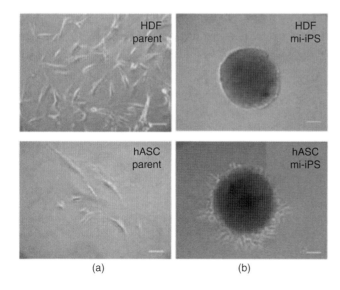

(a)   (b)

Figure 5.39 Brightfield microscopy of microRNA-induced pluripotency in human dermal fibroblasts (HDFs) and murine adipose stromal cells (ASCs). (a) Parental lines. (b) Embryonic stem cell-like colonies after pluripotency induction with microRNAs known to be expressed at high levels in embryonic stem cells. (Photos courtesy Norikatsu Miyoshi, Masaki Mori and *Cell* (Miyoshi et al., 2011); reprinted with permission.)

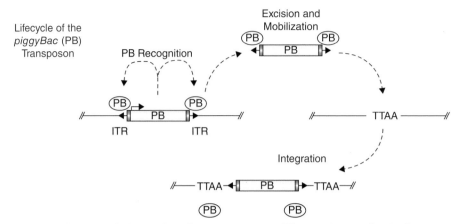

Figure 5.40 Diagrammatic illustration of the *piggyBac* (PB) lifecycle. See text for details. (Diagram courtesy Wikimedia Commons; reprinted with permission.)

of inducing pluripotency by introducing molecules other than those that code for the key inducing factors or the key factors themselves (Miyoshi et al., 2011).

Transposon Delivery **Transposons** are defined as DNA sequences which possess the ability to change positions within the genome. *PiggyBac* **(PB)** is a transposon that can shuttle between target cell chromosomes and extra-chromosomal vectors. This occurs through an efficient cut-and-paste mechanism and has been exploited by numerous researchers for the shuttling of unique nucleic acid sequences in and out of the host genome. The PB transposase recognizes specific inverted terminal repeat (ITR) sequences and transposes internal sequences to be incorporated into genomic regions containing TTAA chromosomal sites (Figure 5.40).

It is an extremely efficient and precise process, thus targeting transgenes to specific sites within the genome of the cell. In addition, re-expression of the transposase can reverse the process, allowing for excision of transgene integrants and leaving behind no deleterious mutations. Two independent groups at the University of Edinburgh in Scotland and the Samuel Lunenfeld Research Institute, Mount Sinai Hospital, Toronto, Canada employed the *piggyBac* system in 2009 for the generation of iPS cells from mouse embryonic fibroblasts. In addition, in each study it was confirmed that re-expression of *piggyBac* allowed for the complete excision of transgene integrants. Efficiency of iPS induction was reported to range between 0.02% and 005%.

Direct Protein Delivery Direct delivery of proteins representing the key transcription factors is another method that would eliminate any possibility of leaving a footprint on reprogrammed cells. This is due to the fact that no transgene integration system is used and exogenous protein introduced into target cells would be diluted over successful cell divisions. Yet delivering protein intracellularly is not a trivial matter, as protein does not readily cross the plasma membrane. In addition, purified protein is often prepared in a denatured format, thus losing valuable biological activity. In 2009, research teams at Harvard Medical School in Boston and the Scripps Research Institute in La Jolla, California independently addressed each of these limitations. Specifically, proteins representing the key factors required for pluripotency induction were synthesized in *E. coli* and subsequently solubilized and renatured to allow for retention of biological activity. In addition, a **poly-arginine leader sequence** was incorporated into each protein. This sequence has been characterized as a potent mediator of protein translocation across the plasma membrane. Multiple rounds

(a)

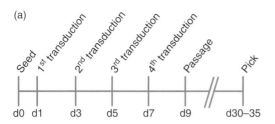

Seed   1ˢᵗ transduction   2ⁿᵈ transduction   3ʳᵈ transduction   4ᵗʰ transduction   Passage   Pick

d0   d1   d3   d5   d7   d9   d30–35

(b)

     Phase   GFP

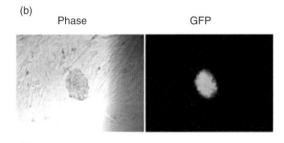

(c)

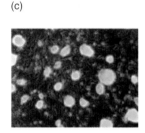

Figure 5.41 Generation of iPS cells by recombinant protein transduction. (a) Timeline for sequential protein transduction. (b) Green fluorescent protein colonies indicating active Oct4 expression, a hallmark of iPS cells. (c) GFP positive iPS colonies. (Figure courtesy Hongyan Zhou, Sheng Ding and *Cell* (Zhou et al., 2009); reprinted with permission.)

of protein transduction over a 1-week period in both mouse and human fibroblasts with these modified proteins resulted in the generation of iPS cells at an efficiency of about 0.006% and 0.001%, respectively (Figure 5.41) (Zhou et al., 2009).

Basic Properties of iPS Cells

iPS cells exhibit many of the same characteristics as embryonic stem cells. These include both long-term self-renewal and pluripotency capabilities. In addition, as discussed in numerous cases above, iPS cells share some of the same morphological, molecular, enzymatic, and potency features. Below is a brief comparison of these properties between ES and iPS cells.

- Morphology—Individual iPS cells are typically round with very little visible cytoplasm but a large nucleolus. Human iPS cell colonies are compact and flat with visible sharp demarcating edges while mouse cell colonies are less flat and more disaggregated. In general, human iPS cells more closely resemble human ES cells and the same is true for their murine counterparts (Figure 5.42).
- Stem cell marker expression—Like embryonic stem cells, both human and mouse iPS cells express Nanog, Tra-1-60, Tra-1-81, and Tra-2-49/6E. While human iPS cells express SSEA-3 and SSEA-4, mouse iPS cells do not. They do, however, express SSEA-1 in a manner similar to that for pluripotent, undifferentiated mouse ES cells. Other stem cell specific genes similarly expressed by iPS and embryonic stem cells include Sox2, GDF3 Rex1, FGF4, DPPA2, DPPA4, and hTERT.

(a)   (b)

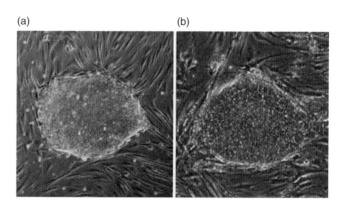

Figure 5.42 Brightfield microscopic comparison of ES and iPS cells. (a) Embryonic stem cells. *Source:* Loring & Peterson, 2012. Reproduced with permission from Elsevier and J. F. Loring. (b) Induced pluripotency cells. *Source:* Courtesy of J. F. Loring.

- Telomerase activity—Both human embryonic stem cells and iPS cells exhibit high levels of telomerase activity in comparison to other cell types, which is critical for driving mitotic activity and nearly unlimited self-renewal (Figure 5.43) (Buseman et al., 2012).
- Pluripotency—Pluripotency is the most definitive characteristic for confirming the iPS cellular identity. As with embryonic stem cells, both human and mouse iPS cells have been demonstrated *in vitro* to be induced toward differentiation into specific terminal lineages. These include neurons and glial cells as well as cardiomyocytes. The derivation of lineages representing the three primary germ layers is even more definitive, and this has now been shown for iPS cells in embryoid body development *in vitro* as well as teratoma formation germline transmission *in vivo* (Figure 5.44).

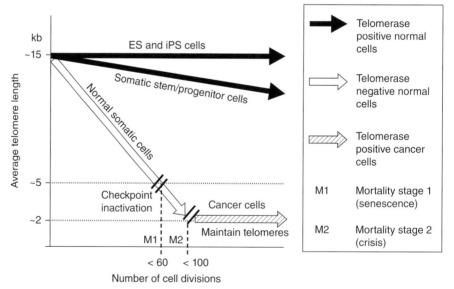

Figure 5.43 Graph of telomere length to number of cell divisions for various cell types. ES cells and iPS cells demonstrate considerably longer average telomeric lengths in comparison to other cell types and this length appears to be unaffected by the number of cell divisions indicating high telomerase activity. *Source:* Courtesy of K. Okita/Center for iPS Cell Research and Application, Kyoto University.

Figure 5.44 Germline transmission of induced pluripotency-generated stem cells. The mouse on the lower right was generated from iPS cells. (Photography courtesy Shinya Yamanaka and the Center for iPS Cell Research and Application, Kyoto University, Kyoto, Japan; reprinted with permission.)

A General Comparison of iPS Cells and Embryonic Stem Cells Despite the numerous similarities between iPS cells and embryonic stem cells described above, beginning in 2009 some researchers began to describe differences between the two cell types. These differences include, for example, global patterns of gene expression. Researchers in the Department of Molecular, Cell and Developmental Biology, University of California, Los Angeles, performed comprehensive gene microarray analyses and compared resulting data between iPS and ES cell clones to reveal hundreds of differentially expressed genes (Figure 5.45) (Chin et al., 2009).

DNA methylation patterns were also observed to be different between ES, iPS, and fibroblastic cell lines. A **CpG island** is defined as a genomic region containing a high frequency of CpG sites. A CpG site is made of cytosine and guanine connected by a ("p") phosphodiester bond. Researchers can observe the methylation status of CPG islands through a process known as bisulfite DNA sequencing. A modified version of this also allows for the monitoring of the global methylation status of a cell's genome. Specifically, both hypo- and hypermethylation of specific CpG islands were observed in iPS lines in contrast to their parental fibroblast donors. The unique methylation patterns were localized

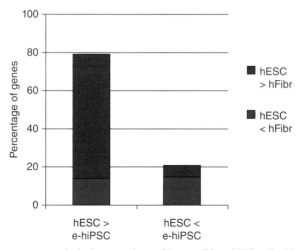

Figure 5.45 Signature gene analytical comparison of human ES and iPS cells. The graph on the left indicates genes expressed at higher levels in ES cells than iPS cells. The graph on the right indicates the converse. Red denotes genes expressed at higher levels in ES cells than fibroblasts and blue denotes the opposite. The minimum threshold to be considered a significant change in expression level was 1.5. (Graph courtesy Mark H. Chin, William E. Lowry and *Cell Stem Cell* (Chin et al., 2009); reprinted with permission.)

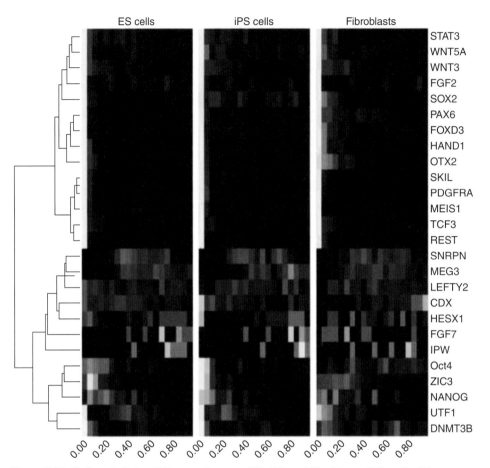

Figure 5.46 CpG methylation differences between ES, iPS, and fibroblast cell lines for 26 genes. Different yellow shading indicates differences in methylation state for each gene. Clustering analysis of CpG methylation distribution was also performed and is noted to the left. (Image courtesy Jie Deng, Kun Zhang and *Nature Biotechnology* (Deng et al., 2009); reprinted with permission.)

to tissue-specific and cancer-specific gene clusters. In addition, iPS cells exhibited higher total CpG island methylation than embryonic stem cells (Figure 5.46) (Deng et al., 2009).

Aside from differences in genomic content and gene expression, iPS cells and ES cells have also been shown to be unique in their capacities to differentiate into terminal lineages. The ability to differentiate into mature neurons has long been a hallmark of the ectodermal potency of embryonic stem cells and even considered by some as a "default" differentiation pathway in the absence of stimuli directing differentiation into other lineages. Su-Chun Zhang's team in collaboration with James Thomson in the Department of Neurology, School of Medicine and Public Health, University of Wisconsin, Madison, compared 12 iPS cell lines to 5 ES cell lines for their ability to differentiate into neurons and consistently observed a 90% efficiency for the generation of neurons from embryonic stem cells, but only ~10–50% when iPS cells were used as the source (Figure 5.47) (Hu et al., 2010).

Thus, while there are obvious differences between embryonic stem cells and induced pluripotency cells, these differences vary from one clonal cell population to another, and in

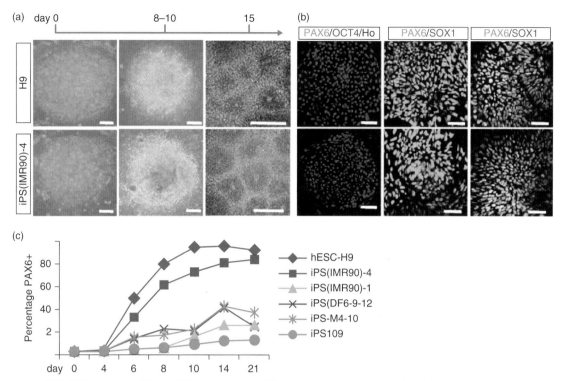

Figure 5.47 Differences in iPS and ES cell neuronal differentiation potential. (a) Brightfield microscopy demonstrating neuronal differentiation of both H9 human ES cells and (IMR90)-4 iPS cells over time. (b) Immunofluorescence confirming the presence Oct4 (Day 0), Pax6 (Day 6–8 and Pax6 plus Sox1 (Day 15). (c) FACS analysis demonstrating different differentiation potential for ES versus iPS cells. hESC line H9 was the most efficient followed by various iPS lines. (*Source:* Hu & Weick, 2008. Reproduced with permission from National Academy of Sciences, USA).

fact there are a number of iPS cell lines that are difficult to distinguish from embryonic stem cells by any of the properties mentioned above. Many researchers speculate that both differences and similarities between iPS and ES cells could be simply due to the different developmental time points for either induction of pluripotency or isolation from blasto-cysts. What has been confirmed is that the more iPS cells differ from ES cells in each of the categories above, the less viable they are with respect to pluripotency capabilities or even true *in vivo* germline competency (Figure 5.48). Additional similarities and differences between iPS cells and embryonic stem cells are described below.

Examples of Derived iPS Cells

As mentioned at the beginning of this chapter, induced pluripotency stem cells have been derived from a number of different and unique somatic cell lineages. This suggests a uni-versal utility for the production of pluripotent cell lines from numerous sources and high-lights the very real possibility of a virtually unlimited supply of patient-specific stem cells and differentiated cells. The following sections describe two high profile examples of unique donor cell sources from separate species (mouse and human) and the resulting iPS cell lines produced. A more detailed comparison to an embryonic stem cell line of the same species is provided to clarify the properties of each derived iPS line.

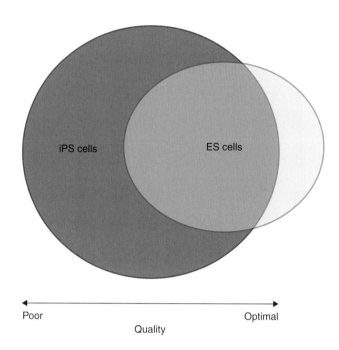

Figure 5.48 Diagrammatic illustration of the relationship between iPS and ES cells and differentiation quality.

Poor   Optimal

Quality

Mouse Embryonic Fibroblast-Derived iPS Cells Although embryonic in origin, mouse embryonic fibroblasts are not considered to be stem cells but are of somatic origin. Unaltered, they do not possess the capacity to differentiate into other lineages. As discussed above (see Focus Box 5.1 and Case Study 5.1) the first example of induced pluripotency using defined key factors was accomplished using MEFs as the donor cell source by Shinya Yamanaka and colleagues in the Department of Stem Cell Biology, Institute for Frontier Medical Sciences, Kyoto University, Kyoto, Japan. Morphologically, MEF-derived iPS cells resemble murine embryonic stem cells individually as well as upon the formation of colonies, with the exception that they appear to be more refractory in light absorbance (Figure 5.49) (Takahashi and Yamanaka, 2006).

Cell division rates for individual clonal MEF-derived iPS cell lines in comparison to embryonic stem cells were slightly slower (Figure 5.50).

Considerable differential gene expression was observed for individual iPS clones in comparison to embryonic stem cells for various stem cell-specific markers. Most striking was the low expression levels of Oct3/4 in each iPS line in comparison to that for embryonic

(a)   ES   (b)   iPS-MEF24-1-9   (c)   MEF

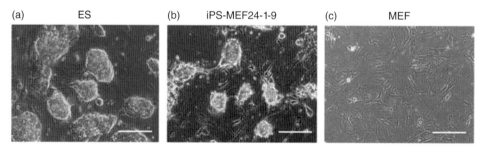

Figure 5.49 Brightfield microscopy of mouse embryonic fibroblast-derived iPS cells. (a) Mouse embryonic stem cells; (b) Mouse embryonic fibroblasts induced to pluripotency and; (c) Mouse embryonic fibroblasts. (Photographs courtesy Kazutoshi Takahashi, Shinya Yamanaka and *Cell* (Takahashi and Yamanaka, 2006); reprinted with permission.)

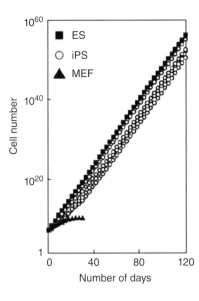

<figure>Figure 5.50 Graphical analysis of mitotic rates for murine embryonic stem cells and four independently isolated MEF-derived iPS cells lines. (Graph courtesy Kazutoshi Takahashi, Shinya Yamanaka and *Cell* (Takahashi and Yamanaka, 2006); reprinted with permission.)</figure>

stem cells as measured by **quantitative RT-PCR (reverse transcriptase polymerase chain reaction)** (Figure 5.51). The lack of Oct3/4 expression in each iPS line compared to that for embryonic stem cells could be due to extensive silencing of the locus via hypermethylation passed on from somatic donor cells (see the section A General Comparison of iPS Cells and Embryonic Stem Cells).

Yet the most striking difference between MEF-derived iPS cells and mouse embryonic stem cells occurred at the genomic level, with extensive CpG island (see the section A General Comparison of iPS Cells and Embryonic Stem Cells) methylation seen in the stem

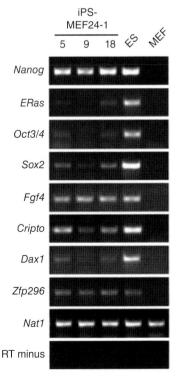

Figure 5.51 Quantitative RT-PCR analysis and comparison of MEF-derived iPS cells and murine embryonic stem cells. (Figure courtesy Kazutoshi Takahashi, Shinya Yamanaka and *Cell* (Takahashi and Yamanaka, 2006); reprinted with permission.)

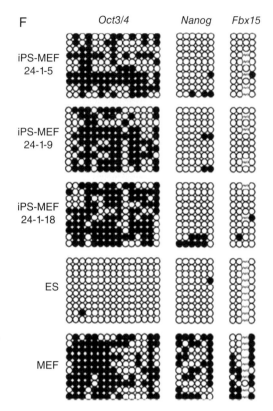

Figure 5.52 Representation of the CpG island methylation pattern for key stem cell-specific marker loci in MEF-derived iPS cells and mES cells. Closed circles represent methylated CpG islands. Open circles represent unmethylated CpG islands. (Figure courtesy Kazutoshi Takahashi, Shinya Yamanaka and *Cell* (Takahashi, Takahashi et al., 2007); reprinted with permission.)

cell-marker locus Oct3/4 in contrast to mES cells. Nanog and Fbx15 yielded no significant methylation pattern difference between the cell lines (Figure 5.52).

Thus, it is clear that there are considerable similarities but also differences between iPS cell lines derived from mouse embryonic fibroblasts and mES cells, but perhaps the most crucial test for commonality is pluripotency potential. In this respect, both types of stem cells have been demonstrated to form teratomas when injected into nude mice and contribute to embryonic development (Figure 5.53) (Takahahi and Yamanaka, 2006). These same cells were later demonstrated to give rise to adult chimeras and transmit through the germline.

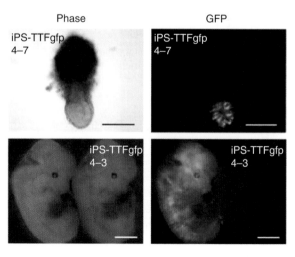

Figure 5.53 Contribution of MEF-derived iPS cells to embryonic development. Individual fluorescence-tagged iPS cells were injected into host blastocysts and observed for contribution to embryonic lineages (Figure courtesy Kazutoshi Takahashi, Shinya Yamanaka and *Cell* (Takahahi and Yamanaka, 2006); reprinted with permission.)

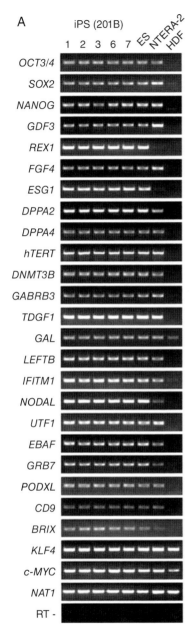

Figure 5.54 Quantitative RT-PCR demonstrating gene expression levels in human dermal fibroblast-derived iPS cells, donor HDFs, and embryonic stem cells. (Figure courtesy Kazutoshi Takahashi, Shinya Yamanaka and *Cell* (Takahahi and Yamanaka, 2006); reprinted with permission.)

Human Adult Skin-Derived iPS Cells An obvious step toward translational value in iPS cell technology is the derivation of pluripotent lines from adult human somatic cells. In 2007, Shinya Yamanaka's team expanded upon their initial success in inducing mouse embryonic fibroblasts to become pluripotent with a focus on human skin-derived cells as the donor source. These adult human dermal fibroblasts were transduced with the key pluripotency induction factors Oct3/4, Sox2, c-Myc, and Klf4 via a modified technique (discussed above) to that implemented for murine fibroblast pluripotency induction. iPS cells generated displayed striking similarities to hES cells with respect to stem cell-specific marker expression (Figure 5.54) (Takahahi and Yamanaka, 2006).

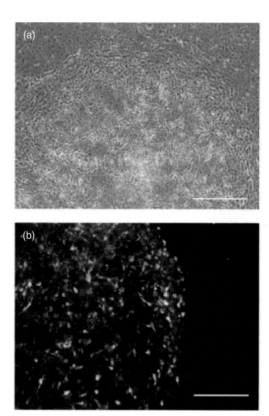

Figure 5.55   Brightfield and immunofluorescence microscopy of HDF-derived iPS cells differentiated toward a dopaminergic phenotype. (Red) βIII-tubulin, (Green) tyrosine hydroxylase, and (Blue) nuclei. (Photographs courtesy Kazutoshi Takahashi, Shinya Yamanaka and *Cell* (Takahashi et al., 2007); reprinted with permission.)

HDF-derived iPS cells could also be directed to differentiate into dopaminergic neurons in tissue culture at high efficiencies (Figure 5.55) (Takahashi et al., 2007). In addition, like their hES cells counterparts, HDF-derived iPS cells formed teratomas representing the three primary germlayers when transplanted into SCID mice.

Other considerable similarities between HDF-derived iPS cells and hES cells observed by Yamanaka's team included high telomerase activity and a marked degree of unmethylated promoter regions for key stem cell marker genes such as Oct3/4, Rex1, and Nanog. Thus, in the first documented case of human iPS cell generated the resulting cell lines appear to resemble human embryonic stem cells to a much greater degree than that observed in the murine system.

ADVANTAGES OF IPS CELLS OVER OTHER CELL TYPES

Numerous inherent advantages exist regarding the utility of induced pluripotency stem cells in comparison to stem cells of other origins. These advantages mostly apply to driving toward translational applications in cell- or tissue-based transplantation initiatives. Most notable is the potential for the development of patient-specific cell-based transplantation strategies. The following sections outline the utility and inherent value of iPS platforms in comparison to stem cells derived by other means.

Origin and Bioethics

Human embryonic stem cells are, by definition, derived from human embryos. In almost all cases with the exception of single cell isolation this derivation results in the destruction of the embryo. Many ethicists and religious leaders consider this to be an unethical and blatant destruction of human life. Thus, alternative strategies for the generation of truly pluripotent stem cell lines would alleviate the controversial nature of stem cells derived from human embryos. Induced pluripotency technology and resulting iPS cell lines effectively address ethical concerns regarding the destruction of human embryos.

Patient Specificity

The possibility of generating stem cell populations with broad differentiation potential using the patient as a somatic cell donor source has been a long sought goal of stem cell biologists worldwide and can be considered as one of the holy grails of stem cell research and stem cell-based translational medicine. The derivation of patient-specific stem cells addresses one primary biological limitation: immunorejection. Current cell and tissue transplantation strategies are limited by the high probability of immunorejection by the patient's own immune system. **Immunorejection** results from recognition of transplanted cells or tissues derived from a unique source as foreign (an allogeneic transplant): thus transplanted material is attacked, and any therapeutic benefit is compromised. The ability to generate stem cells and corresponding differentiated lineages from a patient's own donor cell source, for example, a skin biopsy, results in a genetically matched resource for a virtually unlimited supply (see below) of differentiated cell types valuable for translational applications.

Broad Lineage Differentiation Potential

"Adult" stem cells (ASCs) are derived from non-embryonic sources and thus address bioethics concerns. In addition, ASCs may be derived from a donor patient and therefore serve as a resource for autologous cell or tissue replacement therapeutics. However, the differentiation potential of ASCs is considerably limited in comparison to either embryonic stem cells or induced pluripotent stem cells. Most ASC lines are restricted in differentiation capacity and are considered to be unipotent, multipotent, or, at best, oligopotent (see definitions for cell potency terms in Chapter 2). iPS cell technology has circumvented this limitation. The very definition of "pluripotent"—having the ability to give rise to any fetal or adult cell type—signifies that iPS cells are not limited in differentiation capacity as are ASCs. Numerous studies on iPS cells have confirmed their potential to differentiate into lineages representing the three primary germlayers such as neurons and hepatocytes that are invaluable with respect to cell and tissue transplantation initiatives.

"Unlimited" Supply

Although many studies suggest "indefinite" expansion capabilities, many adult stem cell lines have a limited mitotic lifespan, thus reducing the amount (number of viable cells) and even the integrity of derived, differentiated lineages. For example, adult MSCs have shown much promise in differentiation potential, giving rise to musculoskeletal, urogenital, dermal, and even blood cell types. Yet the number of population doublings inherent in MSCs has been reported in the range of 30–40. This severely limits the utility of MSCs in a therapeutic setting as it directly correlates with a limited supply of terminally differentiated cells. Induced pluripotent cells have been demonstrated, if cultured properly, to yield a virtually unlimited supply of stem cells and resulting progeny. As discussed in the section Basic Properties of ES Cells, this is at least in part due to high telomerase activity inherent in iPS cells compared to other cell types, most notably ASCs (embryonic stem cells of course exhibit similar indefinite replicative properties).

Ease of Generation

Techniques for the generation of iPS cells mostly involve the introduction of key factors required for pluripotency induction. Since Yamanaka's first induction of pluripotency in murine fibroblasts in 2006, numerous methodologies have been developed and streamlined to make the generation of iPS cells from virtually any cell type relatively straightforward. It is anticipated that future technologies including the use of microRNAs and small molecules to induce endogenous key factor expression will even further refine iPS cell generation further and increase efficiencies for the production of stable, clonal iPS cells. This is in contrast to the isolation of embryonic stem cells, which is a tedious process requiring dissection and dispersion of the inner cell mass of donor blastocysts followed by tedious single cell plating/isolation, culture, and expansion procedures. These technical hurdles combined with the limited availability of human embryos and the ethical questions mentioned above make iPS cell generation and appealing technology for the production of stem cells to be used in basic research or translational applications.

CHAPTER SUMMARY

Examples of Nuclear Reprogramming in Nature

1. Nuclear reprogramming occurs naturally in response to environmental influences and even bacterial infection.
2. Anura Rambukkana and colleagues discovered leprosy bacterium's unique ability to reprogram adult Schwann cells to stem cell-like cells.

Cell Fusion

1. Adult somatic cells may be reprogrammed by cell fusion either via methods and technologies implemented in the laboratory or naturally.
2. Cell fusion occurs naturally during embryonic development, an example of which is the fusion of myoblast cells to form a syncytium.
3. Cell fusion is a common occurrence in many non-vertebrate species and is executed for the proper development of tissues and organs.
4. Cell fusion studies can be traced back to 1960 when Georges Barski, Serge Sorieul, and Francine Cornefert of the Institut Gustave Roussy, Laboratorie de Culture de Tissus et de Virologie in Villejuif, France, first observed the *in vitro* fusion of tumor cells derived from two different strains of mice.
5. Hybridomas, first created in 1975 by Cesar Milstein and Georges Kohler, are produced by cell fusion and used to produce monoclonal antibodies.
6. Various proteins, referred to as fusogens, have been identified which play key roles in cell fusion.
7. Cell fusion in a laboratory setting may be accomplished by electrofusion, pegylation, or viral induction.

8. Cell fusion results in cellular reprogramming through alterations in epigenetic status.

9. Helen Blau of Stanford was the first to demonstrate the plastic nature of cells which could be manipulated through cell fusion.

Somatic Cell Nuclear Transfer

1. Somatic cell nuclear transfer (SCNT), also known as "cloning", is accomplished via direct injection of a nucleus into the host egg or by fusion.

2. Environmental cues emanating from the enucleated oocyte drive a reprogramming of the epigenetic status of the cell.

3. Somatic cell nuclear transfer was successfully employed for the generation of the world's first cloned mammal, Dolly the sheep, in 1996 by Sir Ian Wilmut and colleagues at the Roslin Institute.

4. SCNT may also be used for the creation of stem cells from adult somatic cells.

5. SCNT-derived stem cells are sometimes considered embryonic in origin and can be derived from a variety of species.

6. Considerable differences in gross chromosomal makeup have been noted when compared to naturally derived organisms or cell lines of a similar age.

7. Dolly the sheep exhibited telomeric length considerably shorter than normal.

8. Methylation patterns are grossly different in SCNT-derived versus normal embryos.

9. Embryos derived via cloning methodologies fail to consistently regulate the process of X chromosome inactivation.

Induced Pluripotency

1. Walter Gehring and his team at the University of Basel were the first to demonstrate a form of induced pluripotency in their studies on the *Drosophila melanogaster* transcription factor *Antennapedia.*

2. Numerous transcription factors have been identified that have the regulatory power to guide cellular fate. These include the formation of muscle cells (MyoD), neuronal lineages (neurogenin), and even blood (C/EBP).

3. In 2006, Shinya Yamanaka and colleagues at the Institute for Frontier Medical Sciences, Kyoto University in Japan were the first researchers to identify the crucial factors needed for induction of pluripotency in somatic cells in tissue culture.

4. A "cocktail" of four factors: Oct3/4, Sox2, c-Myc, and Klf4, to be necessary and sufficient for cellular reprogramming.

5. Induction of pluripotency has now been accomplished in a variety of adult somatic cells.

6. Methods for introducing the key reprogramming factors include the use of retroviruses/lentiviruses, adenoviruses, sendai viruses, plasmids, mRNA, transposons, and protein.

7. Several microRNAs have been identified which result in the induction of pluripotency.

8. iPS cells exhibit many of the same characteristics as embryonic stem cells.

9. iPS and ES cells differ in global patterns of gene expression.

10. DNA methylation states are different between iPS cells and donor somatic cells.

11. iPS cells and ES cells are unique in their capacities to differentiate into terminal lineages.

12. Morphologically, MEF-derived iPS cells resemble murine embryonic stem cells individually as well as upon the formation of colonies with the exception that they appear to be more refractory in light absorbance.

13. Individual iPS clones exhibit differential gene expression in comparison to embryonic stem cells for various stem cell-specific markers.

14. Extensive CpG island methylation is seen in the stem cell-marker locus Oct3/4 in mouse iPS cells in contrast to mouse ES cells.

15. Human iPS cells display striking similarities to hES cells with respect to stem cell-specific marker expression.

16. Human iPS cells, like hES cells, exhibit high telomerase activity and a marked degree of unmethylated promoter regions for key stem cell marker genes such as Oct3/4, Rex1, and Nanog.

Advantages of iPS Cells Over Other Cell Types

1. Induced pluripotency technology and resulting iPS cell lines effectively address ethical concerns regarding the destruction of human embryos.

2. The derivation of patient-specific stem cells via induced pluripotency addresses risks of allogeneic immunorejection.

3. iPS cells are not limited in differentiation capacity as are ASCs.

4. Induced pluripotent cells have been demonstrated if cultured properly, to yield a virtually unlimited supply of stem cells and resulting progeny.

5. It is now easier to generate iPS cells than embryonic stem cells.

KEY TERMS

(Key terms are listed by order of appearance in the text.)

- **Autologous stem cell-based therapy**—cell or tissue transplantation in which the donor provides the original source of cells as the basis for the transplant.
- **Nuclear reprogramming**—the induction of changes in gene activity that result in a comprehensive transformation of cellular phenotype.
- **Transdifferentiation (metaplasia)**—the change of a differentiated cell into another unique differentiated cell.
- **Schwann cells**—the principle glial cells of the peripheral nervous system that function to support neurons in a myelinating and non-myelinating fashion.
- **Progenitor/stem like cells (pSLC)** —reprogrammed Schwann cells.
- **Cell fusion**—the joining of two unique cell types to form a single entity.
- **Hybrid**—cell fusion which undergoes mitosis.
- **Heterokaryon**—cell fusion which does not divide and consists of multiple nuclei.
- **Syncytium**—Multinucleated cell.

- **Synciotrophoblasts**—cells that make up the epithelial covering of the embryonic placental villi.
- **Adherens junctions**—protein complexes that occur at the contact points of cells.
- **Hybridoma**—a cell line resulting from the fusion of an antibody-producing B cell with a myeloma cancer cell.
- **Splenocytes**—white blood cells consisting of populations of T and B lymphocytes, dendritic cells, and macrophages.
- **Sendai virus(SeV)**—an RNA virus known to induce the formation of eukaryotic syncytia *in vitro*.
- **Hypoxanthine-guanine-phosphoribosyl transferase (HGPRT)**—an enzyme which drives *de novo* nucleotide synthesis and expressed by B cells.
- **Hemifusion**—fusion of only the outer cell membranes between two cells.
- **Cell–cell fusogens**—proteins involved in the fusion process between two cells.
- **Electrofusion**—the use of an electric field or pulse to drive the fusion of cells.
- **Polyethylene glycol (PEG)**—a polyether compound with many applications from industrial manufacturing to medicine.
- **Epigenetics**—gene expression or cellular phenotype patterns resulting from properties other than underlying DNA sequences.
- **Homokaryon**—the fusion between two identical cell types to result in multiple genetically identical nuclei in a common cytoplasm.
- **Synkaryon**—a single cell nucleus formed by the fusion of two pre-existing nuclei.
- **Heterokaryon** - an intermediate stage in the formation of a synkaryon in which a single cell nucleus contains multiple nuclei.
- **Somatic cell nuclear transfer (SCNT)**—as a laboratory technique for the generation of a cloned cell utilizing a donor nucleus.
- **Cloning**—the process of generating similar populations of genetically identical entities.
- **Reproductive Cloning**—cells cloned and grown into an early-stage embryo for transplantation into a surrogate mother.
- **Therapeutic Cloning**—cells cloned and expanded in tissue culture for therapeutic purposes.
- **Quiescence**—the G0 phase of the cell cycle.
- **Maturation Promoting Factor (MPF)**—a heterodimer of cyclin-dependent kinase, CDK1, and cyclin B. It promotes mitosis through the phosphorylation of key proteins involved in mitotic activity.
- **Telomeres**—repetitive nucleotide sequences of noncoding G-rich DNA $(TTAGGG)^n$ in combination with specific binding proteins present on the ends of chromosomes involved in chromosomal protection from degradation or fusion with neighboring chromosomes.
- **Telomerase**—an RNA-dependent DNA polymerase.
- **Epigenetics**—the study of heritable changes in gene expression resulting from environmental factors other than the underlying DNA sequence of said genes.
- **X-chromosome inactivation (XCI)**—an epigenetically regulated process of gene dosage compensation resulting in the silencing of genes on the X chromosome.
- **Xist**—a non-coding RNA molecule which drives X chromosome inactivation.
- **Induced pluripotency**—the artificial promotion of pluripotent properties in non-pluripotent cells via forced expression of specific genes.

- **Induced pluripotent stem (iPS) cells** - cells generated from the induction of pluripotency in normally non-pluripotent cells.
- *Antennapedia*—a transcription factor discovered in *Drosophila melanogaster* that has the ability to transform fly antennae into legs when overexpressed in developing fly embryos.
- **Homeobox**—a DNA regulatory element involved in the control of body plan and the development of anatomical features during embryonic development.
- **Retrovirus**—RNA virus that replicate in a host cell.
- **SV40 large T antigen** - a proto-oncogene derived from the polyoma virus SV40 which has the ability to stimulate replication of viral genomes in host cells.
- **Lentivirus**—a genus of the retroviridae family characterized by the unique ability to infect non-dividing cells and exhibit extended incubation periods.
- **Transgene**—a gene or genetic material transferred from one organism to another.
- **P1 bacteriophage Cre recombinase**—an enzyme derived from the bacteriophage P1 known as a topoisomerase which has the ability to carry out site-specific recombination.
- **Site-specific recombination**—DNA strand exchange occurring between nucleotide sequences possessing only a limited degree of sequence homology.
- **LoxP site**—Homologous DNA sequence recognized by the P1 bacteriophage Cre recombinase in site-specific recombination.
- **Adenovirus**—a virus which only infects non-dividing cells. In adenoviral infection the transgene does not incorporate into the host cell genome.
- **Sendai virus**—a virus that introduces genetic material into the cytoplasm but not nucleus of the infected cell.
- **Multiplicity of infection (MOI)** —the ratio of viral particles to target cells.
- **Self-cleaving peptide sequence** - a sequence of amino acids derived from the foot-and-mouth disease picornavirus 2A allowing for efficient cleavage of polypeptides to generate multiple peptide/protein fragments.
- **Episomal plasmid (a.k.a. Episome)**—closed circular DNA molecule that replicates within the nucleus of cells.
- **MicroRNAs (miRNAs)**—short, non-coding RNA molecules of about 22 nucleotides in length found naturally in both prokaryotic and eukaryotic cells that function to control gene expression, both transcriptionally and post-transcriptionally.
- **mi-iPS cell**—induced pluripotent cell derived via the application of miRNAs.
- **Transposon**—DNA sequence which possesses the ability to change positions within the genome.
- *PiggyBac* **(PB)**—a transposon that can shuttle between target cell chromosomes and extra-chromosomal vectors.
- **Poly-arginine leader sequence**—a peptide rich in arginines that mediates translocation across the plasma membrane.
- **CpG island**—a genomic region containing a high frequency of CpG sites, the methylation state of which plays a key role in transcriptional regulation and gene activity.
- **Quantitative RT-PCR (Reverse Transcriptase Polymerase Chain Reaction)**—a method for quantitatively measuring the presence of mRNA transcripts.
- **Immunorejection**—rejection of foreign materials by the body's innate immune system.

REVIEW QUESTIONS

(Answers to select review questions can be found at www.stemcelltextbook.com.)

1. What are the three groundbreaking technologies developed to reprogram cells?
2. Describe at least two examples of naturally occurring cellular reprogramming.
3. How does leprosy bacterium enhance its infectivity success?
4. Give an example of naturally occurring cell fusion.
5. Who first documented cell fusion?
6. Why are hybridoma cell lines so valuable in research and medicine?
7. Describe the proteins involved in cell–cell fusion.
8. Compare and contrast the three primary methods for laboratory induced cell fusion.
9. What role does epigenetics play in cellular reprogramming?
10. What is the difference between a homokaryon, synkaryon, and heterokaryon?
11. Describe the process of somatic cell nuclear transfer.
12. What is the difference between reproductive cloning and therapeutic cloning?
13. Describe the process implemented by Sir Ian Wilmut and colleagues in the cloning of Dolly the sheep.
14. List at least five species cloned since 1952.
15. What genomic property was different in Dolly the sheep compared to sheep of the same age?
16. How can epigenetics affect nuclear reprogramming efficiency?
17. How could aberrant X chromosome inactivation affect cloning efficiencies?
18. What discovery in flies did Walter Gehring and colleagues at the University of Basel make in 1987 that set the stage for studies on the regulation and manipulation of cell and body plan fate?
19. How did Shinya Yamanaka and colleagues first successfully induce pluripotency in mouse somatic cell fibroblasts?
20. List at least five somatic cell types that have been reprogrammed into iPS cells.
21. What are the key factors involved in pluripotency induction?
22. How did George Q. Daley's team at the Harvard Stem Cell Institute increase the efficiency of using retroviruses for inducing pluripotency in somatic cells?
23. Why would someone use a lentivirus over a retrovirus to induce pluripotency in a somatic cell?
24. What is the major drawback of using retroviruses and lentiviruses to introduce the key factors required for pluripotency into cells?
25. What is the major advantage of using adenoviruses to introduce the key factors required for pluripotency into cells?
26. What are the four major advantages for using Sendai virus to introduce the key factors required for pluripotency into cells?
27. Describe how episomes work.
28. How would one make mRNA transfection less immunogenic?
29. Give an example of inducing pluripotency in somatic cells using molecules other than those directly representing the key factors.

30. Why would someone use the *piggyBac* transposon system to deliver sequences encoding the key factors required for pluripotency induction in somatic cells?

31. What is the strategy employed for efficiently introducing proteins into cells to induce pluripotency?

32. Describe the four basic properties of iPS cells.

33. List at least two differences between iPS and ES cells.

34. What were the first two species from which iPS cells were generated?

35. How was the pluripotency of mouse embryonic fibroblast-derived iPS cells definitively proven?

36. What are the five advantages of iPS cells in comparison to other stem cell types?

THOUGHT QUESTION

Devise an experiment utilizing retroviral technology for the generation of iPS cells from adult somatic cells that does not leave a transgene stably incorporated into the genome of the host cell.

SUGGESTED READINGS

Buseman, C. M., W. E. Wright, et al. (2012). "Is telomerase a viable target in cancer?" *Mutat Res* **730**(1–2): 90–97.

Byrne, J. A., D. A. Pedersen, et al. (2007). "Producing primate embryonic stem cells by somatic cell nuclear transfer." *Nature* **450**(7169): 497–502.

Chin, M. H., M. J. Mason, et al. (2009). "Induced pluripotent stem cells and embryonic stem cells are distinguished by gene expression signatures." *Cell Stem Cell* **5**(1): 111–123.

Deng, J., R. Shoemaker, et al. (2009). "Targeted bisulfite sequencing reveals changes in DNA methylation associated with nuclear reprogramming." *Nat Biotechnol* **27**(4): 353–360.

Hu, B. Y., J. P. Weick, et al. (2010). "Neural differentiation of human induced pluripotent stem cells follows developmental principles but with variable potency." *Proc Natl Acad Sci U S A* **107**(9): 4335–4340.

Kang, Y., C. J. Xu, et al. (2006). "A novel strategy to compensate the disadvantages of live vaccine using suicide-gene system and provide better antitumor immunity." *Vaccine* **24**(12): 2141–2150.

Kohler, G. and C. Milstein (1975). "Continuous cultures of fused cells secreting antibody of predefined specificity." *Nature* **256**(5517): 495–497.

Lengerke C, Grauer M, et al. (2009) " Hematopoietic development from human induced pluripotent stem cells." *Ann N Y Acad Sci* **1176**:219–227.

Masaki, T., J. Qu, et al. (2013). "Reprogramming adult schwann cells to stem cell-like cells by leprosy bacilli promotes dissemination of infection." *Cell* **152**(1–2): 51–67.

Miyoshi, N., H. Ishii, et al. (2011). "Reprogramming of mouse and human cells to pluripotency using mature microRNAs." *Cell Stem Cell* **8**(6): 633–638.

Niemann, H., X. C. Tian, et al. (2008). "Epigenetic reprogramming in embryonic and foetal development upon somatic cell nuclear transfer cloning." *Reproduction* **135**(2): 151–163.

Nolen, L. D., S. Gao, et al. (2005). "X chromosome reactivation and regulation in cloned embryos." *Dev Biol* **279**(2): 525–540.

Okada, Y. (1962). "Analysis of giant polynuclear cell formation caused by HVJ virus from Ehrlich's ascites tumor cells. III. Relationship between cell condition and fusion reaction or cell degeneration reaction." *Exp Cell Res* **26**: 119–128.

Okita, K., M. Nakagawa, et al. (2008). "Generation of mouse induced pluripotent stem cells without viral vectors." *Science* **322**(5903): 949–953.

Sapir, A., O. Avinoam, et al. (2008). "Viral and developmental cell fusion mechanisms: conservation and divergence." *Dev Cell* **14**(1): 11–21.

Schneuwly, S., R. Klemenz, et al. (1987). "Redesigning the body plan of Drosophila by ectopic expression of the homoeotic gene Antennapedia." *Nature* **325**(6107): 816–818.

Seki, T., S. Yuasa, et al. (2010). "Generation of induced pluripotent stem cells from human terminally differentiated circulating T cells." *Cell Stem Cell* **7**(1): 11–14.

Soldner, F., D. Hockemeyer, et al. (2009). "Parkinson's disease patient-derived induced pluripotent stem cells free of viral reprogramming factors." *Cell* **136**(5): 964–977.

Stadtfeld, M., M. Nagaya, et al. (2008). "Induced pluripotent stem cells generated without viral integration." *Science* **322**(5903): 945–949.

Takahashi, K., K. Tanabe, et al. (2007). "Induction of pluripotent stem cells from adult human fibroblasts by defined factors." *Cell* **131**(5): 861–872.

Takahashi, K. and S. Yamanaka (2006). "Induction of pluripotent stem cells from mouse embryonic and adult fibroblast cultures by defined factors." *Cell* **126**(4): 663–676.

Warren, L., P. D. Manos, et al. (2010). "Highly efficient reprogramming to pluripotency and directed differentiation of human cells with synthetic modified mRNA." *Cell Stem Cell* **7**(5): 618–630.

Wilmut, I., A. E. Schnieke, et al. (1997). "Viable offspring derived from fetal and adult mammalian cells." *Nature* **385**(6619): 810–813.

Yu, J., K. Hu, et al. (2009). "Human induced pluripotent stem cells free of vector and transgene sequences." *Science* **324**(5928): 797–801.

Yu, J., M. A. Vodyanik, et al. (2007). "Induced pluripotent stem cell lines derived from human somatic cells." *Science* **318**(5858): 1917–1920.

Zhou, H., S. Wu, et al. (2009). "Generation of induced pluripotent stem cells using recombinant proteins." *Cell Stem Cell* **4**(5): 381–384.

Chapter 6

CANCER STEM CELLS

As it is clearly evidenced from the first five chapters of this book, stem cells and stem cell biology are pervasive in the life cycle of the vast majority of multicellular organisms. Stem cells can be embryonic in origin, adult in origin, or artificially crafted from virtually any cell type via the application of induced pluripotency technologies. Self-renewal, multi and pluripotency, and high proliferative potential give stem cells novel capacities compared to other cell types, yet these properties that make them both essential for embryonic development and postnatal survival may go awry and result in unwanted phenotypes. The third property, extensive, prolonged, and perhaps even unlimited self-renewal (as in the case of embryonic stem cells), is the focus of this chapter. Should stem cell mitotic activity become unregulated or uncontrolled, a tumorigenic and perhaps malignant phenotype may result, hence the term cancer stem cell (CSC). Yet these cells may also act as unique and valuable targets for therapeutic intervention. For example, the targeting of therapies to CSCs may be a potent technique for combating cancer by homing in on its very origins. This chapter outlines the highly controversial field of CSC research, including early discoveries and biological properties. Various examples of characterized CSC lineages are also described.

Focus Box 6.1: Rudolph Virchow and the Embryonal Rest Hypothesis

 Rudolph Virchow (1821–1902), was a German doctor and biologist, and is known as the factor father of modern pathology and is considered to be one of the founders of social medicine. He was the first to discover leukemia cells while acting as the director of the Pathological Institute at Charite Hospital in Berlin. He is widely credited with proposing the *embryonal rest hypothesis* and *cell theory* (see text for details). He has received numerous awards including the Copley Medal in 1892. (Photo courtesy Wikimedia Commons; reprinted with permission.)

BACKGROUND ON THE ORIGINS OF CANCER

The origins of cancer depend upon alterations in one or more of three properties: regulatory control of the cell cycle; cellular senescence; and/or **apoptosis**, or programmed cell death.

Stem Cells: A Short Course, First Edition. Rob Burgess.
© 2016 John Wiley & Sons, Inc. Published 2016 by John Wiley & Sons, Inc.

These alterations may be either the result of somatic (environmental) mutations or due to genetic predisposition (inherited). In either case, gene-specific mutations result in changes in either the expression level of the locus or the resulting protein product. Transforming mutations tend to occur in **proto-oncogenes**, which are defined as normal cellular genes that contribute to the production of cancer when their activity is affected. Said changes may occur in genes known to control or regulate the cell cycle, apoptosis, or senescence, thus resulting in cancer. Many studies have indicated that mutations in proto-oncogenes may occur in a completely random fashion. Thus the **stochastic hypothesis** describes that virtually any cell type could become transformed given the occurrence of a proto-oncogene mutation. Yet, as discussed below, there is growing evidence that a subpopulation of cells within a tumor drives the majority of its growth. Referred to as **cancer stem cells**, these tumorigenic stem cells are described throughout the remainder of this chapter.

DISCOVERY AND ORIGIN OF CANCER STEM CELLS

The hypothesis that CSCs exist as a subpopulation within tumors is not a new concept, and was first proposed in the 1800s by cell biologists Rudolph Virchow (see Focus Box 6.1) and Julius Cohnheim. Virchow developed the **hypothesis of embryonal rest**, which highlighted the striking similarity between cancer cells and cells encompassing an early-stage developing embryo. In fact, in 1858 Virchow coined the now infamous quote "Omnis cellula e cellula" (All cells come from cells). This quote is now widely recognized as **cell theory**. These concepts were later expanded upon by Virchow, driving his hypothesis that mature organs may have rudimentary components that are embryonic-like and will promote tumorigenesis given the right circumstances. Two models evolved regarding the origins of the growth and heterogeneity of tumors. In the first and most widely accepted model, most cells within the tumor are heterogeneic and have the ability to divide, giving rise to new tumor cells. In the second model, only a subset of cells have the capacity to proliferate extensively, driving the growth of the existing tumor and the formation of new tumors (Figure 6.1).

A wealth of information has amassed over the last 40 years regarding the molecular origins of cancer. Many seminal research publications exist suggesting that certain forms of cancer harbor cells which exhibit similarities to stem cells, that is, to a certain degree they possess the same three properties described earlier in this text for normal stem cells.

Two Models for Tumor Heterogeneity

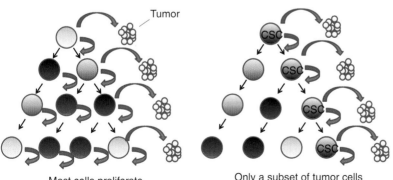

Most cells proliferate
and drive tumor growth

Only a subset of tumor cells
(cancer stem cells) proliferate
and drive tumor growth

Figure 6.1 Diagrammatic illustration of the two models for tumor heterogeneity and growth. See text for details.

These cells have been referred to as **tumor-initiating cells**. **Malignancy** is defined as a tumor tending to invade normal tissues or to recur after removal, and tumor-initiating cells that have malignant properties have been referred to as **CSCs**. The existence of CSCs has been debated over the years due to the fact that clear characterization of these cells as unique and different in comparison to normal tissue stem cells has been difficult. In addition, if it is indeed accepted that CSCs exist, their origin is even more hotly debated. Do CSCs arise from a population of normal stem or progenitor cells or are they the result of somatic cell dedifferentiation?

In 1961 it was discovered that the heterogeneous population of cells present in a tumor might have differing tumorigenic potential and differing cellular potency. During this year Chester M. Southam and Alexander Brunschwig in the Division of Clinical Chemotherapy at Sloan-Kettering Institute in New York performed a series of autologous tumor cell transplantation experiments that revealed a minimum threshold cell number required for tumorigenesis. These findings were expanded upon over 30 years later by Stewart Sell and G. Pierce in the Department of Pathology and Laboratory Medicine, University of Texas Health Science Center Houston, in a comparison of the cellular composition of teratocarcinomas and the corresponding changes in cellular makeup that occur during chemically induced hepatocarcinogenesis. The researchers concluded that many cancers originate from the arrest of tissue stem cell maturation (Figure 6.2) (Sell and Pierce, 1994). In addition, the researchers speculated that cancer could be treated effectively by eliminating this maturation block. In fact, a real world example of this therapeutic strategy is (Gleevec®), which has been demonstrated to remove the signals leading to cellular immortalization in chronic myelocytic leukemia (Druker et al., 2006 and Jordan et al., 2006).

Acute myeloid leukemia (AML) is a cancer defined by abnormal and uncontrolled growth of white blood cells. In 1997, Dominique Bonnet and John E. Dick in the Department of Genetics, University of Toronto, Canada identified a subpopulation of cells in human patients responsible for the initiation of the disease and possessing all the hallmarks

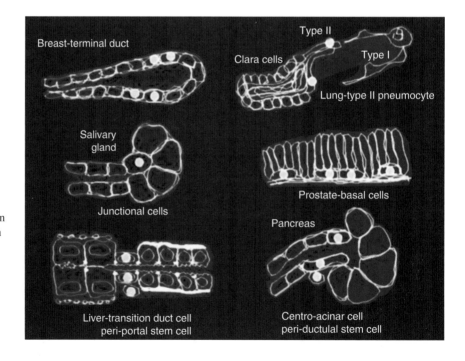

Figure 6.2
Diagrammatic illustration of the origins of CSCs in different organ systems. (Diagram courtesy Stewart Sell, GB Pierce and *Laboratory Investigation* (Sell and Pierce, 1994); reprinted with permission.)

of a stem cell. Case Study 6.1 outlines this seminal paper that set the stage for the development of a new and highly controversial field of stem cell research. More information on CSCs as they pertain to multiple myeloid is provided in the Examples of Cancer Stem Cells section toward the end of the chapter.

Case Study 6.1: Human acute myeloid leukemia is organized as a hierarchy that originates from a primitive hematopoietic cell

Dominique Bonnet and John E. Dick

Researchers Dominique Bonnet and John E. Dick in the Department of Genetics at the University of Toronto in Canada identified a subpopulation of cells within the hematopoietic stem cell population that is responsible for the leukemic phenotype in AML. This was accomplished via cell sorting and transplant of cells capable of initiating AML in nonobese diabetic mice with severe combined immunodeficiency syndrome (NOD/SCID). The subtypes of cells were characterized as CD34^{++} CD38$^-$ and were determined to be normal primitive cells rather than committed progenitor cells. Termed **SL-ICs**, evidence that these cells were indeed the subpopulation responsible for the leukemic phenotype included their ability to proliferate, differentiate, and renew themselves, producing a leukemic phenotype in mice identical to the same disorder observed in humans (Figure 6.3) (Bonnet and Dick, 1997).

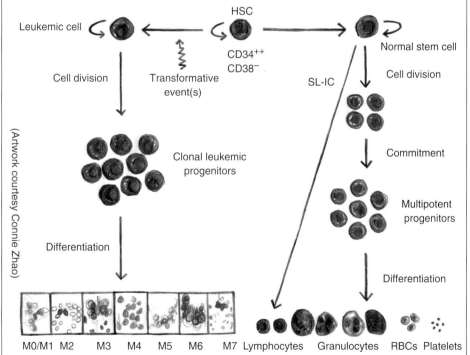

Figure 6.3 Diagrammatic illustration of the model for leukemic CSC development. The CD34^{++} CD38$^-$ hematopoietic stem cell undergoes a transformative event, thus creating an SL-IC subtype that rapidly expands in population size and has the ability to differentiate into all the mature myelo-erythroid cell types found in the blood. (Artwork courtesy Connie Zhao.)

TABLE 6.1 CSCs' markers for various tumor types).

Type of tumor	CSC phenotype (biomarkers expressed on the cell surface)	Percentage of cell fraction
AML	CD34$^+$ CD38$^+$	0.2–1
Brain	CD133$^+$	5–30
Breast	CD44$^+$ CD24$^-$	11–35
Colon	CD133$^+$ or ESAhi; CD44$^+$	1.8–24.5
Head and neck	CD44$^+$	< 10
Hepatocellular	CD133$^+$	1–3
Lung	CD133$^+$	0.3–22
Melanoma	CD20$^+$	~20
Multiple myeloma	CD138$^+$	2–5
Pancreatic	CD44$^+$ CD24$^-$ ESA$^+$	0.2–0.8
Prostate	CD44$^+$ CD133$^+$ a2b1hi or CD44$^+$ CD24$^-$	0.1–3

Source: Adapted from the book *Cancer Stem Cells,* William L. Farrar, Editor.

Thus, Case Study 6.1 outlines the discovery of a subpopulation of cells, CD34^{++}, CD38$^-$ classified as primitive CSCs originating in the hematopoietic system. The authors confirmed the cells' **tumorigenic potential**, which is the ability of cells to drive tumor formation at low cell densities. It is measured by injecting a population of CSC-containing cells into nude or NOD/SCID mice and observing tumor formation. Since this seminal study, a variety of CSCs have been isolated from numerous tumor types, each exhibiting a unique cell surface marker expression profile (Table 6.1).

CSCs may arise from mutations in normal stem or progenitor cells, via refractory effects from chemotherapy or simply escapage from primary tumors (Figure 6.4).

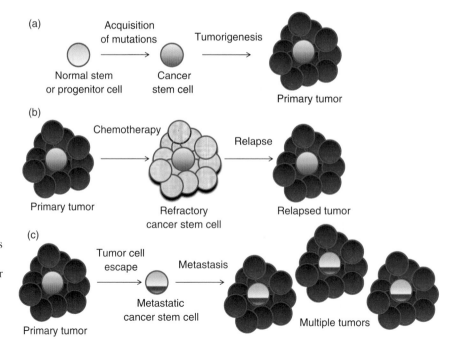

Figure 6.4 Illustration of the origin and types of CSCs. (a) Origin of a CSC via mutations in other stem or progenitor cell populations, (b) CSCs not eradicated by chemotherapy drive tumor regrowth, and (c) CSCs escape a metastasizing tumor resulting in new tumor growth.

BASIC PROPERTIES OF CANCER STEM CELLS

In order to fully understand the concept of a stem cell that promotes the progression of tumorigenesis, malignancy, and even metastasis, it is necessary to compare the similarities and differences between normal adult stem cells and CSCs (also referred to as "cancer-initiating cells" (CICs)). Although these differences will be unique to each tissue or organ system from which cancers arise, nevertheless some common themes exist. Figure 6.5 (a) outlines the marker expression and life cycle of a normal adult stem cell. These cells typically divide asymmetrically, resulting in the "birth" of another stem cell and a committed cell. Numerous loci are involved in the regulation of asymmetric adult stem cell division, including Oct3/4, Nanog, and Stellar. The progenitors resulting from normal adult stem cell division are typically multipotent in nature or, in rare instances, can transdifferentiate into cell types representing different germ layers. Figure 6.5 (b) illustrates some key differences in CICs compared to normal adult stem cells. For example, most CICs do not divide

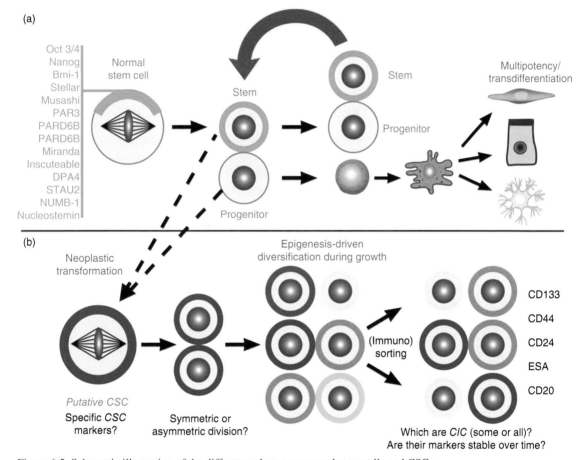

Figure 6.5 Schematic illustration of the differences between normal stem cells and CSCs. (a) Normal stem cells express a variety of stem cell specific markers such as Oct3/4 and Nanog, undergo asymmetric or symmetric division, and differentiate into a variety of terminal lineages, (b) CICs do not necessarily express stemness markers, are not known to undergo asymmetric division, and may or may not differentiate into terminal lineages. (*Source:* Hill & Perris, 2007. Reproduced with permission from Oxford University Press.)

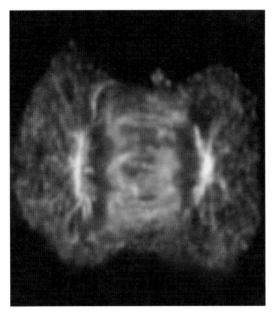

Figure 6.6
Fluorescence
microscopy of a
dividing CSC.
Source:
Courtesy of
Cancer
Research UK.

asymmetrically, but, rather, symmetrically, giving rise to two identical daughter cells from which diversification of these cells is the result of later epigenetic modifications driven by massive, uncontrolled expansion (see Figure 6.6). In addition, only a subset of the genes representing "stemness" expressed in adult stem cells are also expressed in CICs and express unique cell surface markers such as CD20 or CD133 at considerably high levels. It is speculated that CSCs have unique underlying molecular and morphological properties resembling stem cells that enhance their ability to self-renew and drive tumor growth. They differ significantly from "normal" stem cells and have been referred to as **CICs**, given their striking differences from other stem cell types (Hill and Perris, 2007).

A Comparison of Cancer Stem Cells and Normal Stem Cells

As mentioned above, CSCs have the following properties:

- **Ability to Self-renew**

  Self-renewal is perhaps the most critical prerequisite for a CSC to have tumorigenic capability. It is most often measured via serial tumor transplantation *in vivo* or through the application of soft agar sphere formation *in vitro*.

- **Strong Tumorigenic Potential**

  Tumorigenic potential is measured as the number of cells required for tumor induction. Typically, tumorigenic potential can range from as little as 100 cells (high tumorigenic potential) to as many as 50,000 cells (low tumorigenic potential). Tumorigenic potential varies greatly from one CSC to another as illustrated in Figure 6.7 (O'Brien et al., 2007).

Figure 6.7 Graphical analysis of different tumorigenic potentials exhibited by unique CSC populations. CD133+ CSCs clearly exhibit higher tumorigenic potential than CD133− cells. (Data courtesy John E. Dick and *Nature* (O'Brien et al., 2007); reprinted with permission.)

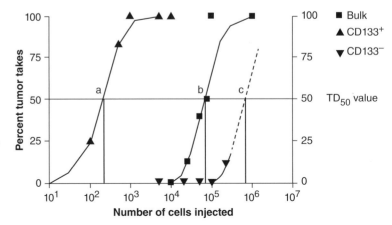

- **Establishment of Tumor Heterogeneity**

  In order to truly be classified as a CSC, the cell must not only be able to self-replicate and result in the formation of tumors, but must also have the capacity to completely recapitulate the phenotype of the original tumor from which it was isolated. Thus the CSC must be able to drive the growth of the heterogenic cell population from which it was isolated, that is, promote **tumor heterogeneity**. Case Study 6.2 illustrates the first example of proof that CSCs can drive tumor heterogeneity.

Case Study 6.2: Prospective identification of tumorigenic breast cancer cells

Muhammad Al-Hajj, Max S. Wicha, Adalberto Benito-Hernandez, Sean J. Morrison, and Michael F. Clarke

Michael Clarke et al., at the Institute for Stem Cell Biology and Regenerative Medicine, Stanford University developed a unique assay to confirm the tumor heterogenic potential of a breast CSC. The assay involved the isolation and serial passaging of a population of breast CSCs marked as CD44$^+$ CD24$^-$/low. Following serial passage, the cells were transplanted into NOD/SCID mice and assayed for both tumorigenic potential and tumor heterogeneity. These studies confirmed that this population of CSCs was capable of driving tumor induction and contributing to the heterogenic makeup of the tumor in a manner similar to that for the original tumor (Figure 6.8) (Al-Hajj et al., 2003).

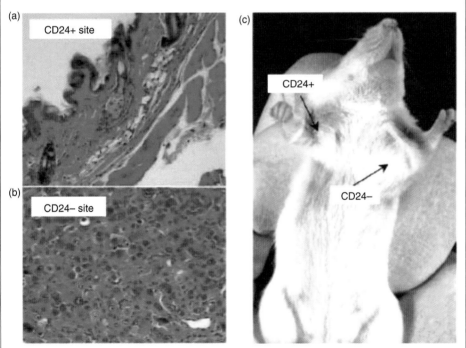

Figure 6.8 Tumorigenic potential of CD24$^-$ breast CSCs. (a) CD24$^+$ transplanted cells exhibiting no tumorigenicity, (b) CD24$^-$ cells demonstrating malignancy, (c) Comparison of CD24$^+$ versus CD24$^-$ injection sites in NOD/SCID mice. (*Source:* Al-Hajj, et al., 2003. Reproduced with permission from National Academy of Sciences, USA.)

SIGNALING PATHWAYS INVOLVED IN CANCER STEM CELL TRANSFORMATION

It has long been speculated that many of the same pathways that regulate normal stem cell self-renewal and maturation may also be involved in regulating the behavior of CSCs. An example of this is the bcl-2 pathway in hematopoietic stem cell development: its forced expression inhibits apoptosis, thus driving increased hematopoietic stem cell numbers. **Bcl-2** is the founding member of the Bcl-2 superfamily of regulatory proteins that control apoptosis in a variety of cell types. Other pathways have long been associated with both embryonic development and oncogenesis and include the Wnt, Shh, and Notch signaling cascades. Each of these pathways has been shown to regulate stem cell self-renewal. For example, Notch activation or the presence of Shh promotes both HSC self-renewal and the self-renewal of other stem cell populations. The Wnt signaling cascade also affects HSC self-renewal, but appears to affect epidermal and gut stem cell self-renewal as well. It has also been widely implicated in promoting oncogenesis in a variety of organ systems (Figure 6.9) (Reya et al., 2001).

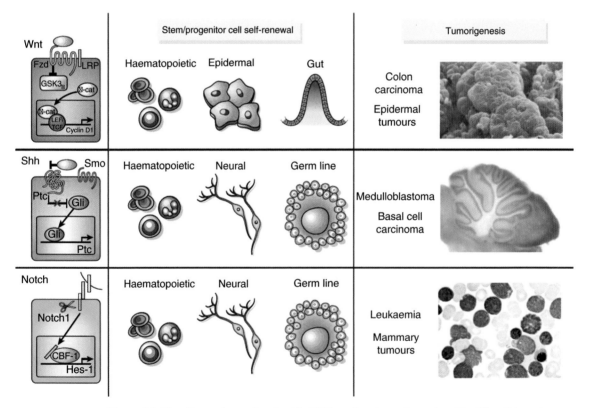

Figure 6.9 Signaling pathways involved in CSC-mediated transformation. The Wnt, Shh, and Notch signaling pathways, known for their important roles in embryonic development, have been shown to be affected in CSC-mediated tumorigenesis and transformation. (*Source:* Reya & Morrison, 2001. Reproduced with permission from Macmillan Publishers.)

EXAMPLES OF CANCER STEM CELLS

CSCs have now been identified and isolated from a variety of different types of cancers. Each type of CSC exhibits unique molecular and morphological phenotypes. In addition, individual CSCs isolated from different tumors or even the same tumor can exhibit drastically different tumorigenic potential due to variations in basic properties such as proliferation rates, symmetric versus asymmetric division, and multipotency. Following are several examples of CSCs or CICs isolated from various tumor types.

Breast

Breast cancer is one of the most devastating forms of cancer yet can be successfully treated with a high survival rate if caught early. As Case Study 6.2 illustrates, breast CSCs have been identified in specimens from patients exhibiting various stages of metastatic breast cancer. These CSCs exhibited a unique pattern of cell surface marker expression, specifically $CD44^+$, $CD24^-$, and could produce tumors in xenograft animal models such as immunocompromised NOD/SCID mice. It was observed that only a rare cell population isolated from the tumor exhibiting the $CD44^+$, $CD24^-$ phenotype could produce tumors *in vivo*. In addition, these purified cells differentiated both *in vitro* and *in vivo* into the various cell types present in metastatic breast cancer, a hallmark of tumorigenic heterogeneity (Galli et al., 2004).

Central Nervous System

Perhaps one of the most significant findings regarding the presence of stem cells in the central nervous system (CNS) occurred in 1992 when researchers in the Department of Pathology, University of Calgary Faculty of Medicine, Alberta, Canada observed the formation of **neurospheres** when culturing CNS cells under nonadherent conditions (Figure 6.10). Neurospheres exhibit robust self-renewal properties and can differentiate into all cell types encompassing the adult brain, including different types of neurons, astrocytes, and oligodendrocytes.

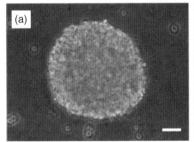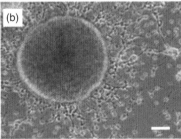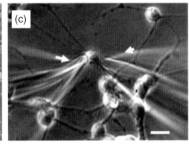

Figure 6.10 Brightfield microscopy of neurospheres. (a) Neurosphere consisting of SVZ cells isolated at E15 that have aggregated in suspension after 2 days in culture. Scale bar: 100 μm, (b) neurosphere of SVZ cells derived at E15 that was attached to the floor of the culture flask after 4 days in culture. Note cells migrating away from the neurosphere. Scale bar: 100 μm, (c) cells at the periphery of neurospheres were chosen for electrophysiological recording. Most of the recorded cells extended processes. Arrows indicate the location of recording (left) and puffer (right) pipettes. Scale bar: 20 μm. (Photographs courtesy D.O. Smith, J.L. Rosenheimer, R.E. Kalil, and Wikimedia Commons; reprinted with permission.)

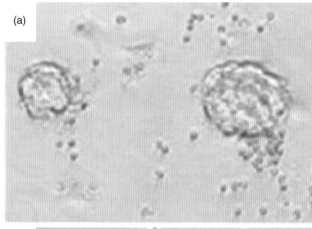

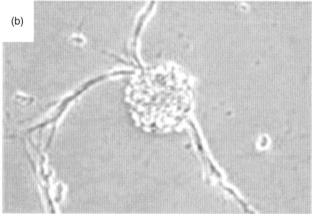

Figure 6.11 Brightfield microscopy of neurospheres isolated from human blastoma tumors. (a) Glioblastoma-derived; (b) medulloblastoma-derived. 20× magnification. (Images courtesy Rossella Galli, Angelo Vescovi and *CancerResearch* ((Galli et al., 2004); reprinted with permission.)

If cultured properly, neurospheres maintain a stem cell compartment that can be isolated and split into single cells, each of which can give rise to a new neurosphere. A full 12 years later, techniques utilized for the generation of neurospheres and related stem cell compartments were implemented for the isolation of CD133+ cells from human gliomas. Researchers at Stem Cell Research Institute and Laboratory of Molecular Diagnostics, H. S. Raffaele, Milan, Italy employed neurosphere cell culture techniques for the isolation of cells from glioblastoma and medulloblastoma tumors and demonstrated that clonal isolates had the capacity to recapitulate the blastoma phenotype *in vitro* and *in vivo,* which had a considerably high tumorigenic potential of 100 cells sufficient for tumor formation upon transplant in NOD/SCID mice (Figure 6.11) (Galli et al., 2004).

It has been suggested that neural progenitor cells contribute to neural cancers, as numerous studies have demonstrated that later-stage committed progenitors can drive the formation of tumors if certain oncogenes such as *ras* and *myc* are dysregulated. This has been demonstrated, for example, in the oligodendrocyte progenitor cell line O-2A which, upon dysregulation of *ras* and/or *myc*, can form tumors when transplanted *in vivo* (Figure 6.12) (Barnett et al., 1998). Therefore, studies on CNS-derived tumors confirm that cancer need not be derived from tissue-specific stem cells, but may indeed arise from later-stage committed progenitors.

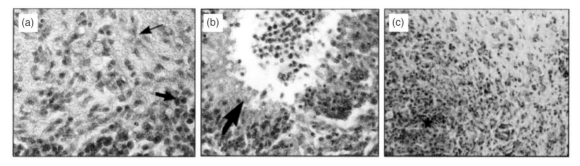

Figure 6.12 Histology of tumors derived from O-2A oligodendrocyte progenitor cells. Hematoxylin and eosin staining of (a) General tumor pathology showing spindle-shaped tumor cells (top arrow), (b) Center of the tumor showing extensive necrosis (large arrow), and (c) Edge of the tumor (star) which extends into the brain. (*Source:* Barnett et al., 1998. Reproduced with permission from Oxford University Press.)

Colon

Focus Box 6.2: John E. Dick and the discovery of cancer stem cells

 John E. Dick is a Canadian scientist and professor at the University of Toronto and is credited with the initial discovery of CSCs in various strains of human leukemia. In 1997 he crafted the "**cancer stem cell hypothesis**," which states that stem cells are unique to each cancer and are key to tumor initiation and growth. He is also responsible for the discovery of human colon CSCs. For his seminal research he has received numerous awards including the 2000 Robert L. Noble Prize for Excellence in Cancer Research, the 2005 William Dameshek Prize, and the 2007 Premier's Summit Award in Medical Research.

(*Source:* Courtesy of the University Health Network.)

Colon cancer and colorectal cancer are most often due to non-genetic environmental influences, with 75–95% of individuals contracting the disorder having no genetic predisposition to the disease. Lifestyle choices such as a high fat content diet or high alcohol intake may promote the initiation and growth of colon cancer. In addition, inflammatory bowel disease, or **Crohn's disease**, results in an increased risk for tumorigenic growth in the colon and/or rectum. Although it is the second most common cause of cancer-related deaths, if caught early, colon cancer can be effectively treated.

As with many cancers, colon cancer originates from the alteration of epithelial cells, in this case those lining the colon or rectal regions of the gastrointestinal tract. The pathways involved in colon cancer tumor progression are now widely understood, and it is known that mutations in loci involved in the Wnt signaling cascade result in epithelial cell transformation and tumorigenesis. In 2007, John E. Dick's research team (see Chapter 1 and the Discovery of Cancer Stem Cells, the section on AML cells and Focus Box 6.2) in the Division of Cell and Molecular Biology, University Health Network, Toronto, Canada sought to determine if each individual colon cancer cell has the capacity to both initiate and sustain tumor growth (stochastic model) or if only a subpopulation of cells present within the tumor possesses these properties (CSC model). The researchers developed a novel xenograft model involving the renal transplantation of a capsule containing human colorectal cancer cells into immunocompromised NOD/SCID mice. Tumors were allowed to grow to a specific size for a total of six different primary colon cancers and cells were harvested for further analysis in immunohistochemistry (IHC) and *in vitro* studies. IHC analysis revealed the presence of a unique set of cells expressing the surface marker CD133, which has been previously demonstrated to be a marker for CSCs in the brain and prostate. Serial xenograft transplants using either diluted cells from the parent tumor or an enriched population of CD133$^+$ cells also from the parent tumor demonstrated a recapitulation of the tumorigenic phenotype originally exhibited in humans (Figure 6.13) ((O'Brien et al., 2007).

Cells were diluted and retransplanted in a range of doses into NOD/SCID mice and tumor growth was observed. The researchers identified approximately one colon cancer-initiating cell (CC-IC) per 5.7×10^4 cells transplanted. These cells were further fractionated based on the CD133 expression. CD133$^+$ cells varied in each tumor sample tested, from 1.8% to 24.5%. Interestingly, CD133$^+$ cells existed in clusters surrounded by CD133$^-$ tumor cells. Thus, Dick's team successfully identified a CD133$^+$ subpopulation of cells,

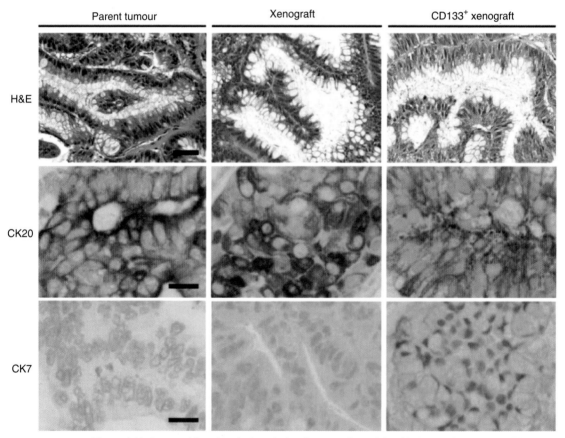

Figure 6.13 Immunohistochemical analysis of xenograft models of human colon cancer. Either unenriched (middle column) or cell populations enriched for the expression of CD133 (right column) recapitulated the original tumorigenic phenotype evident in humans (left column). H&E marked the nuclei and cellular boundaries. The tumors were also positive for cytokeratin-20 (CK20), a marker for human colonic adenocarcinoma. (Figure courtesy Catherine A. O'Brien, John E. Dick, and *Nature* (O'Brien et al., 2007); reprinted with permission.)

which they termed "colon cancer-initiating cells" (CC-ICs), distinct from the vast majority of cells representing the tumor mass. These cells were capable of initiating tumor growth and driving tumor heterogeneity.

Ovary

Ovarian cancer is considered to be the most deadly malignancy found in the female reproductive system. The vast majority of ovarian cancers originate from the ovarian surface epithelium: identifying a subpopulation of epithelial cells driving the growth of ovarian malignancies would allow for possible early therapeutic intervention. Kenneth P. Nephew's team in the Department of Medical Sciences, Indiana University School of Medicine, Bloomington, identified and characterized a subpopulation of ovarian cancer-initiating cells (OCICs) present in ovarian malignancies that possessed all the hallmarks of CSCs including the ability to recapitulate the original tumor phenotype in animal models. The cells exhibited considerably high tumorigenic potential, with as few as 100 cells sufficient for tumor growth in xenograft animal models. The "stemness" properties of the cells were confirmed through cell proliferation assays. The cells exhibited considerable resistance to the ovarian cancer chemotherapies paclitaxel and cisplatin and also showed upregulation of various stem cell markers including Notch-1, Nanog, and nestin. The cells were further characterized for cell surface marker presence and identified as $CD44^+$, $CD117^+$ (c-kit, stem cell factor receptor) (Figure 6.14) (Zhang et al., 2008).

Pancreas

Pancreatic cancer, particularly pancreatic adenocarcinoma, is almost always first detected at advanced stages and as a result is highly lethal. The prognosis typically ranges around 3% reaching a 5-year survival and is the fourth most common cause of cancer-related deaths in the United States. Thus the development of an assay or technology that would allow for the early detection of pancreatic adenocarcinoma would perhaps increase prognoses. Researchers in the Departments of Surgery, Molecular and Integrative Physiology and Internal Medicine at the University of Michigan Medical Center, Ann Arbor, utilized a xenograft model system to identify pancreatic adenocarcinoma CSCs. Specifically, they transplanted human pancreatic adenocarcinomas in immunocompromised mice and identified a $CD44^+$, $CD24^+$, ESA^+ subpopulation of cells that were highly tumorigenic. **Epithelial specific antigen (ESA a.k.a. epithelial specific molecule, Ep-CAM)** is a cell surface marker consisting of two glycoproteins. It is located on the cell surface and in the cyto-

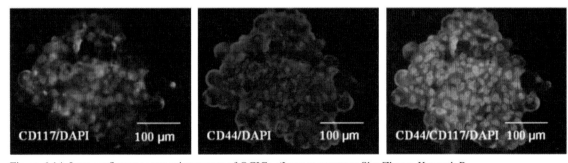

Figure 6.14 Immunofluorescence microscopy of OCICs. (Images courtesy Shu Zhang, Kenneth P. Nephew and *CancerResearch* (Zhang et al., 2008); reprinted with permission.)

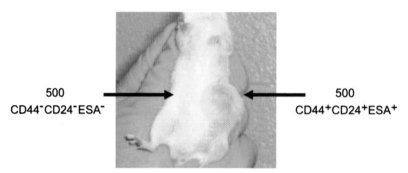

Figure 6.15 Tumor formation in xenograft models of pancreatic cancer. Only CD44+, CD24+, ESA+ cells demonstrated tumorigenic potential when transplanted into NOD/SCID immunocompromised mice. (Figure courtesy Chenwei Li, Diane M. Simeone and *Cancer Research* (Li et al., 2007); reprinted with permission.)

plasm of almost all epithelial cells and has been shown to be upregulated considerably in liver lesions and adenocarcinomas. The subpopulation represented 0.2–0.8% of the entire adenocarcinoma and had an extremely high tumorigenic potential, with 50% of animals developing tumors after transplant of as little as 100 cells (Figure 6.15). The CD44+, CD24+, ESA+ cells also exhibited typical stem cell properties including prolonged self-renewal, the ability to produce differentiated progeny, and an upregulation of sonic hedge-hog expression (Figure 6.16) (Li et al., 2007).

Prostate

Prostate cancer is more frequently diagnosed in men than any other cancer. It is often diag-nosed via the presence of **prostate-specific antigen (PSA)**. PSA is a glycoprotein enzyme secreted by the epithelial cells of the prostate gland and is often upregulated in prostate cancers. Only lung cancer results in more cancer-related deaths than prostate cancer for

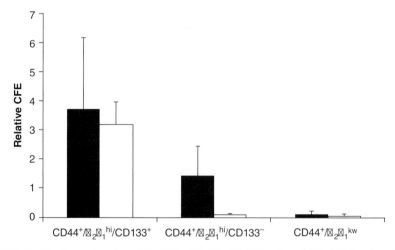

Figure 6.16 Proliferation rates of prostate CSCs. CD44+, CD133+ cells exhibited high colony forming efficiencies (CFE) in comparison to other cell types. Black columns represent enrich selected cells versus unselected cells. White columns represent secondary CFEs for each cell type. (Graph courtesy Anne T. Collins and *CancerResearch* (Collins et al., 2005); reprinted with permission.)

men worldwide. If not caught early, advanced or metastatic prostate cancer can have dev-astating consequences. While prostate cancer responds well to androgen-based ablation treatment, in many cases androgen-resistant cells emerge from the heterogeneous popula-tion. Thus, identifying a new target subpopulation of cells for therapeutic intervention would be of high value. The considerable amount of cell type heterogeneity in prostate cancers has led many researchers to believe that a multipotent stem cell may be driving tumorigenesis. As early as the 1980s, researchers were amassing a wealth of data suggest-ing a stem cell subtype present in prostate cancer that could be a driving force for the can-cerous phenotype. In 2005 Anne T. Collins et al., at the Yorkshire Cancer Research Unit, University of York, York, United Kingdom identified and characterized a population of CSCs from human prostate tumors that exhibited varying Gleason grade metastatic states. The **Gleason grading system** is the most widely used to quantify how far along the path from normal to abnormal (cancerous) a cell has progressed. It is a useful predictor of out-come and has become the standard for prognosing prostate cancer. The researchers identi-fied a subpopulation of cells (0.1%) that were $CD44^+$, $CD133^+$ from both primary and metastatic prostate cancers exhibiting stem cell-like characteristics such as long-term self-renewal, high proliferative capacity, (Figure 6.16) and multipotency. In addition, the cells were demonstrated to have considerable invasion activity potential (Collins et al., 2005). **Invasion activity** is defined as a cell's ability to migrate to other tissues and is an important classification of the potential lethality of a cancerous cell line.

Melanoma

Melanoma is caused by malignant tumorigenic transformation of **melanocytes**, which under normal conditions produce the dark pigment of the skin known as melanin (Figure 6.17). Melanoma is by far the most dangerous of all skin cancers and, if not detected early, is extremely lethal. Melanoma is the cause of 75% of all skin cancer-related deaths. Finding a common cell subpopulation that could be the source of the melanoma tumorigenic phenotype could open new avenues for therapeutic intervention.

Given the growing evidence of the presence of CSCs in other malignancies, in 2008 Markus H. Frank et al., at the Transplantation Research Center, Children's Hospital, Boston decided to focus on the identification of tumor-initiating cells in melanomas. They focused on the expres-sion of the chemoresistance mediator ABCB5 and showed that cells expressing this marker exhibited a primitive phenotype, and their presence as well as growth rates correlated with

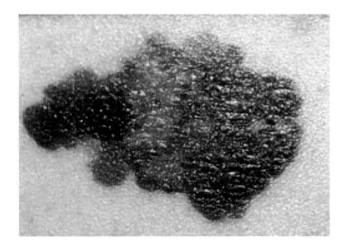

Figure 6.17 Example of melanoma. (Photograph courtesy Wikimedia Commons; reprinted with permission.)

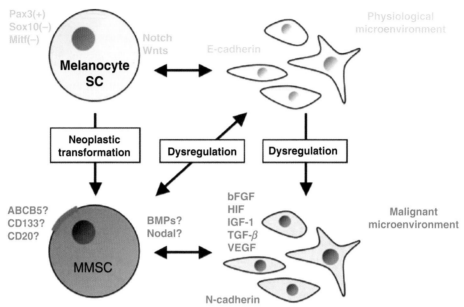

Figure 6.18 Schematic diagram of what goes awry in the formation of melanocyte melanoma stem cells from melanocyte stem cells. See text for details. (*Source:* Schatton & Frank, 2008. Reproduced with permission from John Wiley & Sons.)

clinical melanoma progression. These cells were termed "**malignant melanoma-initiating cells (MMICs, also referred to as MMSCs)**". As defined by serial human-to-mouse xenotransplantation experiments, ABCB5$^+$ melanoma cells have strong tumorigenic potential and possess the ability to recapitulate the heterogeneity of human clinical tumors. The cells were confirmed for long-term self-renewal capacities and can generate both ABCB5$^+$ and ABCB5$^-$ cells. Finally, to confirm that MMICs are required for the growth of established cells, the researchers demonstrated antibody-dependent cytotoxicity and tumor-inhibitory effects when treating xenotransplanted-established tumors with an antibody that recognizes ABCB5 (Schatton and Frank, 2008). Much of the Frank group's research focused on the molecular and physiological differences between melanocyte stem cells and MMSCs. The researchers noted that genetic mutations in melanocyte SCs or local environmentally induced dysregulation via changes in E-cadherin, basic Fibroblast Growth Factor, bone morphogenetic and other factor expression in neighboring cells could create a malignant microenvironment and result in the conversion of melanocyte stem cells to MMSCs (Figure 6.18) (Schatton and Frank, 2008).

Since the seminal discovery of melanoma CSCs by the Frank group, other research teams have also identified similar CSCs. Irving L. Weissman's research team at the Institute for Stem Cell Biology and Regenerative Medicine, Stanford University isolated "**melanoma tumor stem cells" (MTSCs)**" as a highly enriched population of CD271$^+$ cells. These cells were purified from a broad spectrum of melanoma sample representing different sites and stages of tumor progression. Weissman's team confirmed that the CD271$^+$ cells initiated tumors in 90% of the melanomas tested (Figure 6.19). These cells lacked the expression of T cell therapeutic target antigens TYR, MART, and MAGE in the majority of patient melanoma samples tested. The researchers speculated that perhaps the MTSCs are resistant to T cell-based therapies, evidenced by only temporary tumor shrinkage in these patients (Boiko et al., 2010). See section Strategies for Treatment Related to Cancer Stem Cells for an example of melanoma eradication by targeting CSCs.

CD271⁻ CD271⁺

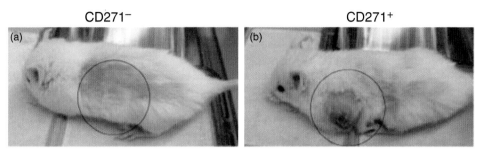

Figure 6.19 Tumor formation in immunocompromised mice driven by MTSCs. Only CD271⁺ cells exhibited tumor formation. Tumorigenic potential was high, with as little as 100 cells driving melanoma when transplanted into NOD/SCID mice. (Figure courtesy Alexander D. Boiko, Irving L. Weissman and *Nature* (Boiko et al., 2010); reprinted with permission.)

Multiple Myeloma

Multiple myeloma (MM), also referred to as plasma cell myeloma, results from the transformation of white blood cells. The term "multiple" refers to the fact that collections of many different types of abnormal plasma cells accumulate in the bone marrow. MM is second only to non-Hodgkin's lymphoma as the most prevalent blood cell malignancy, and, on average, represents 1% of all cancers. Prognosis for survival is typically 5–7 years if conventional treatments are administered; thus, the development of a new method for eradication of these abnormal plasma cells is of high priority. As early as 2004, researchers identified a subpopulation of cells present in MM samples with the ability to both replicate and differentiate into malignant plasma cells. Richard Jones et al., at the Sidney Kimmel Comprehensive Cancer Center, Johns Hopkins University School of Medicine, Baltimore, Maryland identified MM CSCs through an analysis of syndecan-1 positive and negative MM plasma cells. **Syndecan-1 (CD138)** is a well-known transmembrane domain protein

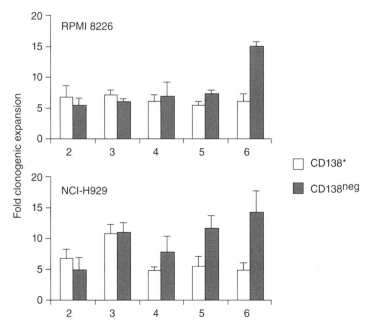

Figure 6.20 Clonogenic potential of MM cell lines. Two independent MM lines, RPMI 8226 and NCI-H929, were analyzed. After serial passaging CD138⁻ cells exhibit high clonogenic potential. (Graphs courtesy William Matsui, Richard J. Jones and *Blood* (Matsui et al., 2004); reprinted with permission.)

and plasma cell marker that has also been detected in cells representing various tumor types. Although the majority of malignant plasma cells express CD138, the Jones research group identified a CD138⁻ subpopulation of cells representing less than 5% of the total population that demonstrated high **clonogenic potential**, which is the degree of a cell's ability to give rise to a clone or colony of cells. It is an *in vitro* measure of tumorigenic potential (Figure 6.20) (Matsui et al., 2004).

These cells exhibited all the properties characteristic of stem cells including the potential to differentiate into malignant CD138⁺ plasma cells. In addition, the growth of these cells could be inhibited in the presence of the anti-CD20 therapeutic monoclonal antibody **rituximab** suggesting that this subpopulation of cells could be effectively targeted through the presence of the CD20 cell surface receptor for therapeutic intervention (see Strategies for Treatment Related to Cancer Stem Cells).

STRATEGIES FOR TREATMENT TARGETING CANCER STEM CELLS

The existence of CSCs and their roles in driving tumor growth and malignancy in numerous tissues provide unique opportunities for therapeutic intervention. Targeting of CSCs or other tumor-initiating cells could allow for the effective treatment of aggressive tumors or those that have become resistant to conventional therapies (Figure 6.21). The following examples illustrate intriguing findings with respect to targeting CSCs and other tumor-initiating cells for the eradication of cancer.

Melanoma Treatment Targeting Chimeric Antigen Receptor

In February 2011, researchers in the Tumorigenetics Department of Internal Medicine and Center for Molecular Medicine, Cologne, Germany led by Hinrich Abken published a key finding on the efficient eradication of melanomas by targeted elimination of a subset of tumor cells, presumed to be CSCs. In these studies, the researchers targeted a subset of tumor cells

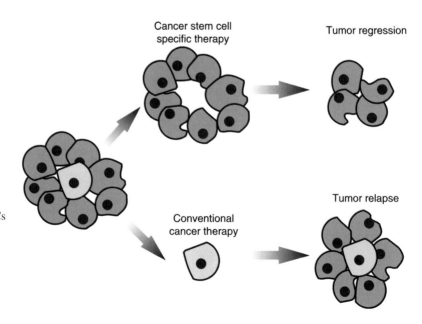

Figure 6.21 Diagrammatic illustration of the effects of CSC-specific therapy versus conventional cancer therapy. Targeting the eradication of CSCs may result in complete tumor regression in contrast to relapse evident from more conventional cancer therapy. (Diagram courtesy Wikimedia Commons; reprinted with permission.)

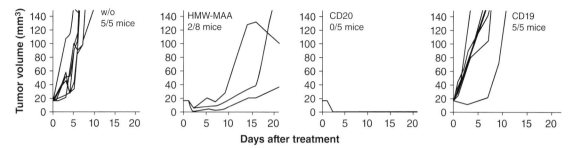

Figure 6.22 Targeted eradication of melanoma through CSC-directed cytotoxicity. Targeting CD20+ cells was effective at tumor volume reduction in comparison to controls w/o (without) HMW-MAA and CD19 targeting. *Source:* Scmidt, et al., 2011. Reproduced with permission from National Academy of Sciences, USA.

and the elimination of this subpopulation was effective in eradicating established melanoma lesions (Figure 6.22). The targeting strategy employed the adoptive transfer of cytotoxic T cells modified to specifically target the CSC subpopulation. Targeting was accomplished through the engineering of T cells with a **chimeric antigen receptor (CAR)**, which is an engineered receptor grafting the specificity of a monoclonal antibody onto a T cell. CAR-engineered T cells targeted a CD20+ subset of melanoma cells represented around 2% of the total melanoma cell population (Figure 6.23) (Schmidt et al., 2011).

Multiple Myeloma Treatment Targeting CD20

As mentioned above, Richard Jones et al., at the Sidney Kimmel Comprehensive Cancer Center identified a subpopulation of CD138+, CD20+ cells capable of driving MM growth and differentiating into mature myeloma lineages. **CD20** is a glycosylated phosphoprotein expressed on the surface of all B cells, beginning early in development and increasing in levels through maturity. CD20 is expressed at abnormally high levels in various lymphomas, leukemias, and on the surface of melanoma CSCs. The Jones team also demonstrated that these cells were sensitive to treatment with the anti-CD20 monoclonal therapeutic antibody rituximab. Specifically, treatment with rituximab significantly decreased the clonogenic recovery rates for these cells, a measurement of tumorigenic potential (Figure 6.23) (Matsui et al., 2004). These studies suggest a possible new therapeutic intervention strategy through the targeting of MM stem cells with an antibody already approved for therapeutic use with respect to cancer treatment.

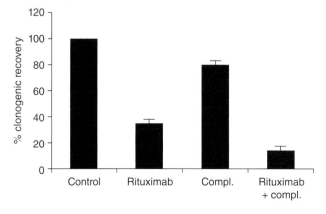

Figure 6.23 Effects of rituximab on clonogenic potential of MM stem cells. Clonogenic recovery was significantly reduced for MM stem cells in the presence of rituximab, evidence of anti-CD20 monoclonal therapeutic antibody sensitivity. (Graph courtesy William Matsui, Richard J. Jones and *Blood* (Matsui et al., 2004); reprinted with permission.)

CHAPTER SUMMARY

Background on the Origins of Cancer

1. The origins of cancer depend upon alterations in regulatory control of the cell cycle, cellular senescence, or apoptosis.
2. Transforming mutations tend to occur in proto-oncogenes.
3. Rudolph Virchow is a German doctor and biologist who was the first to discover leukemia stem cells.

Discovery and Origins of Cancer Stem Cells

1. Rudolph Virchow developed the "embryonal rest" hypothesis and is credited with "cell theory."
2. The existence of CSCs has been debated over the years due to the fact that clear characterization of these cells as unique and different in comparison to normal tissue stem cells has been difficult.
3. Many cancers originate from the arrest of tissue stem cell maturation.
4. Imatinib (Gleevec) has been demonstrated to remove the signals leading to cellular immortalization in chronic myelocytic leukemia.
5. In 1997 John E. Dick's group identified CSCs in AML.
6. Tumorigenic potential is a key property for CSCs.

Basic Properties of Cancer Stem Cells

1. Most CSCs divide symmetrically.
2. A subset of genes representing "stemness" are expressed in CSCs.
3. The basic properties of CSCs are as follows: (1) ability to self-renew; (2) strong tumorigenic potential; and (3) ability to establish tumor heterogeneity.
4. CSCs are sometimes referred to as "cancer-initiating cells."

Signaling Pathways Involved in Cancer Stem Cell Transformation

1. Many of the same pathways that regulate normal stem cell self-renewal and maturation may also be implicated in the behavior of CSCs.
2. Some critical pathways inherent in CSCs include the Wnt, Shh, and Notch signaling cascades.

Examples of Cancer Stem Cells

1. CSCs have now been identified and isolated from a variety of different types of cancers.
2. Each type of CSC exhibits unique molecular and morphological phenotypes.
3. Breast CSCs have been identified in specimens from patients exhibiting various stages of metastatic breast cancer.
4. Techniques utilized for the generation of neurospheres and related stem cell compartments were implemented for the isolation of CD133+ CSCs from human gliomas and medulloblastomas.

5. Late-stage neural progenitor cells have also been suggested to contribute to neural cancers.
6. Colon cancer originates from the alteration of epithelial cells.
7. Altered loci involved in the Wnt signaling cascade result in colon epithelial cell transformation and tumorigenesis.
8. John E. Dick's research team developed a novel xenograft model for colon cancer and employed it for the identification of CC-ICs.
9. Kenneth P. Nephew's group discovered and isolated OCICs.
10. Pancreatic adenocarcinoma CSCs have been identified and characterized as CD44[+], CD24[+], and ESA[+].
11. Anne T. Collins et al. discovered CD44[+], CD133[+] prostate CSCs with considerable invasion activity.
12. Markus H. Frank's team utilized the chemoresistance mediator marker ABCB5 to isolate a primitive cell type with strong tumorigenic potential they referred to as MMICs.
13. Irving L. Weissman's team isolated MTSCs as a highly enriched population of CD271[+] cells.
14. Richard Jones et al. identified MM CSCs through an analysis of syndecan-1 positive and negative MM plasma cells.
15. CD20[+] MM CSCs are sensitive to treatment with the therapeutic antibody rituximab.

Strategies for Treatment Targeting Cancer Stem Cells

1. Targeting of CSCs or other tumor-initiating cells could allow for the effective treatment of aggressive tumors or those that have become refractory to conventional therapies.
2. Hinrich Abken's team eradicated melanomas by targeting cytotoxic T cells to the melanoma CSC population through CAR-CD20 binding.
3. Richard Jones targeted CD20[+] MM CSCs with the therapeutic monoclonal antibody rituximab.

KEY TERMS

(Key terms are listed by order of appearance in the text.)

- **Apoptosis**—programmed cell death.
- **Proto-oncogenes**—normal cellular genes that contribute to the production of cancer when their activity is affected.
- **Stochastic hypothesis**—virtually any cell type could become transformed given the occurrence of a proto-oncogene mutation.
- **Cancer stem cell (CSC)**—tumorigenic cell possessing stem cell-like properties that has the ability to give rise to all cell types of a given cancer.
- **Embryonal rest hypothesis**—Hypothesis of Rudolph Virchow that tumors arise from residual embryonic tissues.

- **Cell theory**—Theory of Rudolph Virchow that all cells come from cells ("Omnis cellula e cellula").
- **Tumor-initiating cells**—cells possessing properties similar to normal stem cells that have the ability to drive the initial formation of a tumor.
- **Malignancy**—a tumor with a tendency to invade normal tissues or to recur after removal.
- **Cancer stem cells** — tumor-initiating cells that exhibit a malignant phenotype. These cells are also referred to as cancer initiating cells.
- **Acute myeloid leukemia (AML)** —cancer of white blood cells.
- **Stem-like-initiating cells (SL-ICs)**—cells identified by John E. Dick's group that are responsible for the leukemic phenotype in acute myeloid leukemia.
- **Tumorigenic potential**—the ability of cells to drive tumor formation at low cell densities.
- **Cancer-initiating cells (CICs)**—cancer stem cells.
- **Tumor heterogeneity**—the diverse makeup of different cell types encompassing a tumor.
- **Bcl-2**—the founding member of the Bcl-2 superfamily of regulatory proteins that control apoptosis in a variety of cell types.
- **Neurosphere**—a free-floating cluster of neural stem cells, usually grown *in vitro*.
- **Crohn's disease**—an inflammatory bowel disease that may affect the entire gastrointestinal tract.
- **Epithelial specific antigen (ESA a.k.a. Ep-CAM)**—a cell surface marker consisting of two glycoproteins located on the cell surface and in the cytoplasm of almost all epithelial cells and has been shown to be upregulated considerably in liver lesions and adenocarcinomas. It is also referred to as epithelial cell adhesion molecule (Ep-CAM).
- **Prostate-specific antigen**—a glycoprotein enzyme secreted by the epithelial cells of the prostate gland and is often upregulated in prostate cancers.
- **Gleason grading system**—the most widely used classification system to quantify how far along the path from normal to abnormal (cancerous) a cell has progressed.
- **Invasion activity**—a cell's ability to migrate to other tissues.
- **Melanocytes**—melanin-producing cells present in the basal stratum layer of the epidermis. These cells are precursors to melanoma.
- **Malignant melanoma-initiating cells (MMICs)**—ABCB5$^+$ cells identified by Markus H. Frank and colleagues which promote clinical melanoma progression. These cells are also referred to as malignant melanoma stem cells (MMSCs).
- **Melanoma tumor stem cells (MTSCs)**—CD271$^+$ cells identified by Irving L. Weissman's research team present in 90% of melanomas tested that promote melanoma progression.
- **Multiple myeloma (MM)**—a collection of many different types of abnormal plasma cells in the bone marrow. It is also referred to as plasma cell myeloma.
- **Syndecan-1 (CD138)**—a transmembrane domain protein and plasma cell marker that has been detected in cells representing various tumor types. It is also referred to as CD138.
- **Clonogenic potential**—the degree of a cell's ability to give rise to a clone or colony of cells.

- **Rituximab**—an anti-CD20 monoclonal therapeutic antibody used to treat cancers positive for CD20 such as multiple myeloma.
- **Chimeric antigen receptor (CAR)**—an engineered receptor grafting the specificity of a monoclonal antibody onto a T cell.
- **CD20**—a glycosylated phosphoprotein expressed on the surface of all B cells, beginning early in development and increasing in levels through maturity.

REVIEW QUESTIONS

(Answers to select review questions can be found at www.stemcelltextbook.com)

1. Alterations in what three processes drive the initiation of cancer?
2. How might a transformation mutation result in cancer?
3. Describe the "stochastic hypothesis" of the origins of cancer.
4. What hypothesis and theory did Rudolph Virchow formulate regarding
5. the origin of cells and cancer?
6. Describe the two models regarding the origin of cancer.
7. Why is the research field of cancer stem cells controversial?
8. What is Stewart Sell's hypothesis on the origin of many cancers?
9. How did John E. Dick's team isolate AML SL-ICs?
10. How do you measure tumorigenic potential?
11. List and describe at least two differences between normal and cancer stem cells.
12. Describe the unique assay developed by Michael Clarke and colleagues for the isolation of breast cancer stem cells.
13. List and describe the three main properties of cancer stem cells.
14. What is an example of a signaling pathway that may go awry and thus be implicated in altering the behavior of cancer stem cells?
15. What is the molecular signature of the breast cancer stem cells isolated by Michael Clarke and colleagues?
16. How was neurosphere culture technique employed to isolate glioblastoma and medulloblastoma cancer-initiating cells?
17. Describe the novel xenograft model developed by John E. Dick's group to identify colon cancer-initiating cells.
18. List the properties exhibited by ovarian cancer initiating cells (OCICs).
19. What markers do pancreatic cancer stem cells express that define them as such?
20. What marker is often upregulated in prostate cancer?
21. How were malignant melanoma-initiating cells (MMICs) isolated by Markus H. Frank's research team?
22. Why did Richard Jones and colleagues focus on syndecan-1 (CD138) in their search for multiple myeloma cancer-initiating cells?
23. Cite and describe an example of targeting cancer stem cells for therapeutic intervention.

THOUGHT QUESTION

Devise a strategy for the eradication of cancer via the targeting of cancer stem cells. Be sure to identify in detail the targeting strategy and therapy employed.

SUGGESTED READINGS

Al-Hajj, M., M. S. Wicha, et al. (2003). "Prospective identification of tumorigenic breast cancer cells." *Proc Natl Acad Sci U S A* **100**(7): 3983–3988.

Barnett, S. C., L. Robertson, et al. (1998). "Oligodendrocyte-type-2 astrocyte (O-2A) progenitor cells transformed with c-myc and H-ras form high-grade glioma after stereotactic injection into the rat brain." *Carcinogenesis* **19**(9): 1529–1537.

Bonnet, D. and J. E. Dick (1997). "Human acute myeloid leukemia is organized as a hierarchy that originates from a primitive hematopoietic cell." Nat Med **3**(7): 730-737.

Boiko, A. D., O. V. Razorenova, et al. (2010). "Human melanoma-initiating cells express neural crest nerve growth factor receptor CD271." *Nature* **466**(7302): 133–137.

Collins, A. T., P. A. Berry, et al. (2005). "Prospective identification of tumorigenic prostate cancer stem cells." *Cancer Res* **65**(23): 10946–10951.

Druker, B. J., F. Guilhot, et al. (2006). "Five-year follow-up of patients receiving imatinib for chronic myeloid leukemia." *N Engl J Med* **355**(23): 2408–2417.

Galli, R., E. Binda, et al. (2004). "Isolation and characterization of tumorigenic, stem-like neural precursors from human glioblastoma." *Cancer Res* **64**(19): 7011–7021.

Hill, R. P. and R. Perris (2007). "Destemming" cancer stem cells." *J Natl Cancer Inst* **99**(19): 1435–1440.

Jordan, C. T., M. L. Guzman, et al. (2006). "Cancer stem cells." *N Engl J Med* **355**(12): 1253–1261.

Li, C., D. G. Heidt, et al. (2007). "Identification of pancreatic cancer stem cells." *Cancer Res* **67**(3): 1030–1037.

Matsui, W., C. A. Huff, et al. (2004). "Characterization of clonogenic multiple myeloma cells." *Blood* **103**(6): 2332–2336.

O'Brien, C. A., A. Pollett, et al. (2007). "A human colon cancer cell capable of initiating tumour growth in immunodeficient mice." *Nature* **445**(7123): 106–110.

Reya, T., S. J. Morrison, et al. (2001). "Stem cells, cancer, and cancer stem cells." *Nature* **414**(6859): 105–111.

Schatton, T. and M. H. Frank (2008). "Cancer stem cells and human malignant melanoma." *Pigment Cell Melanoma Res* **21**(1): 39–55.

Schatton, T., G. F. Murphy, et al. (2008). "Identification of cells initiating human melanomas." *Nature* **451**(7176): 345–349.

Schmidt, P., C. Kopecky, et al. (2011). "Eradication of melanomas by targeted elimination of a minor subset of tumor cells." *Proc Natl Acad Sci U S A* **108**(6): 2474–2479.

Sell, S. and G. B. Pierce (1994). "Maturation arrest of stem cell differentiation is a common pathway for the cellular origin of teratocarcinomas and epithelial cancers." *Lab Invest* **70**(1): 6–22.

Zhang, S., C. Balch, et al. (2008). "Identification and characterization of ovarian cancer-initiating cells from primary human tumors." *Cancer Res* **68**(11): 4311–4320.

Chapter 7

STEM CELLS AS DRUG DISCOVERY PLATFORMS

As evidenced throughout this text, both embryonic and adult stem cells from numerous species have been studied and characterized with inherent morphological, molecular, and behavioral properties. These properties have been studied in the context of *in vitro* and *in vivo* environments with particular emphasis placed on the changes in makeup and behavior that take place within these cells in response to alterations in environmental cues. As a result, a wealth of information and data related to stem cell identity and behavior has been amassed over the last several decades. These data have now been utilized to devise uses for stem cells as platforms for drug or drug target discovery. This chapter illustrates some of the more high-profile examples of the application of stem cells as a discovery platform for defining drug targets and the therapeutics that may bind these targets.

Focus Box 7.1: Gail R. Martin and the discovery of mouse ES cells

Gail R. Martin's pioneering research in embryonic development and discovery of mouse embryonic stem (ES) cells paved the way for a revolution in the study of gene function *in vivo*. She is credited with coining the term "embryonic stem cell" and has been associated with the University of California–San Francisco since 1976, where she is currently in charge of the Program in Developmental Biology. Dr. Martin has received numerous awards for her groundbreaking research including 2002 Conklin Medal from the Society for Developmental Biology and the 2007 Pearl Meister Greengard Prize.

(*Source:* Reproduced with permission from G. R. Martin.)

EMBRYONIC STEM CELLS AND MOUSE MODELS OF GENE FUNCTION

Animal models of human disorders have been studied for decades, and many therapeutic uses have resulted either directly or indirectly from these efforts. Examples include experiments on

---

Stem Cells: A Short Course, First Edition. Rob Burgess.
© 2016 John Wiley & Sons, Inc. Published 2016 by John Wiley & Sons, Inc.

Figure 7.1 Wild-type *Mus musculus.* (Photograph courtesy Wikimedia Commons; reprinted with permission.)

the use of insulin to treat diabetes in dogs and the testing of penicillin for safe use in guinea pigs. Yet over the last 20 years the model system of choice for studying the function of genes has been the mouse (Figure 7.1). Many reasons exist for the convergence of the genetics research field toward mice as an ideal model system. To begin, the genetic similarities between humans and mice are staggering. For example, 99% of mouse genes have human orthologues; 90% of the mouse genome can be aligned in a sequence-specific manner with the human genome; and 75% of the genes in mice are represented as equivalent loci in humans. In addition, a multitude of human disorders are genetically based or have genes for which the corresponding protein products may respond to therapeutics and these genes are represented in the mouse genome. In fact, many current therapies are the result of the study of mouse models of human disease and, more specifically, the study of gene function in mice. Therefore, the study of gene function in the context of mouse physiology, disease, and/or behavior could very well have applicability to what might occur in the context of humans. Mice exhibit characteristics suitable for the study of gene function *in vivo.* These factors include:

- Mammalian phenotype
- Short gestation period
- Reasonably short life cycle
- Relatively large litter size
- Small size
- Ease of handling

It is also clear that the study of genes in *ex vivo* model systems such as cell culture or tissue explants is incomplete and does not take the entire context of mammalian physiology into consideration. So how are genes studied at the functional level in a mammalian organism as complex as the mouse? The most widely used method for defining individual gene function in a mouse model is via **gene targeting**, which is the site-specific alteration of a gene through the targeting of DNA sequences to that gene (see HGPRT targeting below). As this text is focused on the value and utility of ES cells, only technologies involving ES cells for the study of gene function will be described here. As mentioned in Chapters 1 and 2, mouse ES cells were first successfully derived by two independent groups from blastocyst-stage embryos in 1981. Gail R. Martin in the Department of Anatomy at the University of California–San Francisco developed a microdissection technique for the isolation of mouse ES cells from blastocyst-stage embryos. Specifically, Martin employed a whole-embryo culture system in serum on a fibroblast feeder layer followed by microdissection of the **inner cell mass (ICM)**, which is a component of mammalian blastocyst-stage embryos destined to form the embryo proper. The **fibroblast feeder layer** was a crucial component of Martin's stem cell isolation methodology, as it allowed for the provision of both a living support material and growth factors secreted by the fibroblasts essential for growth. Feeder layer technology has since been used in many stem cell culture applications. Microdissected

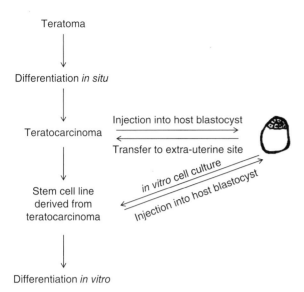

Teratoma

Differentiation *in situ*

Teratocarcinoma

Injection into host blastocyst

Transfer to extra-uterine site

in vitro cell culture

Stem cell line derived from teratocarcinoma

Injection into host blastocyst

Differentiation *in vitro*

Figure 7.2 Schematic diagram of Gail Martin's technology for the isolation of mouse ES cells. See text for details.

cells were clonally expanded into colonies and subsequently differentiated into cell types representing the three primary germ layers endoderm, mesoderm, and ectoderm, thus confirming pluripotency both *in vitro* and in immunocompromised mice (Figure 7.2).

A team led by Martin and Andrew Kaufman in the Department of Genetics at the University of Cambridge instead homed in on the small number of cells present within the ICM. As these cells are known to contribute to the embryo proper, the Evans/Kaufman team sought to increase their numbers, improving the chances of isolating true ES cells. They developed an intrauterine cell culture technique that promoted an increase in the number of cells present within the ICM, thus improving chances for successful murine ES cell isolation. The method involved the administration of the hormone progesterone coupled with ovary removal in order to delay the implantation of embryos into the uterine wall. Delayed implantation allowed for an increase in ICM cell number. In a manner similar to that for the Martin group, whole embryos were cultured on a fibroblast feeder layer and clonal ES cells isolated from these cultures. These cells were also confirmed for pluripotent properties both *in vitro* and *in vivo* (Evans and Kaufman, 1981).

Gene targeting in the context of mouse ES cells has literally revolutionized the study of gene function *in vivo*. Gene targeting is based on the endogenous ability of a genome to recombine with itself between sister chromatids in a sequence-specific manner. This is referred to as **homologous recombination** and is an endogenous property of many mammalian-derived ES cell lines. It was in 1987 and 1989 when three independent research groups led by Mario Capecchi, Martin Evans, and Oliver Smithies employed homologous recombination to sequence specifically mutate and inactivate the **hypoxanthine guanine phosphoribosyl transferase (HPRT)** gene in mouse ES cells. HPRT is a transferase that plays a central role in the generation of purine nucleotides essential for DNA synthesis. Mutations in HPRT have been shown in humans to result in **Lesch–Nyhan syndrome**, which is a rare inherited genetic disorder resulting in the building of uric acids in the body. As an x-linked enzyme, mutation of this locus results in a functional null that can be selected and enriched for in tissue culture. Thus HPRT is an ideal candidate locus to prove site-specific gene-targeting technology. The three independent research groups successfully employed homologous recombination using DNA vectors designed to recombine and mutate the endogenous HPRT locus. Murine ES cells identified with

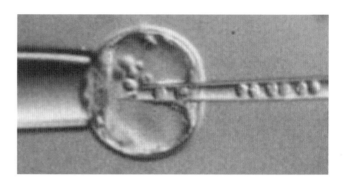

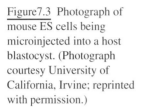

Figure7.3 Photograph of mouse ES cells being microinjected into a host blastocyst. (Photograph courtesy University of California, Irvine; reprinted with permission.)

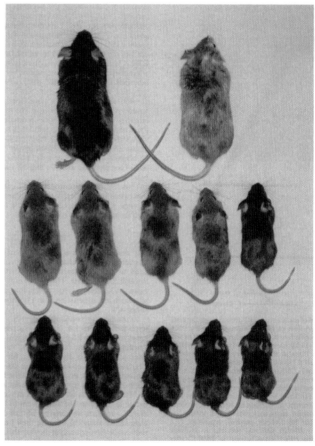

Figure 7.4 Chimeric and gene-targeting mice using mouse ES cells. Upper panel: Chimeras generated from the injection of targeted mouse ES cells into host blastocysts. Lower panel: A transmitting chimeric male (right top) and C57Bl/6J female (left top) along with their litter with germline transmission confirmed by a black coat color. (*Source:* Koller et al., 1989. Reproduced with permission from B. H. Koller.)

correct mutations were subsequently introduced via microinjection techniques into host blastocysts which were implanted into pseudopregnant female mice, resulting in mosaic mice referred to as **chimeras** containing both wild-type and gene-targeted cell populations (Figure 7.3). Breeding of chimeras with wild-type mice resulted in **germline transmission** and the world's first gene-targeted mammals (Figure 7.4) (Koller et al., 1989).

These technologies have since been refined to allow for the production of tissue- and cell-type specific **knockout mice** as well as **knockin mice** whereby entire genes have been replaced with exogenous sequences. These genetically modified animals have provided an invaluable resource for the study of gene function and ultimately the design of new therapeutic strategies (see Case Study 7.1).

Case Study 7.1: Requirement of the paraxis gene for somite formation and musculoskeletal patterning

Rob Burgess, Alan Rawls, Doris Brown, Allan Bradley, and Eric N. Olson

In 1996, researchers at the University of Texas, MD Anderson Cancer Center in Houston and the University of Texas Southwestern Medical Center in Dallas utilized gene-targeting technologies to inactivate the basic helix-loop-helix (bHLH) transcription factor paraxis in mice. A technology referred to as **positive-negative selection**, which is defined as the utilization of a combination of positive and negative selectable markers to enrich gene targeting via homologous recombination, was implemented to mutate the *paraxis* locus. Mutation involved the truncation of the protein and removal of the critical bHLH motif necessary for dimerization and DNA-binding activity. Resulting mutant mice displayed a rostral-to-caudal truncation and defects in lineages derived from the **somites**, which are epithelial spheres present along the axis of the embryo that become patterned to form vertebrae, ribs, skeletal muscle, and dermis. Paraxis-null embryos exhibited a failure in somites to epithelialize, although internal compartmentalization into the dermatome, myotome, and sclerotome was unaffected. Newborn null mice exhibited improperly patterned axial skeleton and skeletal muscle (Figure 7.5). It was demonstrated by the researchers that paraxis plays a key role in somite epithelialization, yet the formation of epithelial somites is not required for the establishment of embryonic segmentation or for the development of somitic lineages (Burgess et al., 1996). Thus, the application of gene targeting in murine ES cells and the production of corresponding mice harboring an inactivating mutation in the paraxis local was a critical tool for defining the function of this transcription factor *in vivo*.

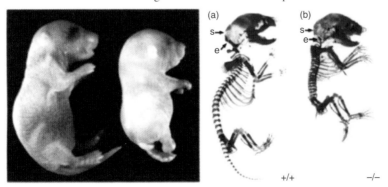

Figure7.5 Gene targeting in mouse ES cells defines the function of the transcription factor *paraxis*. (Left panel) Lateral views of wild-type (left) and paraxis null newborn mice; (Right panel) Lateral view of dissected skeletons isolated from wild-type and paraxis-null embryos. Cartilage is stained red and bone is stained blue. (Photographs courtesy Rob Burgess, Eric N. Olson and *Nature* (Burgess et al., 1996); reprinted with permission.)

STEM CELL-BASED SCREENING ASSAYS

In addition to acting as an excellent living reagent for the study of gene function *in vitro* and *in vivo*, stem cells may provide valuable insight regarding the design or testing of a therapeutic platform. Cell-based **high-throughput screening (HTS), a.k.a. high-content screening (HCS)**, which is defined as a rapid method for identifying a key molecule or pathway, is not a new concept, and while the use of cells and cell-based assays in HTS has significantly increased over the past 10 years due to technological improvements, the success of the system is largely cell-type-dependent. Numerous different types of high-throughput cell-based screens exist, including screens in the following categories:

- *Therapeutic*—Identification of therapeutic molecules
- *Reprogramming*—Identification of agents which drive the reprogramming or the induction of new cellular potencies
- *Directed differentiation*—Identification of compounds which promote differentiation of stem cells into particular lineages

In order to comprehend the benefits of stem cell-based HTS, it is necessary to first understand HTS as applied using other cell sources. In the majority of cases, recombinant immortalized cells are used in various formats, for example, either liquid or in planar culture, to screen libraries of small molecules. A "**hit**" is identified when a small molecule binds or in some other way affects the cell. Hits are confirmed as legitimate through rescreens and enriched libraries of hits are further characterized, often using the same or an optimized cell platform. Many HTS screens are performed on human **primary cells**, which are cells isolated directly from tissue. The major disadvantage of this approach is scarcity. An immortalized or self-renewing cell line, such as an embryonic cell, would provide virtually limitless supplies for screening purposes. In addition, primary cells require much care in handling and often cannot be frozen as stocks without significant cell death observed. Stem cell culture and preservation techniques have been optimized to the point that care, handling, and storage are not issues for HTS applications. In addition, the use and application of stem cells as the data output source is of particular value given their various states of potency. HTS assays designed for the evaluation of directed differentiation of stem cells into specific lineages driven by a "hit" could be powerful for the identification of differentiation inducers. The following sections review recent advances in both reprogramming and directed differentiation high-throughput screens using stem cells as the data output source.

Stem Cells as Lineage Resources for HTS

As is evident throughout this text, there is a wealth of different categories of stem cells representing the ability to generate virtually any cell type in the human body and to amass these cells in large quantities. Thus stem cell technology allows for the production of live, cell-based resources for the HTS of combinatorial chemistry libraries as well as for the validation of early-stage drug candidates. The following sections break down different types of stem cells as resources themselves or as resources for downstream lineages for library and drug candidate screening and validation.

Focus Box 7.2: John McNeish and ES cell applications in drug screening

 Dr. John McNeish is a pioneering researcher in the area of ES cell applications for drug discovery and testing. His research has resulted in the development of several platforms for the testing of the toxic effects of drugs on ES cell-derived differentiated lineages such as neurons, glia, and cardiomyocytes. He is currently in the Regenerative Medicine Program at Glaxo-SmithKline in Boston and previously served as Executive Director of Regenerative Medicine at Pfizer. He is also a Founder of Boston Stem Cells, LLC. (Photo courtesy John McNeish; reprinted with permission.)

Embryonic Stem Cells as a Resource The very biological properties of stem cells that make them an essential component of embryonic development and the maintenance of adult tissues and organs lend them to be a valuable resource for large quantities of differentiated lineages. These properties include:

- Genetic stability
- Scalability
- *En masse* directed differentiation capabilities

Embryonic or adult stem cells thus may act as resources for the large-scale production of desired lineages for use in HTS. Perhaps no differentiated cell type is in more need with respect to HTS experimental design and implementation than that of neurons. As the understanding of the molecular, biochemical, and biological underpinnings inherent in various human neurological disorders has increased over the past 20 years, it has provided opportunities for the use of terminally differentiated lineages such as neurons and glial cells in HTS assays to identify new drug targets or therapeutic entities. For example, the impairment of cognitive function is a hallmark of many neurological diseases such as attention-deficit hyperactivity disorder (ADHD), schizophrenia, and Alzheimer's disease (AD). Yet only a limited number of therapeutic agents have been discovered or developed to treat disease associated with abnormal cognition, which most often is based on dysfunction of the neuronal underpinnings of the central nervous system (CNS). High-throughput screens designed to identify pharmaceutical agents that might impact, or even correct, defects in CNS-based neurobiology would be of enormous value in developing new treatments for CNS disorders. A research team led by John McNeish (see Focus Box 7.2) at the Pfizer Global Research and Development Center utilized ES cells as a source for the *en masse* production of neuronal population for use in HTS. Specifically, the McNeish team developed protocols to efficiently differentiate mouse ES cells into a subtype of neurons that express functional receptors critical for effective excitatory neurotransmission in the CNS. These receptors, referred to as ionotropic α-amino-3-hydroxy-5-methylisoxazole-4-propionic acid (AMPA) receptors, are crucial for the transmission of glutamate-driven post-synaptic signaling in neurons. **AMPA receptors** are tetrameric aggregates composed of four subunits capable of binding glutamate, which subsequently results in the opening of an associated ion pore driving neuronal depolarization and proper neurotransmission. To accomplish large-scale production of AMPA-expressing neurons, the researchers cultured undifferentiated ES cells in the presence of three key factors—transferrin, leukemia inhibitor factor, and Noggin—followed by the formation of embryoid bodies (EBs) (see Chapter 4 and Figure 4.32). EBs were subsequently dissociated in preparation for neuronal precursor selection and cultured in the presence of numerous inducers of neuronal differentiation, including basic fibroblast growth factor (FGF), sonic hedgehog, FGF-8, and Noggin (see Chapter 2). Neuronal precursors were selected from monolayers of cells through the identification of cells expressing the precursor marker Sox-1. Sox-1 expressing cells survived clonal expansion due to an engineered targeted mutation in the Sox-1 locus that allowed for survival of these cells in the presence of the antibiotic G418 (geneticin). Thus, theoretically, only cells expressing the Sox-1 marker gene would survive culture in the presence of G418. Surviving cells were identified, and neuronal precursors expanded and cryopreserved for subsequent use in HTS assays (Figure 7.6) (McNeish et al., 2010).

Neuronal precursor cells were differentiated into AMPA receptor expressing neurons in the presence of cyclic AMP and ascorbic acid, potent neuronal differentiation inducers, and plated in 384-well plates for HTS applications. Roughly 2.4 million compounds were screened against the cells with the identification of novel "hits" accomplished by measuring AMPA potentiation using a Ca^{2+} **flux assay** that could be automated for HTS. From the screen seven compounds were selected for and identified as potent AMPA receptor activity potentiators. Two of these were novel molecules and five were known to promote AMA receptor potentiation. The hits were subsequently confirmed as exhibiting a similar function in human neuronal assays. Thus, in this study, ES cells have been demonstrated to be a valuable resource for the production of neurons that may be used successfully in high-throughput screens to identify potentiators of AMPA receptor activity and these hits may be valuable in the design of therapeutics to treat various cognitive disorders.

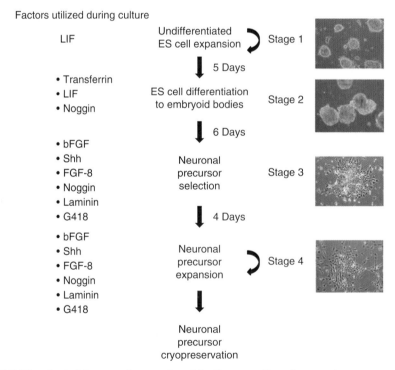

Figure 7.6 Flowchart of the procedure employed for the generation of neuronal precursor cells from embryonic stem cells to be used in high-throughput screening. See text for details. (Flowchart courtesy John McNeish and the *Journal of Biological Chemistry* (McNeish et al., 2010); reprinted with permission.)

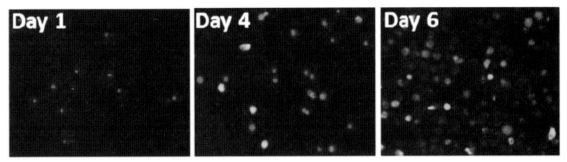

Figure 7.7 Screening hematopoietic stem cells for modulators of megakaryopoiesis. Immunofluorescence was used to discern the presence of megakaryocyte markers CD41 and CD71 at different timepoints of incubation with small molecules. CD41 (red), CD71 (green). (*Source: Boitano et al., 2012. Reproduced with permission from National Academy of Sciences, USA.*)

Adult Stem Cells as a Resource Adult stem cells are multipotent in nature and, depending upon the cell type, may act as a valuable resource for large quantities of cells to be screened for small molecules of therapeutic significance. High-throughput screens may be designed to assess the effects of small molecules on either terminally differentiated lineages derived from adult stem cells or on the stem cells themselves. It is the latter which was the focus of a research study by Peter G. Shultz and colleagues in the Department of Chemistry at the Scripps Research Institute in La Jolla, California. His team developed an assay to screen for modulators of hematopoietic stem cell (HSC) (see Chapters 1 and 2 on discovery and culture) **megakaryopoiesis**, which is the differentiation of HSCs into **megakaryocytes**, cells responsible for the production of **thrombocytes** (platelets). Prior to this study there was no reliable cell line available for use in the identification of megakaryopoiesis modulators. This limitation was overcome by the Schultz team, using advanced automation and imaging technologies which allowed for screens to be conducted directly on human donor purified CD34$^+$ HSCs. A library of 50,000 compounds was screened against CD34$^+$ donor cells in 1,536-well plates and immunostained for CD41 and CD71 expression, which are markers of the megakaryocytic phenotype (Figure 7.7). This screen resulted in the identification of naphthyridinone (MK1) as a compound that increased the number of CD41$^+$ cells in a dose-dependent fashion. Optimal dosing of MK1 resulted in a threefold increase in CD41$^+$ cell production (Boitano et al., 2012). Therefore the Shultz team devised a novel screening strategy that allows for the identification of megakaryogenesis modulators in the conversion of HSCs into megakaryocytes. Implementation of this screen could allow for the identification of small molecules and other compounds that ultimately drive thrombocyte production, enabling enhanced blood clotting when necessary for the treatment of clotting deficiencies.

Human mesenchymal stromal cells (hMSCs) (discussed in more detail in Chapter 3) have been widely regarded as the gold standard for cell-based therapies. This is due to not only their multilineage differentiation capabilities but also their ability to secrete various trophic and immunomodulatory factors. **Osteogenesis** is the formation of bone. A great deal of attention has been focused on the osteogenic differentiation capabilities of human stromal mesenchymal stem cells, as these cells may act as a source for bone regeneration to address injury or disease. Researchers in the Department of Tissue Regeneration, MIRA Institute for Biomedical Technology and Technical Medicine, University of Twente, Enschede, The Netherlands, recently developed and implemented a screen for the identification of compounds regulating the osteogenic differentiation of these cells. Donor hMSCs were plated at low density in 96-well plates and treated with a library of 1,280 pharmacologically active compounds followed by assessment of osteogenesis as measured by the presence of **alkaline phosphatase (ALP)**, which is a marker for bone development (Figure 7.8). ALP presence as well as proliferation status was characterized in a fluorescence plate reader and compounds identified which promoted both. Compounds were selected for further validation as those inducing the highest levels of ALP expression while not affecting the proliferation status of the cells. Numerous compounds were identified as having osteogenic potential while not significantly affecting cellular proliferative properties. These may be valuable for the induction of bone regeneration in various injury- or disease-related states (Alves et al., 2011).

iPS Cells as a Resource Induced pluripotent stem (iPS) cells exhibit many properties that would make them ideal as a source for virtually any cell type to be used in an HTS platform. These include:

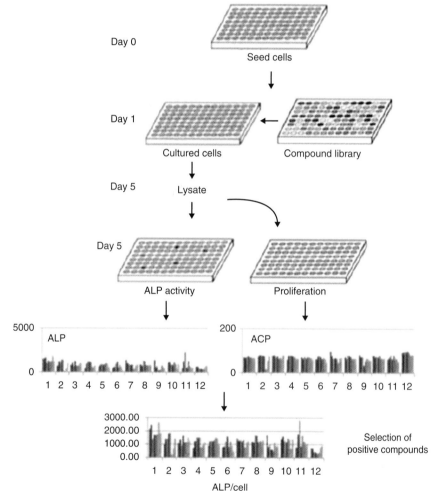

Figure 7.8 Schematic illustration of human stromal mesenchymal stem cell screen for modulators of osteogenesis. See text for details. (Illustration courtesy Hugo Alves, Jan De Boer and *PLOS One* (Alves et al., 2011); reprinted with permission.)

- Pluripotency
- Virtually unlimited proliferative capacities
- Non-embryonically
- Potential for patient specificity

   As a result, many stem cell biologists and researchers have focused on the application of induced pluripotency technology for the production of a variety of cell types for use in HTS as well as drug validation. This is perhaps the common theme of this chapter as it represents the most wide-ranging potential as a resource for almost any cell type to be used in small molecule screening assays. Case Study 7.2 illustrates an example of iPS cell utility in the generation of neuronal cells for the validation of drug candidates effective in the treatment of AD.

Case Study 7.2: Anti-Aβ drug screening platform using human iPS cell-derived neurons for the treatment of Alzheimer's disease

Naoki Yahata et al.

AD is a neurodegenerative disorder characterized by the deposit of **amyloid-β peptide** in the brain. A research team led by Haruhisa Inoue at the Center for iPS Cell Research and Application, Kyoto University, Kyoto, Japan, developed a screening platform based on the use of neurons and glia derived from human iPS cells. These cells were produced by directed differentiation and expressed the amyloid precursor protein. Differentiation was accomplished by a sequential addition of neuronal-inducing factors noggin, BDNF, GDNF, and NT-3 in the context of the neuronal supplementation medium **N2/B27**. These cells were functionally relevant due to the fact that they expressed active enzymes β- and γ-secretase and actively secreted Aβ, a hallmark of neurons in AD patients. When screened against known inhibitors of secretase activity such as BSI (β-secretase inhibitor) and NSAID secretase activity was diminished suggesting that these drugs may act to slow amyloid-β peptide production and amyloid plaque building in AD patients (Figure 7.9) (Yahata et al., 2011).

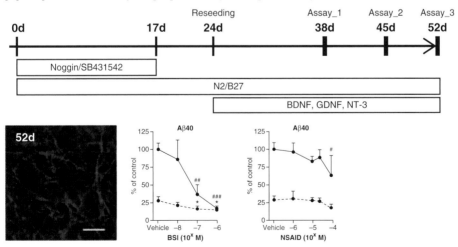

Figure 7.9 Use of iPS cells and descendant differentiated neuronal cells as a platform for drug validation. (Upper) Timeline and strategy for iPS cell differentiation; (Lower left) Neurons expressing the terminal differentiation marker Tuj1; (Lower middle and right) Graphical representation of β-secretase inhibition by BSI and NSAID. (Figures courtesy Naoki Yahata, Haruhisa Inoue and *PLOS One* (Yahata et al., 2011); reprinted with permission.)

Cancer Stem Cell Screens Since the identification and confirmation of the existence of cancer stem cells (CSCs) by John E. Dick's group in 1994 (see Focus Box 6.2), many research groups have sought to therapeutically target these cells, driven by the hypothesis that eradication of CSCs would lead to eradication of the cancer itself. Recently efforts have been directed at the application of HTS strategies to identify compounds that would home in on and kill CSCs and tumor-initiating cells. **Neuroblastoma (NB)** is one of the most devastating forms of cancer. It is the most common and lethal form of extracranial solid tumors in children. In 2011, David Kaplan et al., in the Department of Molecular Genetics, University of Toronto in Canada, developed an HTS assay to screen for small molecule kinase inhibitors affecting NB tumor-initiating cell viability. Kaplan's group previously isolated NB tumor-initiating cells (NB TICs) from bone marrow metastases with high tumorigenic potential. These cells were utilized in a screen to identify key pathways inherent in NB TICs crucial for their survival and self-renewal. Given the wealth of evidence that kinase pathways are critical for the maintenance and promotion of both cellular survival and self-renewal, the Kaplan group performed a comprehensive HTS on NB TICs of a unique collection of pharmacologic kinase inhibitors. In the screen, patient-derived NB TICs were dissociated into single cells and seeded at low density in cell culture plates and compounds were added directly at specific concentrations. Positive hits were classified as those inhibitors that drove greater than 45% NB TIC growth inhibition. Secondary screens were employed to confirm hits (Figure 7.10). The screen allowed for the identification of several small molecules that were either **cytostatic**,

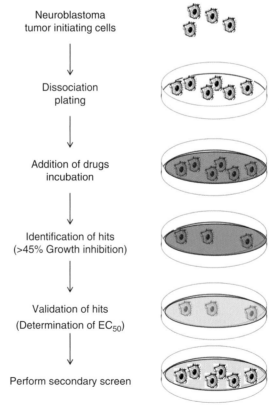

Neuroblastoma
tumor initiating cells

Dissociation
plating

Addition of drugs
incubation

Identification of hits
(>45% Growth inhibition)

Validation of hits
(Determination of EC_{50})

Perform secondary screen

Figure 7.10 Flowchart of a screen for the identification of kinase inhibitors as therapeutics. See text for details.

defined as inhibiting cell proliferation, or **cytotoxic**, defined as causing the death of cells with respect to NB TICs. The cytostatic and cytotoxic nature of these compounds was specific to NB TICs and did not affect normal neural stem cells as therapeutic concentrations. These studies demonstrate the feasibility of using CSCs as a screening tool for the identification of small molecule compounds that may have a therapeutic effect by acting on these cells directly (Grinshtein et al., 2011).

Other researchers have focused on the screening of larger chemical libraries to identify compounds that may inhibit the growth and proliferation of CSCs. Greg Foltz et al., at the Swedish Neuroscience Institute in Seattle, Washington, studied **glioblastoma multiform (GBM), a.k.a. glioma**, which is well documented as the most common and primary aggressive brain tumor in humans. Glioma stem cells (GSCs) have been identified and suggested to be the primary underlying cause for tumor recurrence and resistance to therapy. The Foltz team screened a diverse library of 2,000 chemical compounds, known as the **Micro-Source Spectrum**, which is a library of FDA-approved drugs or drugs in late-phase clinical trials. The screen was designed to assess the antiproliferative capacity of drug candidates in the context of GSCs. It was performed using GSCs isolated from patient samples. GSC isolation was accomplished by confirming the neurosphere formation capability of the cells. Screens were carried out in 384-well plates at a density of 800 cells/well in the presence of the library. Potency of candidate molecules was assessed by the generation of dose–response curves and calculation of the IC_{50}, which is defined as half of the maximum proliferation inhibitory concentration of the compound. These high-throughput screens resulted in the identification of 78 compounds with anti-GSC proliferative capacities. Of these compounds, **disulfiram (DSF)** was identified as one of the most potent inhibitors of GSC proliferation (Figure 7.11) (Hothi et al., 2012).

DSF is a clinically approved drug for alcohol aversion therapy. This provides the unique advantage of well-established pharmacokinetics. In addition, DSF is known to rapidly break down in the bloodstream into an active metabolite, diethyldithiocarbamate (DTTC), which readily crosses the blood–brain barrier (BBB). Efficient passage across the BBB is

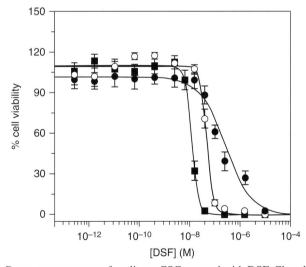

Figure 7.11 Dose–response curve for glioma CSCs treated with DSF. Closed circles represent neural stem cells as a control; Open circles indicate the highest IC50; Closed squares indicate the lowest IC50. (Graph courtesy Parvinder Hothi, Greg Foltz and *Oncotarget* (Hothi et al., 2012); reprinted with permission.)

of paramount importance for effective treatment of brain tumors. So how does DSF inhibit glioma cancer cell proliferation? The researchers noted that **aldehyde dehydrogenase (ALDH)**, which is a known marker for CSCs and hypothesized to play a role in the maintenance of the progenitor cell phenotype, is a known target of DSF. DSF has been characterized to irreversibly inhibit the function of ALDH in a variety of systems. The Foltz team went on to confirm that ALDH function was inhibited in GSCs, specifically when treated with DSF. Thus, not only has a drug candidate for the treatment of GBM been identified, at least one mechanism of action for this drug has been identified using CSCs as a screening platform.

Reprogramming Screens

The ability of certain somatic cell lineages to be induced to reprogram into a pluripotent-state is potentially of great value for the large-scale production of stem cells or even mature, differentiated lineages that may be used in cell or tissue transplantation experiments. Thus, identifying small molecules which drive the reprogramming of cells into a pluripotent phenotype could yield valuable entities for use in therapeutic cell production. Numerous research groups have attempted to identify small molecule compounds that would induce pluripotency based upon knowledge of the molecular and biochemical pathways involved in reprogramming. Yet a more comprehensive approach is to design a phenotypic output screen that would allow for the high-throughput identification of small molecules that induce pluripotency, irrespective of the pathways through which they might act. Researchers in the Department of Chemistry at the Scripps Research Institute in La Jolla, California designed just such a screen. Utilizing mouse embryonic fibroblasts as the preferred parental cell type, a phenotypic output screen was designed and used to examine a collection of 2000 known drugs for their ability to induce pluripotency in a somatic cell type. Induction of pluripotency was assessed by an observance of characteristic ES cell morphology and expression of the pluripotency marker ALP (discussed throughout this text), which is a hydrolase enzyme responsible for removing phosphate groups from a variety of different types of molecules. ALP is highly elevated on the surface of pluripotent stem cells, making it an ideal marker for pluripotency. In addition to the characterization of ALP expression, the researchers monitored the expression of the pluripotency markers Oct4 and **Kruppel-like factor 4 (Klf4)**, for which expression vectors were introduced into the somatic cells via retroviral transduction (O/K transduced). After identification of pluripotency inducers by assessing morphology, ALP, Oct4, and Klf4 expression, other pluripotency markers, including Nanog, SSEA1, and Sox2, were characterized to further confirm the pluripotent phenotype. These studies allowed for the identification of a histone methyltransferase inhibitor (BIX) and an L-channel calcium agonist (Bay K8644) as potential inducers of pluripotency. When combinations of these small molecules were tested, potent pluripotency induction properties were confirmed (Figure 7.12) (Shi et al., 2008). Therefore, a combination of morphology and marker expression characterization has allowed for the efficient identification of small molecules that have the capacity to reprogram somatic cell fibroblasts into iPS cells.

In a manner similar to that of the Scripps researchers, Rudolph Jaenisch's group at the Whitehead Institute for Biomedical Research, Massachusetts Institute of Technology in Cambridge, focused on the role of the reprogramming factor Klf4. **Klf4** is one of four key transcription factors that, when overexpressed in somatic cells, drive reprogramming to induce a pluripotent phenotype. It has been hypothesized that Klf4's role in reprogramming is through **chromatin remodeling**, which is defined as the modification of DNA and

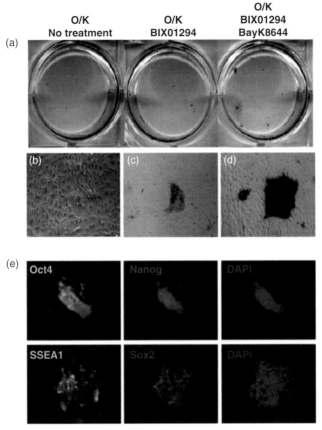

Figure 7.12 Output data for the high-throughput identification of small molecule inducers of pluripotency. (a) Alkaline phosphatase expression in the presence or absence of a combination of small molecules identified in HTS assays as conferring pluripotency characteristics on mouse embryonic fibroblasts; (b) Empty retroviral vector results in no colonies, (c) O/K transduced cells result in several small flattened colonies with weak ALP expression; (d) ES cell-like colonies observed after O/K vector transduction and treatment with BIX/BayK; (e) Confirmation of induced pluripotency as measured by immunofluorescence of pluripotent markers. (Images courtesy Yan Shi, Sheng Ding and *Cell Stem Cell* (Shi et al., 2008); reprinted with permission.)

protein architecture to allow open access of transcriptional control factors to condensed genomic regions for purposes of gene regulation. The researchers designed an HTS method for the identification of small molecules that could act to replace Klf4 in the context of iPS cell formation. The screen was based on the utilization of a Nanog-luciferase reporter mouse somatic cell line. **Luciferase** is a catalytic enzyme that oxidizes a luciferin substrate resulting in an electronically excited state. Light is emitted from the substrate upon return of electrons to the ground state. The gene encoding firefly luciferase was stably introduced into the Nanog locus via site-directed homologous recombination discussed in Chapter 6. This allowed for luciferase expression to be under the control of the Nanog promoter, which marks induction of pluripotency (see Chapter 2). Nanog-luciferase mouse embryonic fibroblasts (NL-MEFs) were transduced with a retroviral vector encoding the other three crucial transcription factors Oct4, Sox2, and c-Myc, expanded and seeded into 1,536-well plates. Cells were subsequently screened against 500,000 small molecule compounds

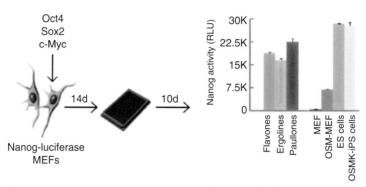

Figure 7.13 Screening strategy and results for the identification of Klf4 inducers in mouse embryonic fibroblasts. Three of the four transcription factors, Oct4, Sox2, and c-Myc, were stably introduced into NL-MEF cells that were subsequently seeded into 1,536-well plates and screened against a 500,000-complexity small molecule library. RLUs were assayed for Klf-4 luciferase activity, an indicator of pluripotency induction, and various types of small molecules identified as Klf4 activators. Induction of Klf4 expression was also compared against control ES cells. (*Source:* Lyssiotis, et al., 2009. Reproduced with permission from National Academy of Sciences, USA.)

representing three structural classes with the desired data output focused on luciferase expression and light emission (relative light units, RLUs) in the presence of the luciferin substrate (Figure 7.13) (Lyssiotis et al., 2009).

Candidate hits were further characterized as potential pluripotency inducers through a secondary screen assessing colony formation and ALP activity. The screen ultimately identified the small molecule kenpaullone as the most potent inducer of pluripotency in NL-MEFs (Figure 7.14) (Lyssiotis et al., 2009). Thus the screening platform designed and implemented by the Jaenisch research team allowed for the selective identification of a small molecule that could replace the requirement of Klf4 as one of the key factors involved

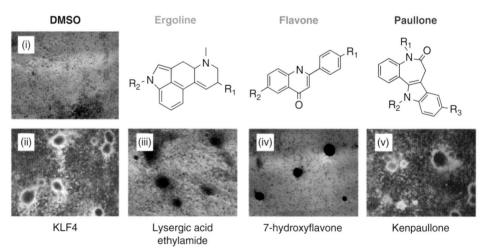

Figure 7.14 Identification of small molecule pluripotency inducers in mouse embryonic fibroblasts. Three structural classes of small molecules were characterized with paullones identified as the most potent inducers of pluripotency as assessed by marker expression and colony morphology (v). (*Source:* Lyssiotis, et al., 2009. Reproduced with permission from National Academy of Sciences, USA.)

in somatic cell pluripotency induction. This screening strategy could potentially allow for the identification of small molecules to effectively replace the requirement for all four key transcription factors, Oct4, Sox2, Klf4, and c-Myc, resulting in a "cocktail" of small molecules that induce pluripotency in somatic cells. Such a cocktail would eliminate the need for the introduction of viral vectors or other genomic modifications, making derived iPS cells potentially safe for cell- or tissue-based transplant therapy.

ANALYSIS OF DISEASE PATHWAYS

It is clear that the characterization of gene function in animal and insect models through gain- or loss-of-function experiments has yielded invaluable information with respect to genetic cascades involved in normal and diseased physiological states. Yet these models have been less successful in the prediction of drug efficacy specifically for human diseases. One high-profile example is that of the murine loss-of-function knockout of the HPRT locus. As mentioned above, in humans, mutations in this gene result in Lesch–Nyhan syndrome, a rare inherited genetic disorder characterized by a buildup of uric acid in most body fluids (discussed above). Surprisingly, the mouse knockout of HPRT has no phenotype. Differences in human versus other animal model gene-specific phenotypes are speculated to be the result of variations in gene dosage, mutation variability, and genetic background. Thus, other models of human disease employing cells and tissues of human origin may be needed in some cases to recapitulate a disease phenotype or a cascade of events. In many cases, sources of patient-specific cell lines include lymphocytes isolated from the blood or skin fibroblasts isolated during routine biopsy procedures. Yet these cells may not truly mimic some pathways due to the fact that they are mature, differentiated lineages. Pluripotent human cells, such as ES cells, may allow for a more accurate representation of the genetic and biochemical pathways which go awry in disease as they are capable of differentiation into virtually any cell type of the body. Disease-specific human pluripotent ES cells exist which contain chromosomal or genetic abnormalities. These are often derived from blastocysts discarded during *in vitro* fertilization procedures. Due to inherent mutations in these cells, they may represent excellent *in vitro* models of human disorders. In addition, with the recent advancements made in induced pluripotency, true patient-specific pluripotent stem cell lines may now be created which mimic cell-based disorders specific to that patient. Table 7.1 lists the number of diseases now modeled using characterized lines of either human ES cells or human iPS cells.

TABLE 7.1 Diseases modeled by either human ES cells, iPS cells, or both.

Disease human ES cells	Molecular defect	Phenotype demonstrated
Alport syndrome	Mutation in *COL4A5*	Not determined
Androgen insensitivity syndrome	Deletion of androgen receptor gene	Not determined
Fabry syndrome	Mutation in *GLA*	Not determined
Fanconi anemia (carrier)	Mutations in *FANCA*	Not determined
Marfan syndrome	Mutation in *FBN1*	Not determined
Multiple endocrine neoplasia type 2A	Mutation in *RET*	Not determined

(continued)

Disease human ES cells	Molecular defect	Phenotype demonstrated
Myotonic dystrophy	Trinucleotide expansion in *DMPK* or tetranucleotide expansion in *CNBP*	Decreased expression of two members of the SLITRK family; altered neurite outgrowth, neuritogenesis, and synaptogenesis in motor neuron and muscle cell co-cultures
Neurofibromatosis type 1	Point mutation in *NF1*	Not determined
Saethre–Chotzen syndrome	Mutation in *TWIST*	Not determined
Spinocerebellar ataxia type 2	Trinucleotide expansion in *ATXN2*	Not determined
X-linked myotubular myopathy **Human embryonic stem cells and iPS cells (iPSCs)**	Mutation in *MTM1*	Not determined
Becker muscular dystrophy	Mutation in dystrophin gene	Not determined
Cystic fibrosis	Mutations in *CFTR*	Not determined
Duchenne muscular dystrophy	Mutation in dystrophin gene	Loss of dystrophin expression in muscle tissue derived from diseased iPSCs; restored by human artificial chromosome-mediated dystrophin expression
Fragile X syndrome	Trinucleotide (CGG) expansion, silencing of *FMR1*	Not determined
Gaucher's disease	Point mutation in β-glucocerebrosidase	Not determined
Huntington's disease	Trinucleotide expansion in Huntington gene	Enhanced caspase activity following growth factor withdrawal in iPSC-derived neurons from patients
X-linked adrenoleukodystrophy **iPS Cells**	Mutation in *ABCD1*	VLCFA levels increased in iPSC-derived oligodendrocytes; reduced after treatment with lovastatin or 4-phenylbutyrate
ADA–SCID	Mutations in *ADA*	Not determined
Atypical Werner syndrome	Mutation in *LMNA*	Nuclear membrane abnormalities, increased senescence and susceptibility to apoptosisobserved in iPSC-derived fibroblasts
β-Thalassemia	Deletion in β-globin gene	Not determined
Crigler–Najjar syndrome	Mutation in *UGT1A1*	Not determined
Type 1 diabetes	Multifactorial; unknown	Not determined
Down syndrome	Trisomy 21	Not determined

Disease human ES cells	Molecular defect	Phenotype demonstrated
Dyskeratosis congenita	Mutations in *DKC1*, *TERT*, or *TCAB1*	Progressive telomere shortening and loss of self-renewal of iPSCs
Dystrophic epidermolysis bullosa	Mutations in *COL7A1*	Lack of expression of type VII collagen, restored following gene correction; no difference between diseased and control formation of three-dimensional skin equivalents
Familial amyotrophic lateral sclerosis	Mutation in *SOD1* or *VAPB*	Reduced levels of VAPB in fibroblasts, iPSCs, and motor neurons derived from patients with VAPB mutation
Familial dysautonomia	Mutation in *IKBKAP*	Decreased expression of genes involved in neurogenesis and neuronal differentiation; defects in neural crest migration
Familial hypercholesterolemia	Mutation in gene encoding LDL receptor	
Glycogen storage disease type 1A	Deficiency in glu-cose-6–phosphate	Hyperaccumulation of glycogen
Gyrate atrophy	Mutation in *OAT*	Not determined
Hereditary tyrosinemia type 1	Mutation in fumarylace-toacetate hydrolase	Not determined
Hutchinson–Gilford progeria syndrome	Mutations in *LMNA*	Accelerated cell senescence, progerin accumulation, DNA damage, nuclear abnormalities, inclusions in VSMCs; phenotype corrected by HDAδ-based gene repair
Inherited dilated cardiomyopathy	Mutation in *LMNA* causing *LMNA* haploinsufficiency	Nuclear membrane abnormalities, increased senescence, and susceptibility to apoptosis in iPSC-derived fibroblasts
Lesch–Nyhan syndrome (carrier)	Heterozygosity of *HPRT1*	Not determined
Long QT syndrome	Mutation in genes encoding KCNQ1 or KCNH2	Arrhythmogenicity in cardiac cells; treatment with ranolazine rescues arrhythmia
MPS type I (Hurler syndrome)	*IDUA* deficiency	Not determined
MPS type IIIB	Defective α-*N*-acetyl-glucosaminidase	iPSCs and differentiated neurons derived from patients show defects in storage vesicles and Golgi apparatus

(continued)

Disease human ES cells	Molecular defect	Phenotype demonstrated
Parkinson's disease	Unknown or mutations in *LRRK2* or *PINK1*	Impaired mitochondrial function in PINK1-mutated dopaminergic neurons, corrected by lentiviral expression of PINK1; sensitivity to oxidative stress in LRRK2-mutant neurons
Polycythemia vera	Heterozygous point mutation in *JAK2*	Enhanced erythropoiesis
Progressive familial hereditary cholestasis	Unknown	Not determined
Retinitis pigmentosa	Mutations in *RP1*, *RP9*, *PRPH2,* or *RHO*	Decreased numbers of differentiated rod cells and expression of cellular stress markers
Rett syndrome	Mutation in *MECP2*	Decreased synapse number, reduced number of spines, and elevated LINE1 retrotransposon mobility
Schizophrenia	Unknown	iPSC-derived neurons from patients show diminished neuronal connectivity and decreased neurite number, PSD95, and glutamate receptor expression; neuronal connectivity is improved following treatment with loxapine
Scleroderma	Unknown	Not determined
Shwachman–Bodian–Diamond syndrome	Mutation in *SBDS*	Not determined
Sickle cell anemia	Mutation in *HBB*	Not determined
Spinal muscular atrophy	Mutation in *SMN1*	Reduced SMN levels in iPSCs, reduced size and number of motor neurons; valproic acid, and tobramycin increases the number of SMN-rich structures (gems) in iPSCs derived from patients
Wilson's disease	Mutations in *ATP7B*	Mislocalization of mutated ATP7B and defective copper transport in iPSC-derived hepatocyte-like cells; rescued by lentiviral gene correction or treatment with the chaperone drug curcumin
X-linked chronic granulomatous disease	*CYBB* deficiency	Lack of ROS production in neutrophils, corrected by insertion of *CYBB* minigene
α_1-Antitrypsin deficiency	Mutation in α_1-antitrypsin	

Source: Adapted from Grskovic et al., 2011.

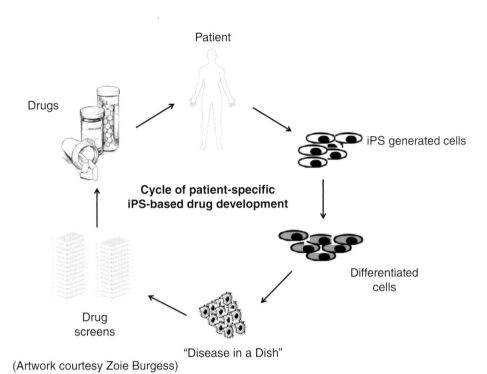

Patient

Drugs

iPS generated cells

**Cycle of patient-specific
iPS-based drug development**

Differentiated
cells

Drug
screens

"Disease in a Dish"

(Artwork courtesy Zoie Burgess)

Figure 7.15 Diagrammatic illustration of the stepwise process for drug discovery using iPS cells as patient-specific disease models. See text for details. (Artwork courtesy Zoie Burgess.)

iPS cells are of particular importance in this respect as they do indeed have the potential to represent a disease or disorder inherent in a specific patient. These cells may be harvested, induced for pluripotency capabilities, and directed to differentiate into mature, disease modeling lineages. They may then be screened in high-content assays to identify "hits" which act to abrogate the disease phenotype, or the "**disease in a dish**" (Figure 7.15).

iPS cells also afford the ability to study and characterize multiple cell types from an individual patient. These studies may be applied to assess either drug efficacy, toxicity, or both, and the analysis may be quite comprehensive, given the fact that effects can be determined on multiple lineages, hence the term "**patient in a dish.**" Yet, as discussed below, it is important to consider that the use of iPS technology to test drug efficacy and toxicity on multiple lineages for individual patients is not cost-effective. Strategies must be developed to implement the use of iPS cells in drug screening and characterization for application to large patient populations and not merely individual patients. A bank of iPS cells isolated from patients with different genetic backgrounds, perhaps representing different races and even inherited disease states, would be an invaluable resource for HCS and drug testing.

As is evident from Table 7.1, much research on pluripotent cell-based modeling of neurological disorders has resulted in the establishment of numerous lines that recapitulate these disorders *in vitro*. For example, familial Parkinson's disease, Huntington's disease, and Rett syndrome are all examples of neurological diseases for which cell line models exist. These are all based on the ultimate model of providing a resource for either neurons or glial cells with disease modeling genetic mutations that are either inherited or sporadic in nature (Figure 7.16) (Mackay-Sim, 2013).

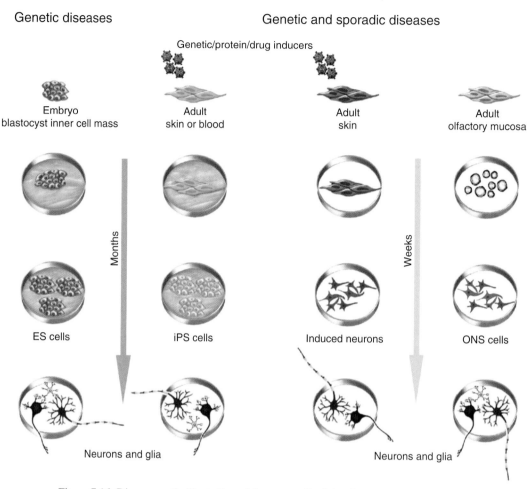

Figure 7.16 Diagrammatic illustration of the stem cell origins for neurons and glia to model human neurological disorders. (Diagram courtesy Alan Mackay-Sim and *Frontiers in Cellular Neuroscience* (Mackay-Sim, 2013); reprinted with permission.)

Yet challenges still exist for the accurate mimicry of neurological and other disorders using either ES or iPS pluripotent cell types as a base system. For example, ES cells are limited by the availability of preimplanted embryos genetically tested to carry disease-specific mutations. In addition, iPS cells have inherent properties unique to their generation such as integrated reprogramming vectors that might compromise the integrity of desired *in vitro* model phenotypes. Current techniques are also somewhat limited in their abilities to drive truly efficient directed differentiation of either ES or iPS cells into homogeneous populations of disease-modeling mature, differentiated cell types. Finally, the production of ES and iPS cells coupled with their expansion and directed differentiation into terminally differentiated lineages on a scale sufficient for disease modeling and drug discovery is considerably expensive and thus will most likely take place in the private sector, given lack of government funding for these initiatives.

STEM CELLS AS A TOXICITY-TESTING PLATFORM

Stem cells not only act as an excellent source for live data output with respect to drug candidate identification through screening or existing drug efficacy evaluation, but they may also yield important information regarding toxicity. Specifically, the testing and evaluation of existing drugs, drug candidates, and even particles or other entities present in the environment are all areas where stem cell testing can yield valuable information. There are three main impacts stem cells may have with respect to toxicity testing:

1. Developmental environmental toxicity
2. "Adult" environmental toxicity
3. Therapeutic toxicity

Stem cells may act as a valuable platform for the characterization of potentially hazardous materials within the environment, in regard to either embryonic development or exposure of individuals postnatally. In addition, stem cells may act as a source for the characterization of a drug or drug candidate's toxicity with respect to certain cell types. The following examples illustrate cases where stem cells acted as a valuable resource for the testing of the toxic effects substances may have on embryonic development and postnatally into adulthood.

Stem Cells as a Resource for Developmental Toxicity Testing

Developmental neurotoxicity (DNT) is defined as the impairment of nervous system development, with resulting structural defects in the central or peripheral nervous system. DNT is considerably difficult to model in animals due to variations in phenotypes resulting from treatment with potentially toxic agents. A more standardized system is needed to characterize toxic effects at the cellular level. ES cells have been demonstrated to accurately represent various stages of early embryonic neural development *in vitro* and thus may act as a good platform for assessing DNT. Researchers in the Department of Biology, University of Konstanz, Konstanz, Germany, recently devised a human ES cell-based three-dimensional *in vitro* model of embryonic neural development to evaluate the potential DNT of chemically inert polyethylene nanoparticles (PE-NPs). The system was based on human ES cell neurosphere generation, which was generated through directed differentiation of **human embryonic stem cells (hESCs)** into neural progenitors followed by aggregation of these cells on low-adhesion plates. Neural differentiation medium was added to the culture in addition to inhibitors of differentiation such as **rho C kinase inhibitor (ROCKi)**. ROCKi is an inhibitor of cellular differentiation acting through the direct inhibition of the enzyme **rho C kinase (ROCK)**. ROCK is a serine-threonine kinase that plays a role in a variety of cellular functions, one of which is the control of cell cycle. Inhibition of ROCK is known to promote cell proliferation, allowing for the expansion of the neurosphere cell population. ROCK was subsequently removed and differentiation was allowed to proceed, with confirmation of neural lineages made by the observance of rosettes, which are radial arrangements of columnar cells that express many of the proteins expressed in neuroepithelial cells in the neural tube(Figure 7.17) (Hoelting et al., 2013).

The three-dimensional nature of the system allowed for a more realistic interaction between living cells and the PE-NPs as well as morphogen gradients involved in embryonic maturation. This system was utilized to screen the toxic effects of

d15	d22	d29

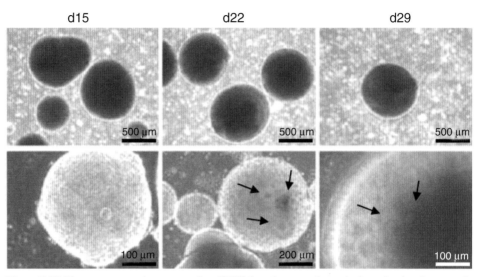

Figure 7.17 Phase contrast microscopy of an hESC-based neurosphere platform for developmental neurotoxicity testing. Days of differentiation are indicated above each panel set. Lower panels are at higher magnification with arrows indicating formation of rosettes. (*Source:* Hoelting et al., 2013.)

nanoparticles. Polyethylene nanoparticles (PE-NPs) were shown to readily incorporate into neurospheres. Intracellular ATP levels are known to decrease in toxic environments. In this study, PE-NPs were acutely toxic as measured by the intracellular ATP content of the cells following treatment with different PE-NP concentrations (Figure 7.18) (Hoelting et al., 2013). Therefore human ES cells have been shown to be a source for a neural screening system to characterize the potential developmental toxicity of nanoparticles. It should be noted that the concentrations of PE-NPs tested in this assay are magnitudes above what an embryo/fetus would be exposed to in the real world. The findings related to nanoparticle toxicity on embryonic development must be considered in this context.

Stem Cells as a Source for Post-Natal Environmental Toxicity Testing

Predictive toxicology is the study of the action of compounds on biological systems to predict a toxic effect. Until recently, a great deal of toxicological testing of compounds that may affect humans postnatally has been primarily based upon the use of laboratory animals, primary cell lines, or immortalized cancer cell lines. Immortalized cell lines do not accurately reflect the normal or even diseased physiological state. Most primary cell lines are heterogeneous in nature and thus produce unreliable toxicity analysis results. The use of animals to test compound toxicity also most often does not reflect human physiology accurately. In addition, animal use in toxicology studies has ethical concerns and is considerably expensive. Stem cells may address each of these limitations and have the potential to not only impact the evaluation of compound toxicity with respect to embryonic development, but they may also act as a valuable resource for mature adult lineages present in humans postnatally and throughout adulthood. Stem cell lineages may be analyzed for phenotypic changes that might occur

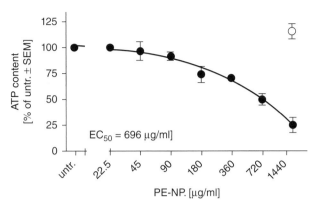

Figure 7.18 Toxicity of nanoparticles on human ES cell-generated neurospheres. Increasing concentrations of polyethylene nanoparticles were demonstrated to be toxic to neural lineages present in neurospheres. (*Source:* Hoelting et al., 2013.)

upon exposure to various compounds, including drug candidates or materials and compounds present in the environment. In this context, numerous stem cell types may be employed as the source for terminally differentiated lineages. These include adult stem cells and iPS cells. The application of iPS cell technology for toxicity testing circumvents the obvious two main issues of ethics and immunological concerns related to ES cell use in this respect. Recent years have seen the advent of directed differentiation technologies that allow for the efficient conversion of iPS cells into various terminal, mature lineages. These lineages may be of vast importance not only for the testing of drug efficacy as mentioned in the previous sections but also for predictive toxicology studies via their use in various *in vitro* toxicity assays (Figure 7.19) (Anson et al., 2011).

Features of iPS cells for the design of *in vitro* predictive toxicology testing include:

- Ability to generate "unlimited" cell populations from both diseased and normal human phenotypes
- Wide-ranging terminal lineage phenotypes is possible
- Patient-specific cell-based testing could become the standard for personalized medicine

Following are several examples of iPS cell sources for differentiated lineage phenotypic characterization to predict compound toxicological properties.

Cardiotoxicity **Cardiotoxicity** is defined as the toxic effect a compound has on heart cells or tissue. Unwanted adverse events affecting the cardiovascular system are the primary causes of treatment withdrawal, limitation of therapeutic regimens, or cessation of clinical trials. **Ventricular tachyarrhythmia**, which is defined as an abnormal heartbeat localized to the ventricles, is the number one cardiovascular disorder caused by drug toxicity. It is also known as drug-induced **torsades de pontes (TdP)** and its cause at the cellular level is due to inhibition of the inward rectifying potassium channel hERG (human ether-a-go-go), resulting in a prolonged **QT interval**, which is the time between depolarization and repolarization. Yet, screening for hERG (human ether-a-go-go) irregularities is not optimal due to the fact that many compounds that

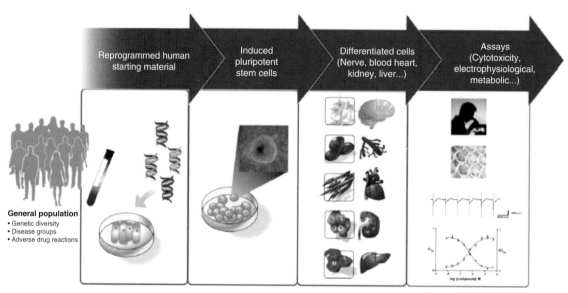

Figure 7.19 Schematic diagram of the application of induced pluripotency stem cell technology in predictive toxicology. Patient-derived somatic cells may be reprogrammed to a pluripotent state and directed to differentiate into different cell types such as hepatocytes or cardiomyocytes. These may be produced *en masse* and applied in various *in vitro* assays for the predictive assessment of compound toxicity. (Schematic diagram courtesy Black D. Anson, Timothy J. Kamp, and *Clin. Pharmacol. Ther.* (Anson et al., 2011); reprinted with permission.)

inhibit hERG (human ether-a-go-go) do not induce TdP, thus a new system for screening drugs that truly addresses prolonged QT and TdP is needed. Researchers at Hoffmann-La Roche in Nutley, New Jersey, developed a unique *in vitro* toxicity screening assay utilizing iPS cells as a resource for cardiomyocytes. A 96-well plate system with interdigitated electrode sensors was developed to assess in real-time changes in cellular behavior during drug treatment. In it the researchers measured **impedance**, which is the measurement of the opposition a circuit imposes on current when a voltage is applied, across the electrodes, which provides an indirect assessment of both cell number and cell interaction with the interface of the dish. This allowed for the detection of the physical movement of contracting cardiomyocytes differentiated from iPS cell sources (iPSC-CMs). Toxic effects of compounds introduced into the culture media produced irregular, nonuniform cardiomyocyte contraction, mimicking proarrhythmia, and thus producing impedance values unique to those cells in sync (Figure 7.20) (Guo et al., 2011). Ultimately the Hoffman-La Roche team tested 28 different compounds utilizing the iPSC-CM impedance measurement system. They noted specific changes in beat rate and/or impedance amplitude consistent with known data previously acquired for each compound. From these measurements the team was able to develop an index of drug-inducing arrhythmias that allowed for the calculation of a particular drug's proarrhythmic potential (Song et al., 2009). Thus the researchers have implemented iPS technology and directed differentiation methodologies combined with impedance analysis to ascertain the potential toxic effects drugs may have on cardiomyocytes. This new *in vitro* cardiac function-based predictive toxicology screen may yield new opportunities for the more timely and accurate prediction of cardiac toxicity.

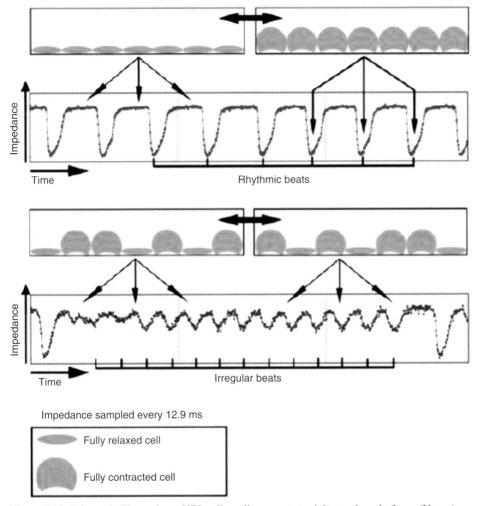

Figure 7.20 Schematic illustration of iPS cell-cardiomyocyte toxicity-testing platform. (Upper) Cardiomyocytes are plated in electrode-lined 96-well plates. Rhythmic contractions of cells in culture result in regular impedance outputs. (Lower) Unsynchronized cardiomyocyte contraction due to drug toxicity results in irregular impedance outputs. (*Source:* Guo et al., 2011. Reproduced with permission from Oxford University Press.)

Focus Box 7.3: Kyle Kolaja and iPS-based cardiotoxicity testing

Dr. Kyle Kolaja has made major contributions to the field of toxicity testing through his studies of iPS cell technology. A scientist at Hoffman-La Roche in Nutley, New Jersey, Dr. Kolaja has developed an iPS-derived cardiomyocyte toxicity-testing platform that detects changes in cellular behavior following drug treatment. Dr. Kolaja is the author of nearly 50 peer-reviewed scientific articles and in 2010 was awarded the prestigious position of Fellow of the Academy of Toxicological Sciences. (Photo courtesy Indiana University School of Medicine.)

Hepatotoxicity **Hepatotoxicity** is defined as toxic effects on either the liver or individual liver cells known as **hepatocytes**. Studies regarding drug-induced hepatotoxicity have often been based on the use of primary human hepatocytes, yet these cells are in many instances unreliable. In many cases they are inaccessible, have high intrinsic variability, and rapidly lose endogenous metabolic activity when placed in an *in vitro* environment. Thus, iPS-derived hepatocytes might provide a more normalized and reproducible hepatocytic cell line for *in vitro* toxicological studies. In fact, iPS-derived hepatocyte-like cells have been demonstrated in tissue culture applications to reliably recapitulate various cellular processes such as CYP450 metabolism, glycogen storage, and the production of albumin. Hongkui Deng et al., in the Department of Cell Biology at Peking University in Beijing, China, developed a systematic methodology for the efficient directed differentiation of human iPS cells into mature, functional hepatocytes. The stepwise protocol involved the addition of various growth and differentiation factors including:

- Activin A
- FGF4 (fibroblast growth factor 4)
- BMP2 (bone morphogenetic protein 2)
- HGF (hepatocyte growth factor)
- KGF (keratinocyte growth factor)
- Dexamethasone

Successive addition of these factors at specific timepoints activated iPS cell endogenous developmental cues to drive hepatocyte cell formation at roughly 60% efficiency as confirmed by both morphology and alpha fetoprotein as well as albumin marker expression (Figure 7.21) (Song et al., 2009). These cells are homogeneous in nature and exhibit key hallmarks of hepatocyte form and function, indicating that they may be ideal for liver predictive toxicology studies.

Chapter 7 focused on the utilization of various types of stem cells as drug discovery platforms. Chapter 8 will address the use of stem cells themselves as therapies for the treatment of a variety of disorders.

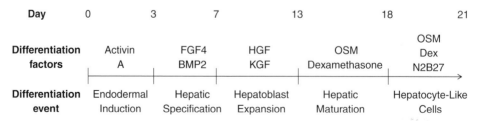

Figure 7.21 Systematic directed differentiation of iPS cells into hepatocyte-like cells. Flowchart of the *in vitro* induction system. See text for details.

CHAPTER SUMMARY

Embryonic Stem Cells and Mouse Models of Gene Function

1. Over the last 25 years the vertebrate model system of choice for studying the function of genes has been the mouse.
2. The most widely used method for defining individual gene function in a mouse model is via gene targeting.
3. Gail R. Martin's pioneering research in embryonic development and discovery of mouse ES cells paved the way for a revolution in the study of gene function *in vivo*.
4. Gene targeting in the context of mouse ES cells has literally revolutionized the study of gene function *in vivo*.
5. HPRT is an ideal candidate locus to prove site-specific gene-targeting technology.

Stem Cell-Based Screening Assays

1. Stem cells may provide valuable insight regarding the design or testing of a therapeutic platform.
2. Numerous different types of high-throughput cell-based drug screens exist.
3. Stem cell culture and preservation techniques have been optimized to the point that care, handling, and storage are not issues for HTS applications.
4. ES cells make an excellent source for differentiated neurons to be used in high-throughput screens.
5. High-throughput screens designed to identify pharmaceutical agents that might impact, or even correct, defects in CNS-based neurobiology would be of enormous value in developing new treatments for CNS disorders.
6. Adult stem cells are multipotent in nature and, depending upon the cell type, may act as a valuable resource for large quantities of cells to be screened for small molecules of therapeutic significance.
7. iPS cells exhibit many properties that would make them ideal as a source for virtually any cell type to be used in an HTS platform.
8. High-throughput screens have now been developed and implemented to identify therapeutic candidates for the eradication of CSCs.
9. Identifying small molecules that drive the reprogramming of cells into a pluripotent phenotype could yield valuable entities for use in therapeutic cell production.
10. A number of research groups have developed screening assays to identify molecules that may drive the reprogramming of adult somatic cells.

Analysis of Disease Pathways

1. Models other than those that are animal-based are needed to assess or recapitulate a human disease phenotype.
2. Pluripotent human cells, such as ES cells, may allow for a more accurate representation of the genetic and biochemical pathways which go awry in disease than that of animal models.

3. iPS cells are of particular importance here as they do indeed have the potential to represent a disease or disorder inherent in a specific patient.

4. iPS cells afford the ability to study and characterize multiple cell types from an individual patient.

5. Current techniques are limited in their abilities to drive truly efficient directed differentiation of either ES or iPS cells into homogeneous populations of disease-modeling mature, differentiated cell types.

6. The production of ES and iPS cells coupled with their expansion and directed differentiation into terminally differentiated lineages on a scale sufficient for disease modeling and drug discovery is expensive.

Stem Cells as a Toxicity-Testing Platform

1. The testing and evaluation of existing drugs, drug candidates, and even particles or other entities present in the environment are all areas where stem cell testing can yield valuable information.

2. ES cells have been demonstrated to accurately represent *in vitro* various stages of early embryonic neural development and thus may act as a good platform for assessing DNT.

3. Most primary cell lines are heterogeneous in nature and thus produce unreliable toxicity analysis results.

4. Stem cells have the potential to not only impact the evaluation of compound toxicity with respect to embryonic development, but they may also act as a valuable resource for mature adult lineages present in humans postnatally and throughout adulthood.

5. The application of iPS cell technology for toxicity testing circumvents the two main issues of ethics and immunological concerns related to ES cell use in this respect.

6. Dr. Kyle Kolaja has made major impacts in the field of toxicity testing through his studies of iPS cell technology and the development of cell-based assays to study drug cardiotoxicity.

KEY TERMS

(Key terms are listed by order of appearance in the text.)

- **Gene targeting**—the site-specific alteration of a gene through the targeting of DNA sequences to that gene.
- **Inner cell mass (ICM)**—a component of mammalian blastocyst-stage embryos destined to form the embryo proper.
- **Fibroblast feeder layer**—a layer of fibroblasts plated upon which stem cells are cultured that provides both support and nutrients.
- **Homologous recombination**—exchange of genetic material between two identical or near-identical nucleotide sequences.
- **Hypoxanthine guanine phosphoribosyl transferase (HPRT)**—a transferase that plays a central role in the generation of purine nucleotides essential for DNA synthesis. Mutations in HPRT have been shown in humans to result in Lesch–Nyhan syndrome.

- **Lesch–Nyhan syndrome**—a rare inherited genetic disorder resulting in the building of uric acids in the body.
- **Chimera**—a mouse composed of two or more genetically distinct populations of cells.
- **Germline transmission**—transmission of a particular genetic background to progeny.
- **Knockout mouse**—a mouse that has had a specific gene or nucleotide sequence deleted, often as a result of gene targeting.
- **Knockin mouse**—a mouse that has had a specific gene or nucleotide sequence introduced into its genome, often to replace a similar sequence or gene.
- **Positive-negative selection**—the utilization of a combination of positive and negative selectable markers to enrich gene targeting via homologous recombination
- **Somite**—epithelial sphere present along the axis of the embryo that becomes patterned to form vertebrae, ribs, skeletal muscle, and dermis.
- **High-throughput screening (HTS), a.k.a. high-content screening (HCS)**—a rapid method for identifying a key molecule or pathway.
- **"Hit"**—a small molecule or compound identified in an HTS or HCS that may represent a therapeutic or modulatory candidate.
- **Primary cell**—cell isolated directly from tissue.
- **AMPA receptor**—tetrameric aggregate receptor composed of four subunits capable of binding glutamate, which subsequently results in the opening of an associated ion pore driving neuronal depolarization and proper neurotransmission.
- **Ca^{2+} flux assay**—an assay for the in-cell measurement of agonist-stimulated and antagonist-inhibited calcium signaling through G protein-coupled receptors (GPCRs).
- **Megakaryopoiesis**—the differentiation of hematopoietic stem cells into megakaryocytes.
- **Megakaryocyte**—cell that is responsible for the production of thrombocytes (platelets).
- **Thrombocyte**—platelet, essential for blood clotting.
- **Osteogenesis**—the formation of bone.
- **Alkaline phosphatase (ALP)**—a hydrolase enzyme responsible for removing phosphate groups from a variety of different types of molecules. It is also a marker for both bone development and pluripotency in stem cells.
- **Amyloid-β peptide**—a 36-43 amino acid peptide processed from amyloid precursor protein that is a hallmark of amyloid plaques in association with Alzheimer's disease.
- **N2/B27**—a neuronal supplement medium which promotes and supports neuronal growth in tissue culture.
- **Neuroblastoma (NB)**—a neuroendocrine tumor that is the most common form of extracranial cancer in children.
- **Cytostatic**—inhibiting cell proliferation.
- **Cytotoxic**—causing the death of cells.
- **Glioblastoma multiform (GBM), a.k.a. glioma**—the most common and primary aggressive brain tumor in humans, it is a cancer of the glial cells of the brain most often caused by sporadic and noninherited genetic mutations.
- **MicroSource Spectrum**—a library of FDA-approved drugs or drugs in late-phase clinical trials.
- **Disulfiram (DSF)**—a clinically-approved drug for alcohol aversion therapy.

- **Aldehyde dehydrogenase (ALDH)**—a known marker for cancer stem cells and hypothesized to play a role in the maintenance of the progenitor cell phenotype.
- **Klf4 (Kruppel-like factor 4)**—one of four key transcription factors that, when overexpressed in somatic cells, drive reprogramming to induce a pluripotent phenotype.
- **Chromatin remodeling**—the modification of DNA and protein architecture to allow open access of transcriptional control factors to condensed genomic regions for purposes of gene regulation.
- **Luciferase**—a catalytic enzyme that oxidizes a luciferin substrate resulting in an electronically excited state. Light is emitted from the substrate upon return of electrons to the ground state.
- **"Disease in a dish"**—recapitulation of human disease phenotypes for purposes of high-content screening.
- **"Patient in a dish"**—the application of iPS cell technology in high-content screening assays for the characterization of compound effects on multiple cell types.
- **Developmental neurotoxicity (DNT)**—the impairment of nervous system development, with resulting structural defects in the central or peripheral nervous system.
- **Rho C kinase inhibitor (ROCKi)**—an inhibitor of cellular differentiation acting through the direct inhibition of rho C kinase.
- **Rho C kinase (ROCK)**—a serine-threonine kinase that plays a role in a variety of cellular functions, one of which is the control of cell cycle.
- **Predictive toxicology**—the study of the action of compounds on biological systems to predict a toxic effect.
- **Cardiotoxicity**—the toxic effect a compound has on heart cells or tissue.
- **Ventricular tachyarrhythmia**—an abnormal heartbeat localized to the ventricles.
- **Torsades de pontes (TdP)**—see Ventricular tachyarrhythmia.
- **QT interval**—the time between cardiomyocyte depolarization and repolarization.
- **Impedance**—the measurement of the opposition a circuit imposes on current when a voltage is applied.
- **Hepatotoxicity**—toxic effects on either the liver or individual liver cells known as hepatocytes.
- **Hepatocytes**—individual mature liver cells.

REVIEW QUESTIONS

(Answers to select review questions can be found at www.stemcelltextbook.com)

1. Name two examples of animal models of human disorders.
2. Why is the mouse an ideal model for human gene function and disease?
3. List three characteristics of mice that make them suitable for the study of gene function *in vivo*.
4. Describe Gail R. Martin's original method for isolating mouse embryonic stem cells.
5. How did Martin Evans and Andrew Kaufman improve their chances for successful murine ES cell isolation?

6. Why is the HPRT locus an ideal candidate gene for proving site-specific gene-targeting-methods?

7. What was the phenotype of paraxis-null mice produced by Burgess et al.?

8. List and describe three types of high-throughput cell-based screens.

9. What is the major disadvantage to performing HTS on primary cells?

10. List the three properties of embryonic stem cells that make them valuable for HTS.

11. Describe John McNeish's high-throughput screening assay for identifying compounds modulating neuronal signaling.

12. Cite two examples of adult stem cells used in high-content screening assays.

13. What properties do iPS cells possess that make them ideal as a source for virtually any cell type to be used in high-throughput screening?

14. Describe Haruhisa Inoue's Alzheimer's disease compound screening platform.

15. How did David Kaplan identify kinase inhibitors that negatively affect the growth and survival of cancer stem cells?

16. What was the data output for Doug Foltz's assay to identify therapeutics for glioblastoma?

17. Why did Rudolph Jaenisch's group choose to seek out replacement compounds for Klf4 activity?

18. List at least five examples of human diseases now modeled by stem cells.

19. What are some disadvantages of using patient-specific iPS cells or embryonic stem cells in HCS assays?

20. What are the three main impacts stem cells may have regarding toxicity-testing strategies?

21. How might neurospheres be used to screen for compound toxicity?

22. What features do iPS cells possess that make them suitable for predictive toxicology testing?

23. How did Kyle Kolaja's group assess changes in cardiomyocyte QT intervals in response to exposure to different compounds?

24. Why would someone use iPS cells to study hepatotoxic effects of compounds?

THOUGHT QUESTION

In your opinion, what is the most ideal stem cell type for HTS and HCS? List the reasons why and cite an example of how this stem cell type could be used to identify a candidate therapeutic.

SUGGESTED READINGS

Alves, H., K. Dechering, et al. (2011). "High-throughput assay for the identification of compounds regulating osteogenic differentiation of human mesenchymal stromal cells." *PLoS One* **6**(10): e26678.

Anson, B. D., K. L. Kolaja, et al. (2011). "Opportunities for use of human iPS cells in predictive toxicology." *Clin Pharmacol Ther* **89**(5): 754–758.

Boitano, A. E., L. de Lichtervelde, et al. (2012). "An image-based screen identifies a small molecule regulator of megakaryopoiesis." *Proc Natl Acad Sci U S A* **109**(35): 14019–14023.

Burgess, R., A. Rawls, et al. (1996). "Requirement of the paraxis gene for somite formation and musculoskeletal patterning." *Nature* **384**(6609): 570–573.

Evans, M. J. and M. H. Kaufman (1981). "Establishment in culture of pluripotential cells from mouse embryos." *Nature* **292**(5819): 154–156.

Grinshtein, N., A. Datti, et al. (2011). "Small molecule kinase inhibitor screen identifies polo-like kinase 1 as a target for neuroblastoma tumor-initiating cells." *Cancer Res* **71**(4): 1385–1395.

Grskovic, M., A. Javaherian, et al. (2011). "Induced pluripotent stem cells—opportunities for disease modelling and drug discovery." *Nat Rev Drug Discov* **10**(12): 915–929.

Guo, L., R. M. Abrams, et al. (2011). "Estimating the risk of drug-induced proarrhythmia using human induced pluripotent stem cell-derived cardiomyocytes." *Toxicol Sci* **123**(1): 281–289.

Hoelting, L., B. Scheinhardt, et al. (2013). "A 3-dimensional human embryonic stem cell (hESC)-derived model to detect developmental neurotoxicity of nanoparticles." *Arch Toxicol* **87**(4): 721–733.

Hothi, P., T. J. Martins, et al. (2012). "High-throughput chemical screens identify disulfiram as an inhibitor of human glioblastoma stem cells." *Oncotarget* **3**(10): 1124–1136.

Koller, B. H., L. J. Hagemann, et al. (1989). "Germ-line transmission of a planned alteration made in a hypoxanthine phosphoribosyltransferase gene by homologous recombination in embryonic stem cells." *Proc Natl Acad Sci U S A* **86**(22): 8927–8931.

Lyssiotis, C. A., R. K. Foreman, et al. (2009). "Reprogramming of murine fibroblasts to induced pluripotent stem cells with chemical complementation of Klf4." *Proc Natl Acad Sci U S A* **106**(22): 8912–8917.

Mackay-Sim, A. (2013). "Patient-derived stem cells: pathways to drug discovery for brain diseases." *Front Cell Neurosci* **7**: 29.

Martin, G. R. (1981). "Isolation of a pluripotent cell line from early mouse embryos cultured in medium conditioned by teratocarcinoma stem cells." *Proc Natl Acad Sci U S A* **78**(12): 7634–7638.

McNeish, J., M. Roach, et al. (2010). "High-throughput screening in embryonic stem cell-derived neurons identifies potentiators of alpha-amino-3-hydroxyl-5-methyl-4-isoxazolepropionate-type glutamate receptors." *J Biol Chem* **285**(22): 17209–17217.

Shi, Y., C. Desponts, et al. (2008). "Induction of pluripotent stem cells from mouse embryonic fibroblasts by Oct4 and Klf4 with small-molecule compounds." *Cell Stem Cell* **3**(5): 568–574.

Song, Z., J. Cai, et al. (2009). "Efficient generation of hepatocyte-like cells from human induced pluripotent stem cells." *Cell Res* **19**(11): 1233–1242.

Yahata, N., M. Asai, et al. (2011). "Anti-Abeta drug screening platform using human iPS cell-derived neurons for the treatment of Alzheimer's disease." *PLoS One* **6**(9): e25788.

Chapter 8

THERAPEUTIC APPLICATIONS OF STEM CELLS

As the field of stem cell research has matured over the last several decades, there has been much speculation that stem cells may act directly as therapeutic agents for the treatment of a multitude of cell- and tissue-based disorders. This is primarily due to the pluri-, multi-, and even unipotency of various stem cell types. The generation of a virtually unlimited supply of mature, differentiated lineages such as neurons represents an opportunity to treat many human disorders. In addition, stem cells themselves may act as carriers of other therapeutic agents, thus affecting treatment indirectly *in vivo*. This final chapter explores stem cells as therapeutic agents, with a focus on disease-specific treatment and recent advances in the field that are taking the application of stem cells as therapeutic agents a step closer to reality.

HISTORY OF STEM CELLS AS THERAPEUTICS

History of Tissue Engineering

The idea and concept of replacing diseased or dead tissue with healthy living material has been contemplated for greater than two thousand years and can be traced back to direct references in the Book of Genesis: "So the LORD God caused a deep sleep to fall upon the man, and he slept, then He took one of his ribs and closed up the flesh at that place. The LORD God fashioned into a woman the rib which He had taken from the man, and brought her to the man" (Genesis 2:21–22). Many scholars believe this to be the first written contemplation of tissue regeneration. However, it is only within the last 30 years or so that **regenerative medicine**—the process of replacing or regenerating human cells, tissues, or organs to restore or establish normal function—has come of age and begun to provide real promise for cell-based replacement therapeutics. Researchers struggled for years to develop cell culture and implant-based platforms that would sustain cellular viability long-term *in vitro* and *in vivo*. For example, **cell seeding**, which is

Stem Cells: A Short Course, First Edition. Rob Burgess.
© 2016 John Wiley & Sons, Inc. Published 2016 by John Wiley & Sons, Inc.

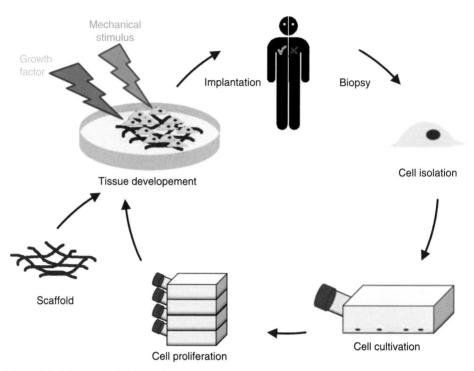

Figure 8.1 Diagrammatic illustration of the principles behind, and process of tissue engineering for therapeutics. See text for details. (Diagram courtesy Wikimedia Commons; reprinted with permission.)

defined as the plating and culture of cells into an artificial matrix, yielded little success initially due to a focus on two-dimensional platforms, which in most cases do not provide the optimal environment conducive to cell viability, growth, and function. It was in the 1970s when pioneering researcher and orthopedic surgeon W.T. Green developed a three-dimensional (3D) system for the propagation of chondrocytes. Green's system employed the seeding of chondrocytes onto bone spicules followed by implantation into nude mice. Although Green's studies were largely unsuccessful, they paved the way for other 3D-based systems for cell culture and cell- or tissue-replacement initiatives. It was Green who first speculated that advancements in biomaterials science would eventually allow for the production of new, healthy tissues through the seeding of cells onto **"smart" scaffolds**. These platforms may be implemented for **tissue engineering**, which is defined as the combination of cells, materials, biochemical or biophysical factors to generate living tissue; this term was coined in 1985 by Y.C. Fung, a biomechanics researcher who regarded living tissue as a unique entity that could be manipulated and even exploited for therapeutic gain. Tissue engineering involves the isolation, culture, expansion, and manipulation of cells with or without scaffolds to develop therapeutic tissues (Figure 8.1).

A few years later in 1988, Robert Langer of the Massachusetts Institute of Technology and Joseph Vacanti of Harvard Children's Hospital, Boston performed the first successful selective cell transplantation experiment using bioabsorbable artificial polymers as 3D matrices. In 1991, a patient with **Poland's Syndrome**, a congenital malformation

Figure 8.2 A human ear grown on the back of a mouse via tissue engineering methods. This is known as the Vacanti mouse after Joseph Vacanti who developed it. (Image courtesy Wikipedia; reprinted with permission.)

of the rib cage and absence of the sternum, was the first human patient to receive an engineered tissue implant composed of autologous chondrocytes seeded onto a synthetic polymer scaffold. The procedure was performed at Harvard Children's Hospital, Boston by Joseph and Charles Vacanti and J. Upton. It was Vacanti who, a full 24 years later, developed the **Vacanti mouse,** a.k.a. "**auriculosaurus**", a laboratory mouse with ear-shaped cartilage engineered onto its back by seeding cartilage cells into a biodegradable ear-shaped mold followed by subcutaneous transplantation (Figure 8.2) (Cao et al., 1997).

This study confirmed the premise that *de novo*, fully functional replacement tissue could be generated from appropriate cells seeded onto 3D scaffolding. Tissue engineering via the use and application of stem cells is not limited to cartilage expansion from chondrocytes and has now been demonstrated for use in a variety of cell- and organ-based systems. For example, the vasculature may be reconstructed through the combination of 3D scaffolding and cell seeding. A second example of tissue engineering involving the engineering of a functional pulmonary artery is outlined in Case Study 8.1.

Thus, early concepts in tissue engineering involving lineages more mature than stem cells have now come to fruition and been demonstrated to translate directly into therapeutic benefits. These advances have now been combined with the utility of various stem cell platforms for the engineering of a variety of tissues and the regeneration of both tissues and specific cells within the body. Case Study 8.1 is but one example of the real-world applications of cell-based tissue engineering. The following sections focus on stem cells as platforms for the treatment of specific diseases, either as straightforward tissue replacement platforms or via state-of-the-art cell and tissue regeneration strategies.

DISEASE-SPECIFIC TREATMENT AND PATIENT TRIALS

The first seven chapters of this book set the foundation for the utility inherent in stem cells as therapeutic platforms for the treatment and a variety of disorders. It is now clear that stem cells have the potential to impact medicine on a number of fronts (Figure 8.4). This section focuses on the potential of stem cells to acts as therapeutic entities in real-world applications.

Case Study 8.1: Transplantation of a tissue-engineered pulmonary artery

Toshiharu Shin'oka and Yoshito Ikada

Researchers at the Tokyo Women's Medical University and Suzuka University of Medical Science in Japan have developed a technique for the generation of a pulmonary artery using a combination of venous cells and a biodegradable 3D matrix. In this real-world application of stem cell-based tissue engineering, a segment of peripheral vein was isolated from a 4-year-old girl with pulmonary atresia. Cells from this biopsy were expanded and seeded onto a polycaprolactone-polylactic acid copolymer tube re-enfoced with polyglycolic acid. Ten days after seeding the graft was transplanted and the pulmonary artery reconstructed. No complications were observed 7 months post implantation (Figure 8.3) (Shin'oka, Ikada, 2001).

Cells from the venous wall were isolated, expanded, and seeded onto a biodegradable polymer scaffold and subwtly implanted as autologous tissue (Diagram redrawn from Shin'oka, Imai et al., 2001).

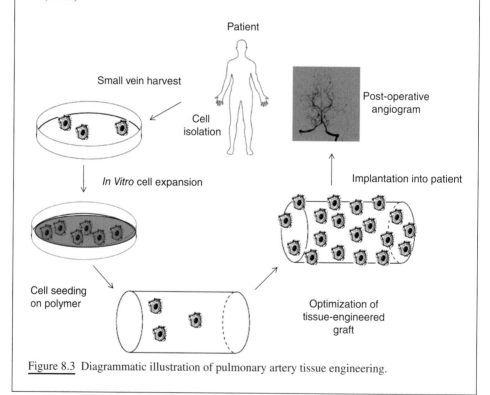

Figure 8.3 Diagrammatic illustration of pulmonary artery tissue engineering.

Stem Cell-Based Patient Trials: An Overview

As has been illustrated throughout this book, a revolution has occurred with respect to advancements in stem cell research. The discovery of different types of stem cells exhibiting different levels of potency combined with advancements in the knowledge of key molecular and biochemical cascades defining potency has allowed for the use of stem cells in the clinic to become a reality. For example, the discovery by Shinya Yamanaka and colleagues of induced pluripotency will certainly have a significant impact on patient-specific cell and

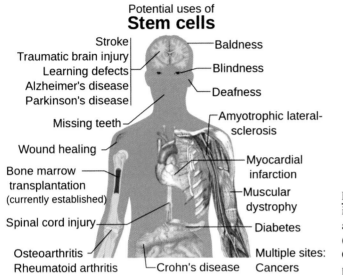

Figure 8.4 Diagrammatic illustration of stem cell applications in medicine. (Courtesy Wikimedia Commons; reprinted with permission.)

tissue transplants. It is important to emphasize that the result of significant advances in stem cell research is the real-world application of these cells to address medical disorders and disease. As of 2012, a number of clinical trials involving stem cells are occurring worldwide, with the predominant effort being undertaken in the United States (Figure 8.5) (Daley, 2012).

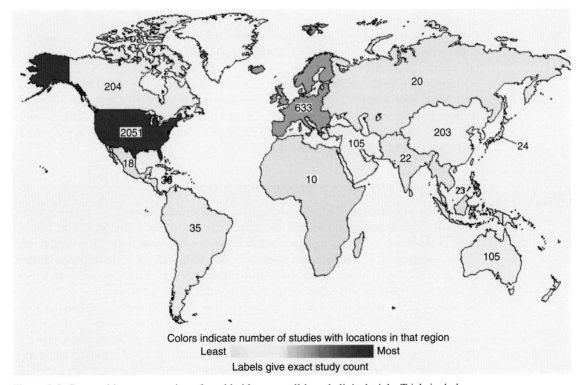

Figure 8.5 Geographic representation of worldwide stem cell-based clinical trials. Trials include those active, pending, and closed. (Figure courtesy George Q. Daley, www.Clinicaltrials.gov and *Cell Stem Cell* (Daley, 2012); reprinted with permission.)

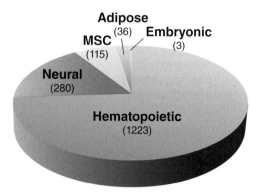

Stem cell types employed in interventional clinical trials (www.clinicaltrials.gov)

Figure 8.6 Graph of stem cell types most often used in clinical trials. The graph illustrates the number of open US clinical trials involving different types of stem cells. (Graph courtesy George Q. Daley, www. Clinicaltrials.gov and *Cell Stem Cell* (Daley, 2012); reprinted with permission.)

What types of stem cells are most often used in clinical trials? There is a tendency toward stem cell phenotypes that are the most understood and well characterized. In addition, the mode of administration, safety, potential efficacy, and disease type all play roles in defining the clinical potential of a particular cell type. Therefore, an obvious choice for most stem cell-based clinical trial efforts is in the hematopoietic area. Over 50 years of research in this area has made hematopoietic stem cell (HSC) transplantation almost routine, and has allowed for the effective treatment and even cure of a number of inherited blood-based genetic disorders possible (see the section on Hematopoietic Disorders below). Other "newcomer" types of stem cells that have shown promise in the clinic include adipose-derived stem cells, mesenchymal stem cells (MSCs), and neural stem cells. Numerous clinical trials are now underway or have been completed using these cell types as platforms to address a variety of disorders. In fact, as of 2013, over 4000 stem cell-based clinical trials have been executed or are underway (see below Figure 8.6). The following sections outline and describe some of the human patient trials completed involving stem cell-based therapeutics platforms, which have demonstrated much promise in patients for the treatment and perhaps even cure of disease.

Cardiomyopathy and Cardiovascular Disease (CV)

Cardiovascular disease (CV), also known as heart disease, is defined as one or more disorders of the heart and blood vessels resulting in decreased blood flow. CV is the leading cause of death and disability in Americans. The death toll arising from CV is higher than cancer, diabetes, HIV, and accidents combined. **Cardiomyopathy** is defined as a chronic disease of the heart muscle (myocardium) in which the muscle is abnormally enlarged, thickened, and/or stiffened; **ischemic cardiomyopathy** is defined as poor oxygen supply to the heart muscle and weakness of the cardiac muscle. Enhancements of the cellular components of cardiac tissue may reduce muscle weakness and improve ventricular contraction. Within the past 10 years, the concept of utilizing stem cells to treat various types of CV has come to fruition and emerged as a realistic treatment option. Indeed, numerous different stem cell types have been shown to have positive therapeutic benefits in preclinical animal models of CV. These cell types include embryonic stem cells, induced pluripotent stem cells, cardiomyocytes, skeletal muscle myoblasts, bone marrow derived stem cells, MSCs, and endothelial progenitor cells. Several criteria have been generally accepted as defining the ideal cell type for use in the treatment of CV disease. These include:

- Safe to use with no possibility of tumor formation or arrythmias
- Result in improvement of cardiovascular function
- Amenable to minimally invasive delivery methods
- Available as standardized "off-the-shelf" reagents
- No chance for immunorejection
- Ethically compliant

Although there is no perfect stem cell meeting all these criteria, cardiac stem cells may be the most ideal platform studied to date for treating patients with CV. **Cardiac stem cells (CSCs),** a.k.a. **cardioblasts**, are defined as a population of cells specified to the cardiac myo-cyte cell lineage prior to their (terminal) differentiation into a fully differentiated cardiac myocytes. These cells meet many of the criteria above and act through a mechanism to pop-ulate the local cardiac tissue with new cardiomyocytes. Certain CSC lineages have shown promise in this respect. **c-kit**, also known as **stem cell factor receptor**, is a cell surface receptor tyrosine kinase that marks the surface of some multipotent stem cell lineages. **c-kit positive, lineage-negative CSCs** are derived from the vasculature and located specifically within vessel walls. They commit to vascular endothelial or smooth muscle lineages upon differentiation *in vivo*. Researchers in the Division of Cardiovascular Medicine at the Univer-sity of Louisville in Kentucky have performed a 16-patient clinical trial in which autologous c-kit positive, lineage-negative CSCs were administered by intracoronary infusion into patients experiencing left ventricular (LV) ejection fractions of <40% normal volume. The researchers observed an 8% improvement in LV ejection fraction at 4 months post-infusion and 13% at 12 months post-infusion with no adverse side effects (Figure 8.7). These studies suggest that autologous infusion of c-kit positive, lineage-negative CSCs is considerably effective in improving LV systolic function in patients with heart failure (Bolli et al., 2011). This is yet one of many examples of stem cells now being utilized as regenerative medicine-platforms in clinical trials for the treatment of various cardiac disorders.

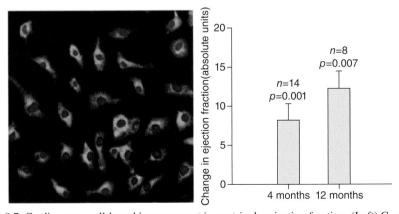

Figure 8.7 Cardiac stem cell-based improvement in ventricular ejection fraction. (Left) Confocal fluorescence microscopy showing the localization of c-kit (green). (Right) graphic representation of the change in ejection fraction percent following CSC infusion as a function of time. (*Source:* Bolli et al., 2011. Reproduced with permission from Elsevier.)

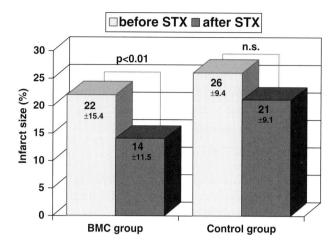

Figure 8.8 Treatment of acute myocardial infarction with bone marrow-derived stem cells. STX = stem cell transplantation. (*Source:* Schannwell et al., 2009. Reproduced with permission from Elsevier.)

Other research groups have focused on the application of bone marrow-derived stem cells (BMSCs) in the treatment of patients with acute myocardial infarction (AMI). AMI most often results in both structural and histopathological changes in the left ventricle decreasing its performance and ejection efficiency. A research team led by Muhammad Yousef in the Department of Medicine, Division of Cardiology, Pneumology and Angiology, Heinrich-Heine-University of Dusseldorf in Germany utilized BMSCs in a clinical trial to treat AMI. They employed autologous BMSCs in a patient population of 124. All patients underwent emergency coronary angiography and either **angioplasty**, which is defined as a procedure implemented to widen blood vessels, or stent implant. Sixty-two patients received a transplant of autologous BMSCs isolated based on the expression of cell surface markers CD34$^+$, CD133$^+$, and CD34$^-$. The cells were directly infused into the damaged artery utilizing an angioplasty balloon catheter. Rather than measuring LV ejection efficiency, the researchers instead studied the repair of the infarcted cardiac tissue. Three months after intracoronary BMSC implantation transplanted patients showed a significant improvement in the repair of damaged cardiac tissue through a reduction in infarct size (Figure 8.8) (Yousef et al., 2009).

Neuropathies and Neurodegenerative Diseases

Perhaps the area that has received the most attention with respect to the use and application of stem cells to address pathology has been in neuropathy and neurodegenerative disease. Potential areas of treatment include brain damage caused by stroke, Parkinson's and Alzheimer's disease, and spinal cord injury (SCI), just to name a few. The following sections describe actual patient trials in each of these areas using stem cell-based therapeutics platforms.

Spinal Cord Injury **Spinal cord injury (SCI)** most often results in considerable loss of nervous tissue and thus corresponding loss of both motor and sensory function. The degree of functional loss is dependent upon the extent of the injury. Unlike the peripheral nervous system, the central nervous system (CNS), specifically the spinal cord, does not allow for repair once damaged. Yes, there is some spontaneous self-repair immediately after injury in some cases, yet the damage done is usually permanent. To date there has been no therapeutic strategy developed which restores spinal cord functionality sufficiently. There are a

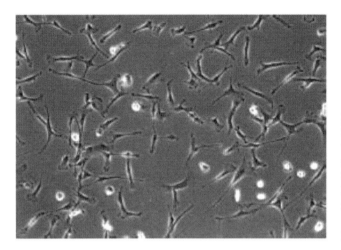

Figure 8.9 Brightfield microscopic image of multipotent stem cells isolated from human umbilical cord blood in preparation for patient spinal cord injury treatment. The cells exhibit a fibroblastic and spindle-shaped morphology and have the potency to differentiate into various neuronal lineages. (Image courtesy K-S Kang and *Cytotherapy* (Kang et al., 2005); reprinted with permission.)

number of reasons why SCI is irreversible. In many cases the injury damages the cell bodies or neuronal processes which often results in cell death. These cells are not typically replaced. In addition, SCI often results in axonal damage, which is the root cause of disability, including voluntary movement, tactile sensibility, and chronic pain. Stem cells show promise in the possible replacement of dead or damaged neurons and perhaps even the repair of damaged axons and encouraging results of SCI repair using stem cells have been demonstrated in preclinical animal models. Specifically, some stem cell types can be differentiated into glial cells or neurons that would allow for at least partial restoration of nervous system functionality if these cells were successfully transplanted and differentiated accordingly. In addition, the presence of undifferentiated stem cells themselves at the site of injury may provide neuroprotective or regeneration-inducing effects on local endogenous cells. Stem cell researchers have focused a great deal of effort at attempting to correct some of the damage done in cases of SCI. Case Study 8.2 outlines some success in this area in a patient with SCI. In this study, South Korean researchers isolated multipotent MSCs from umbilical cord blood (UCB), expanded these cells in tissue culture and prepped them for transplant and directly injected them into the injured spinal cord (Figure 8.9) (Kang et al., 2005).

SCI is not the only anomaly for which MSCs have been tested as a therapeutic platform. Over 120 active clinical trials have been conducted using MSCs to treat various disorders that have been conducted from Phase 1 through Phase III (Figure 8.11). The vast majority of these trials are in Phase I, Phase II or a combination of the two phases with a gravitation toward either bone and cartilage or autoimmune disease treatment due to the differentiation potential of MSCs toward lineages that support these systems (Trounson et al., 2011).

Brain Damage Stroke is defined as the sudden death of brain cells due to inadequate blood flow. Loss of brain cells is indiscriminate and includes the death of both neurons and oligodendrocytes. It often results in significant brain damage affecting both cognition and motility. Stroke is the leading cause of death and disability worldwide; therefore, new avenues of therapy to treat damage caused by stroke are of paramount importance. The ability to replace damaged or lost brain cells, including glial cells and neurons, would be of considerable therapeutic benefit. It is now known that certain regions of the adult brain

Case Study 8.2: A 37-year-old spinal cord-injured female patient, transplanted of multipotent stem cells from human UC blood, with improved sensory perception and mobility, both functionally and morphologically: A case study.

K.-S. Kang et al.

A 37-year-old SCI paraplegic patient exhibiting multiple compression fractures of the 11th and 12th thoracic vertebrae resulting in a severed spinal cord was treated with multipotent MSCs isolated from human UCB. These cells are known to differentiate into a variety of neuronal lineages and were confirmed to have this potential prior to administration. MSCs were harvested from UCB and isolated via the aggregating reagent Ficoll-Paque PLUS (Amersham Bioscience). Cells were expanded in tissue culture in preparation for injection. Patient treatment involved a laminectomy and direct injection of 1 million MSCs into the injured dura of the spinal cord. An additional 1 million MSCs were injected diffusely into the intradural and extradural spaces of the injured area of the spinal cord. On post-operative day (POD) 7, motor activity was observed in the lumbar paravertebral and hip muscles. By POD 15, the patient could move his/her hips and feet and exhibited skin sensitivity in the hip region. By POD 25, the patient's feet responded to stimulation. Spinal cord repair was clearly visible by magnetic resonance imaging (MRI) (Figure 8.10) (Kang et al., 2005).

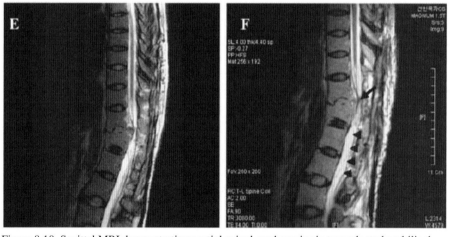

Figure 8.10 Sagittal MRI demonstrating partial spinal cord repair via transplanted umbilical cord-derived multipotent stem cells. (Left) Pre-treatment. (Right) Post-treatment. The arrow and arrowheads denote a thickening of the injured spinal cord following MSC injection and recovery. (Figures courtesy K-S Kang and *Cytotherapy* (Kang et al., 2005); reprinted with permission.)

are capable of neuronal lineage regeneration, a.k.a. neurogenesis. Neurogenesis has in fact been observed in certain regions of the brain including the dentate nucleus of the hippocampus as well as the subventricular zone (SVZ) following stroke. This fact yields credence to the possibility of ectopic stem cell-based repair of brain damage caused by stroke. What stem cell types should be used to treat brain damage caused by stroke? Stem

MSC clinical trials by disease classification (*n* = 123)

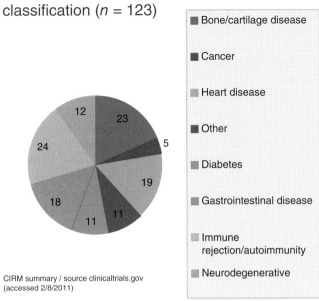

Legend:
- Bone/cartilage disease
- Cancer
- Heart disease
- Other
- Diabetes
- Gastrointestinal disease
- Immune rejection/autoimmunity
- Neurodegenerative

CIRM summary / source clinicaltrials.gov (accessed 2/8/2011)

Figure 8.11 Breakdown of the number of clinical trials conducted using mesenchymal stem cells as a therapeutic platform. (Figure courtesy Alan Trounson and *BMSC Medicine* (Trounson et al., 2011); reprinted with permission.)

cells derived from a vast array of sources, from embryonic to adult, have been considered as candidates for stroke treatment. This is due to the fact that each of these lineages ultimately has the capacity to efficiently generate terminally differentiated neuronal lineages such as glial cells and neurons needed to restore cognitive and locomotor capacities (Figure 8.12).

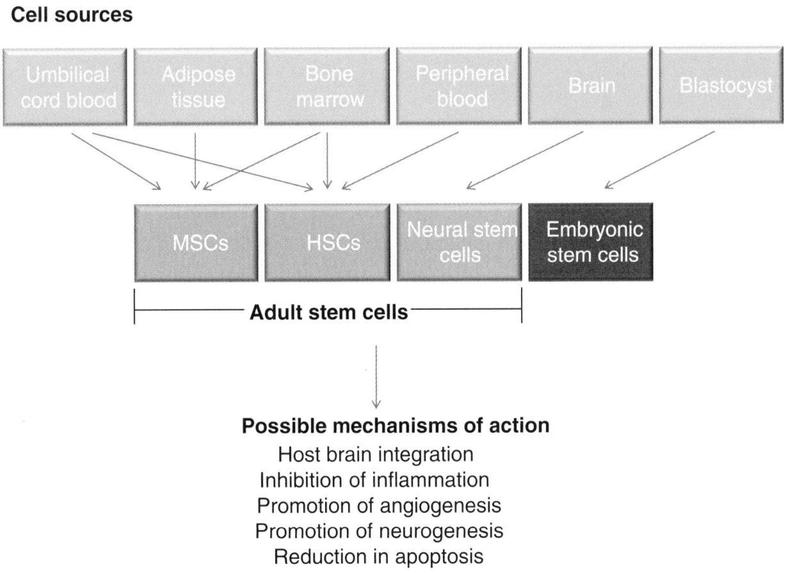

Cell sources

| Umbilical cord blood | Adipose tissue | Bone marrow | Peripheral blood | Brain | Blastocyst |

| MSCs | HSCs | Neural stem cells | Embryonic stem cells |

├──────── **Adult stem cells** ────────┤

Possible mechanisms of action
Host brain integration
Inhibition of inflammation
Promotion of angiogenesis
Promotion of neurogenesis
Reduction in apoptosis

Figure 8.12 Various stem cell sources and types used in clinical trials for the treatment of stroke-related brain damage. See text for details.

Numerous different types of stem cells have been shown in preclinical studies to reverse some of the effects of stroke-induced brain damage. In order to choose the appropriate cell type, both efficacy as well as availability should be taken into account. Allogeneic cell sources currently offer the greatest advantage with respect to availability as they exist broadly and are not derived from the patient's own tissue. As described below, allogeneic stem cell sources can be effective in the treatment of brain damage caused by stroke.

A number of clinical trials have now been conducted utilizing stem cell-based cell replacement platforms to treat brain damage resulting from stroke. In 2013, Glasgow University Professor Keith Muir conducted a small clinical trial involving nine patients aged from 60 to 80 years old, who exhibited brain damage due to stroke. In the study, stem cells isolated 10 years prior from fetal nerve tissue samples were injected directly into the damaged brain tissue. Thus far, over half of the patients have exhibited improvements in neurological function and none of the patients have experienced adverse side effects. Improvements included patient's renewed ability to move fingers and independent upright motility and balance where previously assistance was needed. These findings are striking given that all of the stroke victims in the study had experienced the stroke between 6 months to 5 years prior to treatment—it is generally accepted that any recovery from damage must occur within the first 6 months of the stroke.

A second example of efforts at stem cell-based recovery from ischemic stroke in patients involved the transplantation of autologous MSCs, specifically a research team led by Osamu Honmou in the Department of Neural Repair and Therapeutics at Sapporo Medical University in Japan. The group studied 12 patients in an unblinded clinical trial with ischemic grey and white matter, assessing the safety and feasibility of using MSCs to treat stroke. Patients were chosen who had stroke onset within 6 months prior to treatment and were exhibiting severely disabling defects. Autologous human MSCs were obtained from the posterior iliac crest and cultured for expansion for a period of approximately 3 weeks, harvested and analyzed via flow cytometry for the presence of the markers CD24, CD34, and CD105. Cells were stored at $-150°C$ prior to transplantation. Cryopreserved cells were thawed at patient bedsides on the day of transplant and delivered intravenously. Magnetic resonance angiography (MRA) and MRI were performed, and neurological status was evaluated for each patient using the **modified Rankin score**, which is a scale used for measuring the degrees of disability or dependence of individuals in daily activities which have suffered stroke or other forms of neurological damage. In addition, neurological scores were assessed on the **National Institutes of Health Stroke Scale (NIHSS)**, which is a measurement of impairment caused by stroke. In case study no. 3 of the trial, a 52-year-old male suffering from atherosclerotic stroke as a result of right internal carotid artery occlusion exhibited a significant lesion in the right frontal lobe of the brain. Physically the patient also exhibited paralysis of the left side of his body. Forty-three days after stroke, the patient received an infusion of 1.6×10^8 autologous MSCs. As Figure 8.13 illustrates, the delivery of MSCs reduced lesion size and improved NIHSS scores. Specifically, mean lesion volumes were reduced greater than 20% and NIHSS scores improved 60% following stem cell treatment. (Honmou et al., 2011). Although this degree of improvement was not observed in all 12 patients studied, it is indicative that MSCs may be a cell line of choice for the treatment of neurological disorder such as lesions caused by stroke. The fact that these cells can simply be administered intravenously is a considerable advantage over surgical cell or tissue transplantation methods.

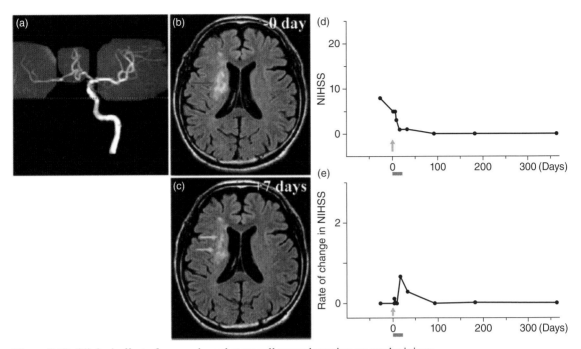

Figure 8.13 Clinical effect of mesenchymal stem cell transplantation on stroke injury. (a) Magnetic resonance angiography (MRA) showing occluded right internal carotid artery. (b) Magnetic resonance imaging (MRI) immediately prior to cell injection. The red arrow denotes the infarcted region of the brain. (c) MRI 7 days-post injection. The blue arrows reveal dissipation in the lesion (less signal intensity). (d) NIHSS score is reduced up to 1 year post cell injection. (e) The rate of NIHHS score change. (*Source:* Honmou et al., 2011. Reproduced with permission from Oxford University Press.)

MECHANISM OF ACTION OF STEM CELL BRAIN DAMAGE REPAIR
Given the great deal of focus on stem cell-based repair of stroke-induced brain damage, a separate section of this text focused on possible repairs mechanisms is worthy. Although not yet confirmed in the patient, there are a number of possible modes by which stem cells exert their effects to repair brain damage caused by stroke. These include:

• Generation of new neurons and neuronal circuitry. It is possible that stem cells grafted into the brain form new neurons and new neuronal connections, although to date there is very little evidence to support this hypothesis.
• Decreased apoptosis. Stroke results in an inadequate supply of blood and therefore nutrients, including growth factors, to the brain, causing both necrotic and apoptotic cell death. Stem cells transplanted at local sites of damage may play a neuroprotective role through both the upregulation and secretion of much needed growth factors such as vascular endothelial growth factor (VEGF), brain-derived neurotrophic factor (BDNF), fibroblast growth factor (FGF), nerve growth factor (NGF), etc. The supply of these factors may rescue proapoptotic cells. Some preclinical research models support this theory indicating a reduction in apoptosis in areas of direct cell injection that also showed extensive repair and some neurological recovery.

- Promotion of angiogenesis. As stroke damage is due to restricted blood flow to the brain, any recovery strategies should be focused not only on the restoration of neuronal function but also local blood supply. Some stem cells have been shown to promote **angiogenesis**, which is defined as the production and growth of new blood vessels, at local sites of engraftment in the brain. These include neural stem cells, CD34$^+$ cells, and MSCs. It is speculated that the primary mechanism of action of stem cell stimulated angiogenesis is through the secretion of VEGF.

Parkinson's Disease **Parkinson's disease** is a progressive nervous system disorder resulting from destruction of the brain cells producing dopamine and characterized by muscular tremor, slowing of locomotion, partial facial paralysis, and general weakness. Some medications such as Levodopa$^{(R)}$ provide for a management of the symptoms of the disease, but do not address the root causes. There have also been some successes and clinical benefits resulting from the transplant of fetal mesencephalic tissue directly into the striatum, but these tissues may often be rejected by the patient's immune system. Thus new strategies that are focused on a cure rather than symptom management would be of great benefit for Parkinson's patients. As Case Study 8.3 illustrates, a research team led by Michel F. Levesque at the UCLA School of Medicine and Brain Research Institute demonstrated the benefits of microinjected autologous adult human neural stem cells and differentiated neurons in Parkinson's disease patients.

Autoimmune Disorders Autoimmune disorders are the result of the body's immune system attacking its own cells and tissues. As the immune system is inherently derived from the hematopoietic system in general and HSCs in particular, the application of MSCs and HSCs to address autoimmune diseases may show great promise. The following sections illustrate several successful clinical trials using autologous stem cells to treat sclerotic and arthritic disease.

MULTIPLE SCLEROSIS (MS) AND AMYOTROPHIC LATERAL SCLEROSIS (ALS)
Multiple sclerosis (MS) is defined as a chronic demyelinating autoimmune disease of the CNS. It is driven by autoimmune dysregulation resulting in inflammation that ultimately drives demyelination and neuronal axon damage. **Amyotrophic lateral sclerosis (ALS)** is a neurodegenerative disease specifically affecting brain motor neurons and leading to respiratory and limb weakness. Neither disorder has been effectively treated with only marginal, limited success observed in the use of immunosuppressive agents to treat MS. It has been speculated that the introduction of multipotent stem cells into the CNS may represent a new avenue of treatment. Certain stem cell lineages, for example, have been shown in animal models to migrate to damaged neural regions and secrete neuroprotective trophic factors or differentiate directly into viable neuronal lineages. Researchers at the Agnes Ginges Center for Neurogenetics and Multiple Sclerosis Center and Department of Neurology, Hadassah-Hebrew University Hospital, Ein Karem, Jerusalem choose to use MSCs as the lineage of choice for a clinical trial involving 15 patients with MS and 19 with ALS. Interestingly, MSCs predominantly give rise to osteocytes, adipocytes, and chondrocytes, not neuronal lineages, yet they do possess immunomodulating properties, including the ability to suppress the activity of certain immune cell populations. In addition, if cultured properly, in some cases MSCs may differentiate into neural- and glial-like cells *in vitro*

Case Study 8.3: Therapeutic microinjection of autologous adult human neural stem cells and differentiated neurons for Parkinson's Disease: Five-Year Post-Operative Outcome

Michel F. Levesque, Toomas Neuman, and Michael Rezak

A 57-year-old Parkinson's disease patient was treated for the disorder via the direct injection of both autologous neural stem cells and differentiated neurons isolated from the patient's own cortical and subcortical tissue samples. Neural stem cells were isolated, expanded, and differentiated in tissue culture for 9 months prior to transplantation. Cell suspensions confirmed as having populations of differentiated dopaminergic and GABAergic neurons were unilaterally injected into the striatum of the patient. Clinical benefits of the transplantation were assessed over the next 3 years using the **Unified Parkinson's Disease Rating Scale (UPDRS)**, which measures nervous system motor functionality, scaling such symptoms as tremors. A low score in this instance is suggestive of improvements in disease treatment. Transplant success was compared to the use of a previously inserted thalamic stimulator used to control tremors. A significant improvement of on average 80% in the UPDRS score was observed through 36 months post treatment with cells with a rapid loss of improvement observed at 4 and 5 years post treatment. This phenotype was evident either with or without thalamic stimulation. The rapid loss in improvement after 4 years is attributed to the death of the transplanted cells. (Figure 8.14) (Levesque et al., 2009).

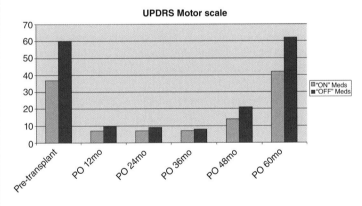

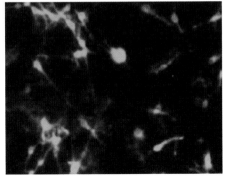

Figure 8.14 Differentiated, patient-derived neurons and clinical improvement in a Parkinsons disease patient based on neuronal transplantation. (Left) Fluorescence microscopy of beta tubulin-positive neurons differentiated from a Parkinson's disease patient-derived neural stem cells. (Right) 5-year time-course of UPDRS measured motor activity following autologous cell transplant. (Figures courtesy Michel F. Levesque and the *Open Stem Cell Journal* (Levesque et al., 2009); reprinted with permission.)

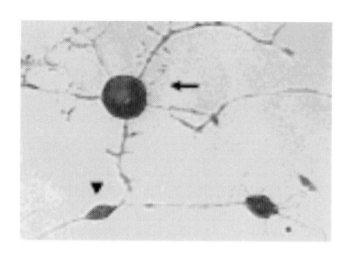

Figure 8.15 Neurons derived from rat mesenchymal stem cells. The cells express the neuronal marker NSE (brown staining). (Figure courtesy Dale Woodbury and the *Journal of Neuroscience Research* (Woodbury et al., 2000); reprinted with permission.)

(Figure 8.15). MSCs have also been shown in animal models to migrate into the CNS and subsequently promote BDNF production and limited proliferation of oligodendrocyte precursors.

Previous clinical trials have demonstrated the feasibility and safety of MSCs and have demonstrated some efficacy in the treatment of disorders unrelated to ALS or MLS. These include myocardial infarction, SCI (see Case Study 8.2 above), and osteogenesis imperfecta, which is a congenital bone connective tissue disorder. In this study focused on MS, roughly 60 million autologous MSCs isolated and cultured from bone marrow aspirates were administered either **intrathecally**, defined as direct injection under the arachnoid membrane of the brain or spinal cord, or intravenously into patients. Clinical effects in MS patients were measured on the **Kurtzke Expanded Disability Status Scale**, which is a method based on a functional system score to quantify disability in MS patients. Time course studies of MS patients post treatment revealed a gradual decrease in the ECSS Score indicating improvement in ambulatory function (Figure 8.16) (Karussis et al., 2010). Similar results, although to a lesser degree, were seen in

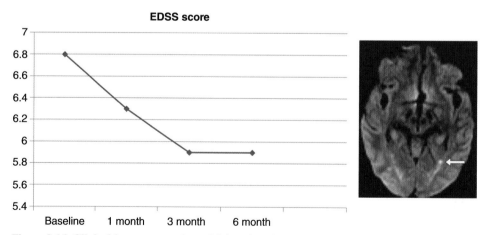

Figure 8.16 Clinical improvement in multiple sclerosis patients treated with mesenchymal stem cells. (Left) Graph of the Expanded Disability Status Scale as a function of time after treatment. (Right) Magnetic resonance imaging demonstrating the presence of transplanted MSCs in the brain (arrow). (Figures courtesy Dimitrios Karussis and *Archives of Neurology* (Karussis et al., 2010); reprinted with permission.)

ALS patients. Thus, this clinical trial suggests that both ALS and MS may be treated via stem cell therapy and that MSCs show promise as the lineage of choice in this respect.

JUVENILE IDIOPATHIC ARTHRITIS

Juvenile idiopathic arthritis (JIA) is a group of illnesses present in children exhibiting the common symptoms of chronic joint inflammation. Current therapies for JIA involve the use of inhibitors of proinflammatory cytokines and their corresponding receptors. Yet in a large population of children with JIA, the disease is refractory to this treatment and becomes progressive and debilitating over time. In the most severe cases, normal growth is retarded, joints become deformed and patients become fully disabled. In addition, the inherent toxicity of antirheumatic medications becomes cumulative, thus further compromising the health of the patient. Therefore new treatment strategies that go beyond near-term inflammatory suppression would be of great value. Studies on the molecular origins of JIA have suggested that dysregulation of T cell function may be a factor in promoting the disease. Therefore, if endogenous T cells were to be depleted from the patient, JIA symptoms might be reduced. A wealth of data from animal models has suggested that either allogeneic or autologous T cell-depleted transplantation of bone marrow-derived stem cells results in the effective remission of arthritis, with the transplanted stem cells ultimately reconstituting the depleted T cell population with new, properly functioning T cells. These findings led to a Phase II clinical trial conducted by researchers at Leiden University Medical Center in Leiden, Netherlands. Twenty-two patients with JIA refractory to conventional treatments were given a combination of immunosuppressive medications and T cell-depleted autologous stem cell transplants (ASCT). In this study, HSCs isolated from bone marrow aspirates of patients were depleted of T cells by negative selection via the application of immunorosette-sedimentation or CD34+ selection. The depleted T cell graft was transplanted into 22 patients and both overall survival as well as disease-free survival was assessed over a period of 100 months. Immune system recovery was noted as delayed as expected and took greater than 6 months to overcome innate deficiencies. This led to a number of complications including viral infections and death in 2 patients. Yet, 8 of the 20 patients undergoing the clinical trial exhibited complete clinical remission. Seven patients

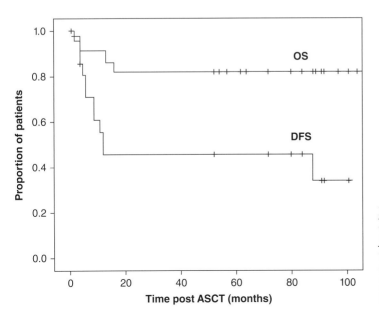

Figure 8.17 Kaplan Meier curve demonstrating the improvement in juvenile idiopathic arthritis patients following autologous stem cell transplant. (*Source:* Brinkman et al., 2007. Reproduced with permission from Wiley.)

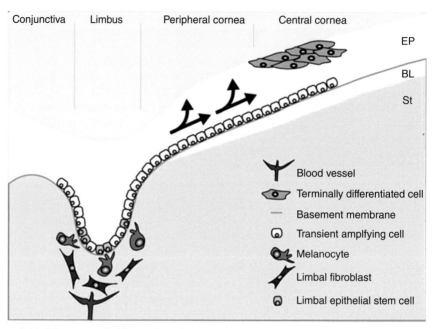

Figure 8.18 Diagrammatic illustration of corneal epithelial stem cells and their contribution to the cornea. (*Source:* Secker & Daniels, 2009. Licensed under Creative Commons Attribution 3.0 Unported License.)

demonstrated partial remission and 5 relapsed. Overall survival of patients from the clinical trial was 82% and more than 1/3 of patients survived disease-free for more than 100 months after ASCT (Figure 8.17) (Brinkman, de et al., 2007). This clinical trial illustrates the power of a combination of immunosuppressive therapies with active T cell depletion and T cell population reconstitution via ASCT to treat JIA.

Corneal Defects The **cornea** is a dome-shaped structure covering the front of the eye. Its clarity and surface properties are crucial for the proper transmission and focusing of light onto the retina at the back of the eye. Diseases of the cornea are the second leading cause of worldwide blindness. One particular corneal disorder involves the deficiency of **limbal stem cells (LSCs)**, a.k.a. **corneal epithelial stem cells (CESCs)**, these cells are normally present in the basal layer of the corneal epithelium and act to replenish corneal epithelial cells which are continuously lost from the surface of the cornea (Figure 8.18).

   Limbal stem cell deficiency (LSCD) is a disorder caused by trauma in which the limbal stem cells become deficient and thus the corneal epithelium cannot be renewed. The result is chronic inflammation, scarring, and loss of vision due to corneal over-vascularization and opacity. Previous attempts at rectifying this disorder have been based on corneal transplants, which do not address the root cause—lack of LSCs. A research team lead by Majlinda Lako at the North East Institute for Stem Cell Research in Newcastle, United Kingdom sought to take a different approach at correcting LSCD through the direct transplantation of LSCs isolated from biopsies of the limbus. In this study, eight patients with unilateral total LSCD received transplants of *ex vivo* expanded autologous LSCs. Specifically, cells were isolated from limbal biopsies and expanded in tissue culture as explants on human amniotic membrane (HAM). A combination of HAM and explanted tissue was transplanted within 12–14 days of the initial biopsy. Three months following transplant, patients were assessed for clinical outcome. Each patient exhibited stable reconstruction of

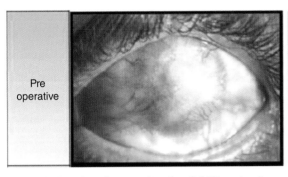
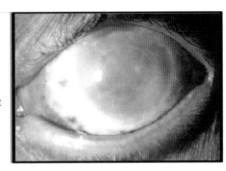

Figure 8.19 Microphotographs of an LSCD patient's eye before and after transplantation treatment with limbal stem cells. (*Source:* Kolli et al., 2010. Reproduced with permission from Wiley.)

the cornea and improvements in visual acuity as well as reduction in eye pain. This was partially due to a decrease in inflammation, neovascularization, and opacity post treatment (Figure 8.19) (Kolli et al., 2010). Therefore, LSCs appear to be an attractive platform for the treatment of injury-induced LSCD.

Hematopoietic Disorders

Hematopoietic disorders are perhaps some of the most amenable anomalies for stem cell-based therapy, given ease of access to the hematopoietic system. In fact, HSC transplantation has become routine for certain types of blood cell disorders. The following examples illustrate the utility of HSC transplantation for treating and even curing hematopoietic disorders.

Sickle Cell Disease **Sickle cell disease** is an inherited blood disorder in which red blood cells exhibit an abnormal rigid, sickle shape. It is the result of a single point mutation in the β-globin chain of hemoglobin and drastically reduces the red blood cell's elasticity. Thus sickle-shaped red blood cells do not conform as needed to the networking of arteries, veins, and capillaries of the circulatory system. This results in red blood cell blockages throughout the circulatory system, primarily at the level of the capillaries, thus leading to occlusion, ischemia, and ultimately death (Figure 8.20).

Sickle cell disease affects millions of individuals worldwide. In the United States, about 1 out of every 500 African–American newborns carries the mutation and is affected by the sickle cell disease. Given its red blood cell localization and restricted phenotype, it is an ideal candidate disorder for stem cell-based therapy. The replacement of diseased red blood cells with stem cell-derived healthy red blood cells has been at the forefront of research in the treatment of this disorder. In fact, it is generally accepted that the only realistic method of curing sickle cell disease is by allogeneic HSC transplant. As of December 2009, approximately 200 children have undergone the procedure. After transplant the donor's hematopoietic cells replace those of the recipient completely in most patients, while some exhibit a mixture of both donor and recipient cells in the blood (**chimerism**). In either case the presence of the donor cells is sufficient to reverse the sickle cell disease phenotype. In the most recent clinical trials a disease-free survival rate of 95% was observed. In many cases, however, **graft vs. host disease (GVHD)** morbidity and mortality were high. GVHD is a complication that can occur following a stem cell transplant in which the newly transplanted cells attack the recipient's body. Researchers at the Molecular and Clinical Hematology Branch of the National Institute of Diabetes and Digestive and Kidney Diseases sought to address GVHD in adult sickle cell patients caused by allogeneic donor T cells

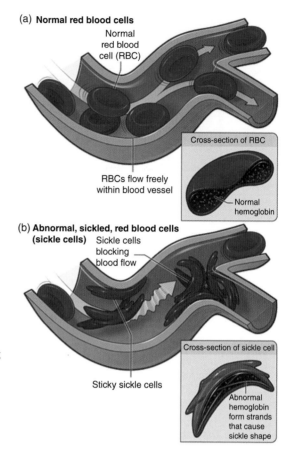

(a) **Normal red blood cells**

Normal red blood cell (RBC)

Cross-section of RBC

RBCs flow freely within blood vessel

Normal hemoglobin

(b) **Abnormal, sickled, red blood cells (sickle cells)** Sickle cells blocking blood flow

Cross-section of sickle cell

Sticky sickle cells

Abnormal hemoglobin form strands that cause sickle shape

Figure 8.20 Diagrammatic illustration of the phenotype resulting from sickle cell disease. (a) Normal red blood cells and blood flow. (b) Sickle cells impeding blood flow into capillaries. (Diagram courtesy Wikimedia Commons; reprinted with permission.)

through the manipulation of a novel mechanism for inducing immunologic tolerance. In a Phase 1–2 study, the researchers applied low-dose radiation plus the drug sirolimus (rapamycin) that inhibits T cell proliferation to 10 adult patients with severe sickle cell disease. Activated T cells unable to proliferate become **anergic**, which is defined as the inability to mount a complete immune response to a foreign antigen. As such, sirolimus was proposed to promote T-cell tolerance. The combination of low-dose radiation with sirolimus allowed for the development of high-level chimerism in patients. In the study, patients aged from 16 to 45 years with severe sickle cell disease received a transplant of CD34+ allogeneic peripheral blood stem cells (PBSCs) mobilized in donors through the administration of **granulocyte colony stimulating factor (G-CSF)**, which is a glycoprotein that stimulates the bone marrow to produce both granulocytes and stem cells and release them into the bloodstream. A minimum of 10 million cells per kilogram of recipient's body weight were isolated and cryopreserved prior to transplantation. Patients were conditioned in preparation for transplantation via the administration of **alemtuzumab**, which is a humanized monoclonal antibody that functions to deplete T cells and B cells from the patient's immune system, thus allowing the recipient to tolerate the introduction of foreign cells into the blood stream. It should be noted that alemtuzumab does not affect HSC maturation and has been used previously to treat GVHD. One year following stem cell transplantation, hemoglobin levels were observed to be on average 30% higher than before the procedure, indicating the presence of normal red blood cells and a reversal of the sickle cell phenotype (Figure 8.21)

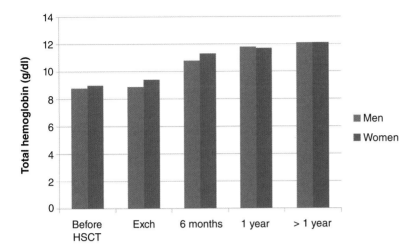

Figure 8.21 Graph of total hemoglobin before, during, and after hematopoietic stem cell transplantation to treat sickle cell disease. Hemoglobin counts significantly increased over the course of time. (Graph courtesy Matthew M. Hsieh and the *New England Journal of Medicine* (Hsieh et al., 2009); reprinted with permission.)

(Hsieh et al., 2009). These studies reveal a striking real-world application of stem cell-based transplantation therapeutics that results in a complete reversal and cure of a devastating and deadly disorder.

Wiskott–Aldrich Syndrome **Wiskott–Aldrich syndrome (WAS)** is an X-linked inherited disorder involving deficiency in the immune system characterized by autoimmunity, eczema, and recurrent infections. In addition, WAS patients exhibit thrombocytopenia, which is a decrease in the number of platelets present in blood. WAS affects roughly 4 out of 1 million live births and is the result of a mutation in the WAS gene, for which the gene product, the **WAS protein (WASP)**, functions to regulate the polymerization of actin in hematopoietic cells. This polymerization defect results in multiple dysfunctions in daughter T and B cells and impaired migratory response by leukocytes. In the most severe cases of WAS, death results most often due to infection or excessive bleeding. As of 2013, the only curative therapeutic strategy has been the allogeneic transplantation of HSCs, yet immunorejection of these cells has often resulted in death or severe complications. Thus new stem cell-based therapeutic strategies are needed to overcome these complications. In 2010 a research team lead by Christoph Klein in the Department of Pediatric Hematology-Oncology at Hannover Medical School in Germany sought to devise a system for the use of the patients own HSCs to treat WAS. The Klein team designed genetically modified HSCs in which the WAS gene was expressed in CD34$^+$ patient-specific HSCs through transduction with a wild-type WASP retroviral expression vector. Cells were isolated either directly from the bone marrow or via **leukapheresis**, which is a laboratory procedure implemented to isolate white blood cells from whole blood. These cells were subsequently transduced with the WASP retroviral expression vector and reinfused into patients 4 days post harvesting. Two patients with WAS, yet exhibiting different symptoms and complications from the disorder, were given transplants of autologous, genetically-modified CD34$^+$ HSCs and assessed for therapeutic benefits. The presence of normal WASP was first characterized and determined to be stable at 7–28% following transplantation. Strikingly, the T-cell proliferative response (stimulated by CD3) was completely normalized 2 years after transplant and eczema symptoms disappeared (Figure 8.22) (Boztug et al., 2010). These studies illustrate a novel autologous gene therapy-based stem cell therapeutics strategy for the treatment of WAS.

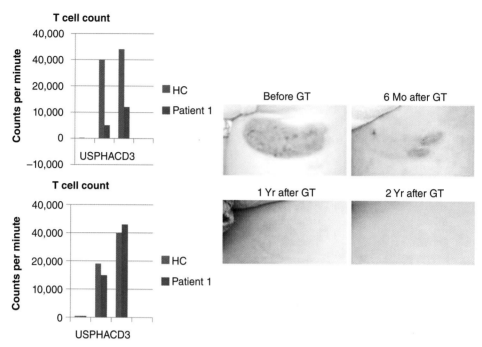

Figure 8.22 Analysis of T cell counts and eczema improvement following autologous stem cell-based gene therapy in Wiskott–Aldrich syndrome patients. (Left) Improvement in T cell numbers following cell transplantation and gene therapy (GT). (Right) Eczema improvement following cell transplantation and gene therapy. US – unstimulated; HC – healthy control patient; PHA – phytohemaglutinin. (Figures and images courtesy Kaan Boztug and the *New England Journal of Medicine* (Boztug et al., 2010); reprinted with permission.)

Focus Box 8.1: Harold Weintraub and the discovery of MyoD

 Dr. Harold (Hal) Weintraub (1946–1995) was a research biologist at the Fred Hutchinson Cancer Research Center in Seattle, Washington from 1978 until his death. He obtained his MD and PhD degrees from the University of Pennsylvania. His contributions to the study of both cell fate and gene function are enormous. He was the first to utilize anti-sense RNA to study gene function and discovered the transcription factor MyoD, which has the capacity to drive the entire muscle development program when introduced into non-muscle cell types. It is his discovery and study of MyoD that resulted in a revolution in research on the transcriptional control of cell fate.

(*Source*: Courtesy of Fred Hutchinson Cancer Research Center.)

Cancer

Stem cells have also been used in clinical trials for the treatment of cancer. In many instances PBSCs, or HSCs have been used in combination with chemotherapy for the treatment of high-risk breast cancer, yet this therapeutic strategy is controversial given the conflicting results regarding efficacy. In addition, recent data gleaned from clinical trials involving high-dose chemotherapy and bone marrow transplantation suggest only limited success. Thus, new strategies for high-risk breast cancer treatment are needed. For example, a combination of

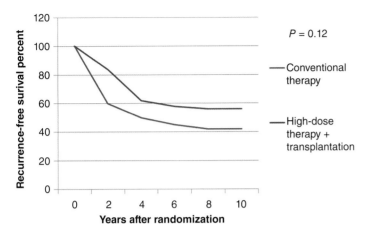

Figure 8.23 Graph of recurrence-free (relapse) survival of high-risk breast cancer patients treated with chemotherapy and stem cells. (Solid line) Conventional therapy with no stem cell transplantation. (Dotted line) High-dose chemotherapy combined with stem cell transplantation. (Graph courtesy Martin S. Tallman and the *New England Journal of Medicine* (Tallman et al., 2003); reprinted with permission.)

stem cell-based immunotherapy following chemotherapy may enhance tumor cell killing. Researchers in the Division of Hematology at Northwestern University and Feinberg School of Medicine in Chicago performed a clinical trial to test just this concept in 540 female patients with primary breast cancer. In this study, patients received cycled chemotherapy with cyclophosphamide doxorubicin, which is a chemotherapeutic agent designed to stop cancer cells from dividing. In addition, the patients received autologous HSC transplants. HSCs used were PBSCs derived from the individual patient's bone marrow. Approximately 100 million cells per kilogram of body weight were transplanted. Some patients also received stem cells isolated from peripheral blood. The researchers studied long-term disease-free survival, overall survival and recurrence-free survival among 511 patients participating in the study over a period of 10 years. It was concluded that the combination of high-dose chemotherapy with stem cell transplantation resulted in an increased recurrence-free survival rate (Figure 8.23) (Tallman et al., 2003). These studies suggest that combination therapies involving both conventional chemotherapy coupled with stem cell transplantation may be effective at reducing disease relapse and increasing survival rates among high-risk breast cancer patients.

Muscular Dystrophy

Muscle loss may occur by injury or through the effects of degenerative diseases such as **muscular dystrophy (MD)**, which is defined as a group of genetic diseases resulting in damage to muscle fibers and related weakness. **Duchenne muscular dystrophy (DMD)** is a severe form of the disease that occurs in boys and is often fatal. It is X-linked and is the result of mutations in the gene **dystrophin**. In patients with MD or DMD, the muscle tissue has become disorganized and the concentration of the protein dystrophin, important in the functioning of muscle, is greatly reduced (Figure 8.24).

Unlike many other organ systems, skeletal muscle has the capacity to regenerate new muscle fibers and thus at least partially repair the injured musculature in some cases. The generation of new muscle tissue is dependent upon the presence of **satellite cells**, which are precursor cells capable of **myogenesis**, the generation of muscle cells, fibers, and tissue. Satellite cells are most often located proximally between the basal lamina and the muscle fiber membrane and act to replenish skeletal muscle locally upon injury or degenerative disease. Over the past 25 years, a vast transcriptional control network has been outlined in tissue culture and in functional genomics animal models such as knockout mice indicating that numerous transcription factors tightly regulate the myogenic phenotype. Of these, perhaps

Muscular dystrophy

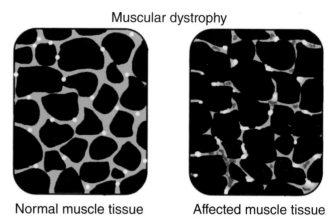

Figure 8.24 Fluorescence microscopy comparing the organization of muscle tissue and presence of dystrophin in normal and muscular dystrophy-based tissue biopsies. (Figure courtesy Wikimedia Commons; reprinted with permission.)

Normal muscle tissue   Affected muscle tissue

the discovery and characterization of the basic helix-loop-helix transcription factor **MyoD, a.k.a. myogenic differentiation factor 1,** has been the most intriguing, as it was the first myogenic factor discovered with the capacity to convert fibroblasts into myoblasts in tissue culture (Figure 8.25 and see Focus Box 8.1).

It is the inherent nature of satellite cells and their ability to generate muscle fibers that makes them amenable and valuable for clinical usage in this respect. For example, it was in 1990 that the first application of satellite cells to treat DMD in a clinical setting was conducted. A 9-year-old boy with DMD received a transplant of satellite cells. The transplant demonstrated that the

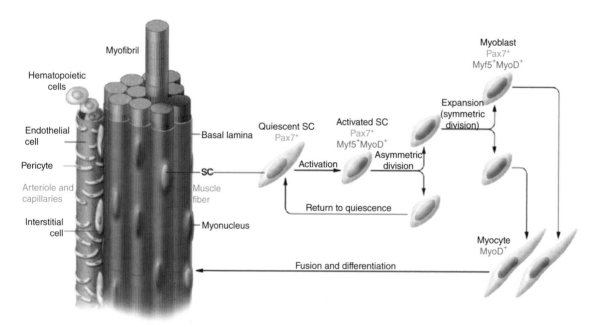

Figure 8.25 Diagrammatic illustration of muscle fiber anatomy in the context of satellite cells and regulatory proteins. Satellite cells result from the asymmetric division of myoblasts activated by the presence of the transcription factors Myf5 and MyoD. These cells reside between the basal lamina and muscle fibers in a quiescent state until extrinsic signaling events promote their division and differentiation into multinucleated muscle fibers. (Diagram courtesy Francesco Saverio Tedesco and the *Journal of Clinical Investigation* (Tedesco et al., 2010); reprinted with permission.)

cells were safe and that they resulted in the production of normal dystrophin protein in the boy's muscles. This study was followed by 11 other clinical trials also involving the injection of satellite cells intramuscularly. Yet, although these results are encouraging, limitations of this system exist. For example, allogeneic sources of satellite cells and myoblasts must be used due to the inherent dystrophin-deficient nature of the patient's own cells. This may result in immunorejection issues. Second, intramuscular injection results only in the presence of new satellite cells locally, thus widespread injection using large numbers of cells must be considered in order to significantly benefit the patient. Finally, rapid cell death of most injected satellite or myoblast cells has been reported, with only a small percentage of injected cells surviving long term. Much experimentation has now been focused on addressing these issues. For example, Helen Blau and colleagues in the Department of Pharmacology at Stanford University School of Medicine in Stanford, California conducted a Phase I clinical trial on 10 patients with DMD to study the effects of using immunosuppressive drugs in combination with myoblast transplantation. **Cyclosporine (CSA)** is an immunosuppressant drug widely used in organ transplant surgeries to prevent organ rejection by the patient. It suppresses the immune system by interfering with T-cell activity. Blau's team administered cyclosporine in combination with allogeneic myoblast transplantation in patients with DMD and assessed the success of the trial through an analysis of muscle force generation. Specifically, anterior tibial muscle strength was assessed using a force transducer. The researchers measured both the **maximum voluntary isometric contraction (MVC)**, which is a measure of strength typically reported in units of force such as Newtons, and tetanic force. **Tetanic force (TF)** is a measurement of the force required to result in a maximal stimulation of a motor unit by its corresponding motor neuron. In both cases, administration of cyclosporine significantly improved muscle strength. This effect was lost upon withdrawal of CS treatment (Figure 8.26) (Miller et al., 1997). Thus Blau's group has demonstrated in clinical trials that a combination of myoblast transplantation with immunosuppression therapy may benefit patients with DMD.

Liver Disorders

The liver is a crucial organ in the body that functions to both synthesize and metabolize endogenous and exogenous substrates such as hormones, drugs, and steroids. Interestingly, the liver is one of the few organs with a capacity for self-regeneration, yet below a critical level of **future liver remnant volume (FLRV)**, this regenerative capacity is not sufficient to regenerate the minimum amount of functioning liver tissue. A vast array of inherited and acquired liver disorders have been successfully treated via liver transplants, a.k.a. **orthotopic liver transplantation (OLT)**, yet a shortage of donors for this purpose has resulted in a crisis situation for those in need of liver therapy. In fact, approximately 15% of patients die while on the waiting list for liver transplants. Therefore techniques not involving OLT must be developed to address donor shortages and other complications such as immunorejection which may arise from liver transplant surgeries. One concept has been the introduction of $CD133^+/CD34^+$ HSCs into the liver. These cells are known to give rise to all lineages of blood cell differentiation, but may also contribute to liver cell proliferation by mechanisms that are not yet well understood. In addition, these cells have been postulated to transdifferentiate into both hepatocytes and bile duct cells. As such bone marrow-derived stem cells (BMSCs) may be a good cell transplantation therapeutic platform for increasing liver mass in patients that require the removal of liver tumors, for example. Although numerous clinical trials have now been conducted in this area, in 2005 a research team lead by Wolfram Trudo Knoefel in the Department of General Surgery at Heinrich-Heine University of Düsseldorf in Germany published a study with surprising results regarding

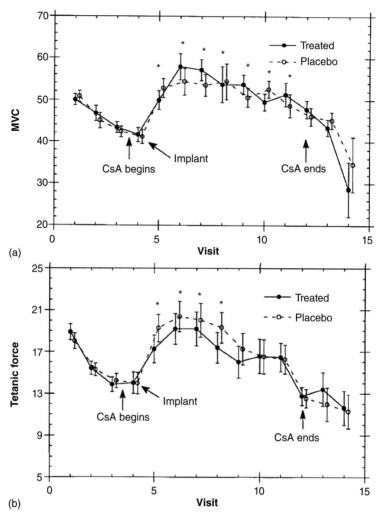

Figure 8.26   Myoblast transplantation and cyclosporine effects on muscle strength in Duchenne muscular dystrophy patients. (a) Graph of maximum voluntary isometric contraction (MVC) as a function of time and patient visits. (b) Graph of tetanic force (TF) as a function of time and visits. (*Source*: Miller et al., 1997. Reproduced with permission from Wiley.)

BMSC-based liver regeneration in patients. In this study patients with liver tumors who required a **hepatectomy**, defined as surgical removal of a portion of the liver, were treated with autologous CD133+ BMSCs with the goal of increasing liver remnant volume to compensate for liver tissue lost by surgical tumor removal. Specifically, CD133+ cells enriched from autologous bone marrow samples were implanted into the left-lateral portal branches with the goal of expanding liver tissue in preparation for tumor removal. Cells were isolated via aspiration from the posterior iliac crest and filtered to remove bone spicule. They were subsequently processed through cell selection to enrich for CD133+ populations. No culture or expansion was conducted prior to transplant. These cells were selectively administered into the portal branches of liver segments II and III (Figure 8.27).

Three weeks after stem cell transplantation but prior to removal of hepatic tumors, the researchers observed significant increases in the volumes of the left lateral ventricle lobe (Figure 8.28) (am Esch et al., 2005). These studies and other similar clinical trials have now

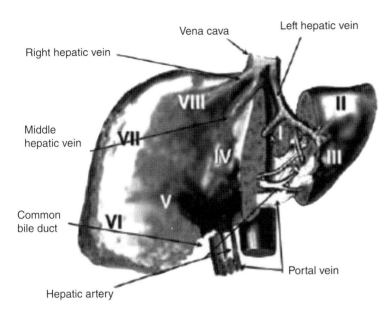

Figure 8.27 Diagrammatic illustration of the hepatic segments. CD133+ cells were introduced into segments II and III. (*Source*: am Esch et al., 2005. Reproduced with permission from John Wiley & Sons.)

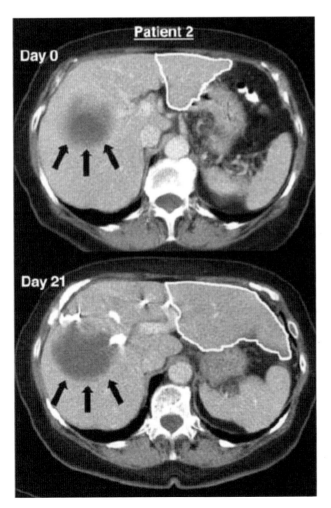

Figure 8.28 Transplantation of bone marrow derived CD133+ stem cells and improvement in liver function. Transverse computed tomography (CT) scans of a patient at day 0 and day 21 after introduction of CD133+ stem cells demonstrate an increase in the size of the left ventricular lobe (white outline). Arrows denote tumor growth. (*Source:* am Esch et al., 2005. Reproduced with permission from John Wiley & Sons.)

confirmed the utility of BMSCs as a cell transplantation platform for increasing liver volume and may become accepted as a standard method for liver regeneration and repair without the need for OLT.

VETERINARY APPLICATIONS

Stem cells may have important therapeutic benefits not only for humans but also for animals as well. In fact, stem cells have been actively used to treat various disorders in horses (Figure 8.29), dogs, and cats. Stem cells were even injected directly into the spine of a Bengal tiger to treat a lumbar fracture. Disorders currently treatable via stem cell transplantations in animals include:

- Myocardial infarction
- Muscular dystrophy
- Osteochondrosis
- Osteoarthritis
- Stroke
- Tendon and ligament damage

   Autologous MSCs are by far the most widely used cell type with respect to veterinary applications. They are typically isolated from adipose tissue or bone marrow and, as discussed above, are multipotent, exhibiting the capacity to differentiate into a wide range of cell types for the repair of cartilage, ligaments, tendons, bone, skeletal muscle, and even neural tissue. Isolates are most often cultured for expansion, purification, and characterization prior to transplant. It is generally accepted that the ease of MSC isolation from adipose tissue makes this source preferable to bone marrow aspirates. The following sections illustrate examples of clinical stem cell applications for different species.

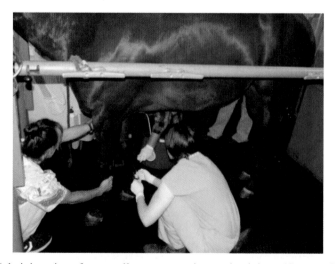

Figure 8.29 Administration of stem cells to treat equine tendon injury. (*Source:* Cornell University College of Veterinary Medicine. Courtesy of L. A. Fortier.)

Equine

Given the nature of their activity, racehorses are particularly susceptible to injuries involving damage to both ligaments and tendons. Most conventional therapies to treat these injuries are only marginally successful. In addition, the scar tissue formed upon healing most often results in stiffness and reduced flexure. More often than not these side effects end the competitiveness of the injured racehorse, forcing an early retirement. Yet the application of MSCs to treat ligament and tendon damage has shown much promise, both in the repair of the damaged tissue and in the minimization of scarring. As outlined in Case Study 8.4 below, researchers at the Comparative Orthopaedics Laboratory, College of Veterinary Medicine, Cornell University conducted a double-blind, placebo-controlled clinical study to evaluate the efficacy of adipose-derived stem cell transplantation for the treatment of tendinitis in horses.

Case Study 8.4: Effect of adipose-derived nucleated cell fractions on tendon repair in horses with collagenase-induced tendinitis

Alan J. Nixon et al.

Tendinitis is a devastating condition for competitive horses and in most cases is career-ending. Historical data have revealed that roughly 40–60% of horses return to athletic competition, but only for a short period of time, as reinjury is extremely common. Thus, strategies for effective treatment and possible reversal of the disease are of great value. In this study, **collagenase**, which is a naturally occurring enzyme that breaks down collagen, was used to induce tendinitis in the superficial flexor tendon (SDFT) of 1 forelimb in 8 horses, aged 2–6 years. For treatment, four horses were utilized as controls and injected with phosphate-buffered saline and four were injected with autologous **adipose-derived nucleated fractions (ADNC)**, which are collections of nucleated adipose cells that can quickly be used to inject into sites of injury. Following injection, histologic evaluations of tendon structure and the alignment of fibers were performed to assess tendon architecture. A significant improvement in the organization of tendon fibers was observed in ADNC-injected horses in contrast to control horses (Figure 8.30). In addition, the expression of **collagen oligomeric matrix protein (COMP)** mRNA, a key protein involved in tendon structure and function, was increased at sites of injury in ADNC-treated horses compared to controls (Nixon et al., 2008).

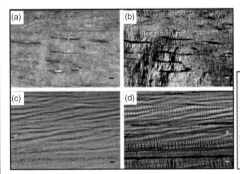

Treatment	Collagen type I	Collagen type III	Decorin	COMP
Control	11.78 ± 3.73	14.70 ± 4.62	9.57 ± 2.30	1.10 ± 0.48
ADNC	11.98 ± 1.95	15.38 ± 2.79	7.62 ± 3.27	2.33 ± 0.64*

Results represent mean ± SD absolute copy number X 10⁴; copy numbers were adjusted on the basis of expression of 18S RNA used as a housekeeping gene.
*Within a column, value differs significantly (P < 0.05) from the value for the control group.

Figure 8.30 Tendon repair in horses using adipose-derived nuclear fractions. (Far left) Photomicrographs of tendon tissue sections after injection with PBS (A and B) or adipose-derived nuclear fractions (C and D). Left photos reveal tissue organization by H&E staining. Right photos display tendon structure by polarized light microscopy. (Far right) gene expression levels of tendon matrix molecules. (*Source:* Nixon, 2008. Reproduced with permission from American Journal of Veterinary Research.)

   As of 2013, the Poway, California-based company Vet-Stem, Inc. has treated over 4000 horses using its proprietary Vet-Stem Regenerative Cell Therapy (VSRC) platform, which is based on the use of adipose-derived stem cells and the ADNF isolate technology described in Case Study 8.4. Vet-Stem cites numerous advantages to the application of adipose-derived regenerative cells and emphasizes their utility in contrast to regenerative cells derived from bone marrow. Some advantages over bone marrow-derived regenerative cells include:

• Readily available source
• Can be collected in higher concentrations than cells from bone marrow
• Wider range of therapeutically valuable cell types including (Figure 8.31):
   ◦ Mesenchymal stem cells
   ◦ Endothelial progenitor cells
   ◦ Pericytes
   ◦ Immune cells
   ◦ Fibroblasts

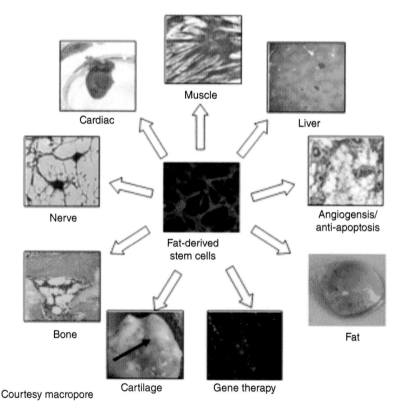

Figure 8.31 Examples of therapeutic uses for adipose-derived stem cells. (*Source:* Macropore, courtesy of Cytori Theraputics.)

Canine

Osteoarthritis (OA) Osteoarthritis (OA) is inflammation due to degeneration of joint cartilage and bone and affects more than 20% of dogs in the United States and is the most frequent cause of chronic pain in these animals. OA involves the degeneration of **articular cartilage**, the cartilage covering the surfaces of bone in synovial joints, and in some cases loss of the entire cartilage surface. Secondary inflammation at sites of degeneration is the result of local chondrocyte release of tissue damaging mediators such as cytokines, free radicals, and certain proteases. Current treatments such as the use of non-steroidal anti-inflammatory drugs (NSAIDs) do not completely alleviate the chronic pain nor promote significant healing. Therefore, the development of alternative strategies for the treatment of canine OA is warranted. Researchers at the company Vet-Stem, Inc. in Poway, California in collaboration with veterinary scientists at The Animal Anesthesia & Pain Management Center in Colorado Springs, Colorado and the San Carlos Veterinary Hospital in San Diego, California conducted a blinded, randomized, placebo-controlled multicenter clinical study to assess the therapeutic benefit of administering adipose-derived stem and regenerative cells in dogs exhibiting lameness associated with osteoarthritis. The focus was on the characterization of improvements in OA of the coxofemoral joints. Twenty-one dogs of multiple breeds ranging in age from 1 to 11 years exhibiting coxofemoral joint OA for at least 6 months were participants in the trial. Each dog exhibited obvious lameness either at a walk or trot. Autologous adipose-derived mesenchymal stem cells (AD-MSCs) were isolated from individual animals from a minimum of 23 grams of lipoaspirated fat collected from the abdominal, inguinal, or thoracic wall regions. After mincing and enzymatic digestion of the adipose tissue, cells were isolated via centrifugation into the stromal vascular fraction (SVF),which typically represents a heterogeneous mixture of fibroblasts, endothelial cells, blood cells, pericytes, and AD-MSCs. A single intra-articular injection of either phosphate buffered saline (PBS) as a control or a suspension of 4.2–5 million viable cells was performed. Following surgery, the evaluation included characterization of the following traits:

- Lameness on walk
- Lameness on trot
- Pain on joint manipulation
- Range of motion
- Total functional disability

Assessment of each of these parameters occurred at 30, 60, and 90 days post surgery and specific scores were assigned according to Table 8.1 below.

A total of 18 dogs completed the study and a final composite score was calculated based on individual scores obtained for the physical evaluation categories listed in Table 8.1. There was a considerable statistically significant improvement in individual scores. Figure 8.32 denotes the composite score improvement as a function of time in comparison to placebo PBS-based controls. On average a 30% improvement in total composite score was observed with AD-MSC treatment.

The positive effect of autologousAD-MSC transplants on osteoarthritic symptoms in canines was profound and is thought to occur through the trophic effects of the cells on local damaged tissues. MSCs are thought to communicate with neighboring cells to suppress immunoreactions and inhibit apoptosisthus reducing the amount of free radicals that promote inflammatory reactions. In addition, it has been demonstrated that bone

TABLE 8.1 Scoring values for the assessment of adipose-derived mesenchymal stem cell treatment of canine osteoarthritis.

	Non-detectable	Intermittent	Persistent	Persistent non-weight bearing	Ambulatory only with assistance	Non-ambulatory
Lameness-walk	1	2	3	4	5	6
Lameness-trot	1	2	3	4	5	6

	No pain		Mild pain (attempts to withdraw limb)		Severe pain (immediate limb withdrawal)	
Pain on manipulation	1		2		3	

	No limitation		Pain only at full range of motion	Pain at less than full range of motion	Pain at any attempt to manipulate joint	
Range of motion	1		2	3	4	

	Normal activity		Slightly stiff gait, only noticeable during running	Stiff, noticeable difficulty walking or running	Very stiff, no desire to walk or run unless coaxed	No desire to walk, must be assisted, cannot run
Functional disability	1		2	3	4	5

(*Source*: Adapted from Black et al., 2007.)

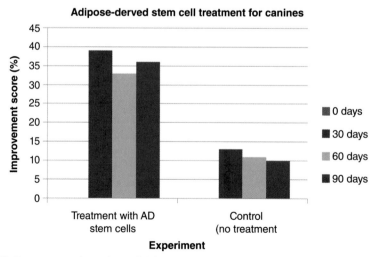

Figure 8.32 Improvement in canine arthritic symptoms upon treatment with adipose-derived stem cells. (*Source:* Black et al., 2007. Reproduced with permission from Vet-Stem and Dr R. Harman.)

marrow-derived MSCs actually deliver new mitochondria to damaged cells, thus restoring aerobic metabolism. This may occur with AD-MSCs as well to advance healing and reduce inflammation.

Myocardial Infarction Treatment of canine disorders utilizing stem cells is not limited to bone or cartilage disorders nor to adipose derived stem cells. Myocardial infarction may also be therapeutically addressed via stem cell transplantation. In fact, numerous strategies and different cell types including endothelial progenitor cells, mononuclear bone marrow cells, and skeletal muscle myoblasts have been used to treat heart attack in dogs and other animals. In 2005, Piero Anversa and colleagues at the Cardiovascular Research Institute, New York Medical College in Valhalla isolated and characterized a CSC in the canine heart and demonstrated the endogenous cell's utility in myocardium repair following heart attack. Specifically, the researchers isolated CSCs from enzymatically digested cardiac tissue based on the presence of stem cell-related antigens c-kit, MDR1, and Sca-1-like, which are noted markers of cardiac stem cells. Cells were analyzed by FACS (fluorescence activated cell sorting) cell sorting and confirmed as negative for hematopoietic cell markers and markers for differentiated cardiac muscle. Only 5% of the total cell population met these criteria. Single cell cloning was implemented to obtain clonogenic populations of cardiac progenitor cells and these cells reconfirmed for CSC marker presence. Thirty-eight individual lines were demonstrated to differentiate *in vitro* into cardiomyocytes, smooth muscle cells, endothelial cells, and fibroblasts. In addition, the researchers confirmed the response of these cells to hepatocyte growth factor (HGF) and insulin-like growth factor (IGF) treatment, which drove the cells toward to both migrate and divide in tissue culture given their expression of the HGF-c-Met and IGF-IGF-1 cell surface receptors. Thus, the Anversa team speculated that HGF and IGF treatment following myocardial infarction might both mobilize and expand the endogenous CSC population and induce cardiac tissue repair following hydraulic occluder-mediated infarction. This was indeed the case as evidenced both at the cellular level and via a characterization of cardiac functional output following HGF and IGF administration (Figure 8.33). This is a profound example of the utility of growth factors directly administered into damaged tissue to induce endogenous stem cell migration, expansion, and functional tissue repair mechanisms (Linke et al., 2005).

BONE FRACTURES

Bone fractures in canines are an extremely common occurrence, yet in most cases do not warrant euthanasia. However, they must be repaired and current methods including simple bone-setting do not allow for sufficient closure of segmental bone defects caused by fracture. Proper healing and repair of bone requires:

- A source of stem cells capable of differentiating into osteoblasts, which are defined as bone-forming cells.
- Critical growth and differentiation factors necessary to induce cells to migrate, proliferate, and differentiate at sites of injury.
- Scaffolding to support cellular attachment and migration.
- Angiogenesis to provide a vascular network for newly formed bone.

Researchers at Tufts University School of Veterinary Medicine in North Grafton, Massachusetts and Osiris Therapeutics in Baltimore, Maryland developed a comprehensive system to address each of the above criteria for bone remodeling and repair by combining stem cells with ceramic scaffolding. Specifically, the team loaded autologous MSCs onto a

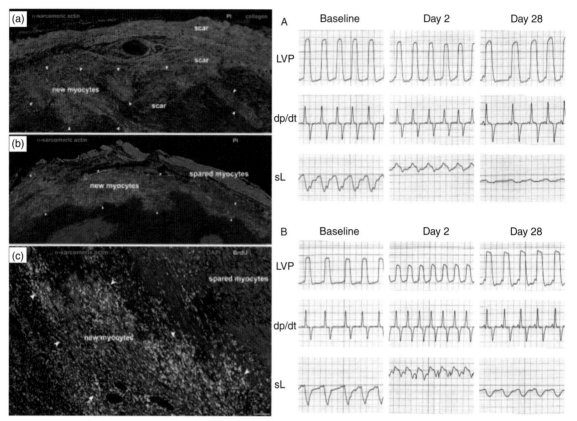

Figure 8.33 New myocyte formation and improvement in cardiac function following growth factor-based cardiac stem cell mobilization in canines with myocardial infarction. (Left) a, b, and c illustrate new myocyte formation in cardiac stem cell-treated post-infarcted cardiac tissue by fluorescence microscopy. New myocytes are stained for α-sarcomeric actin. (Right) Improvement in cardiac output following stem cell transplant. (A) Control untreated post-infarcted cardiac output. (B) CSC-treated. The lower right panel illustrates improvement in paradoxical motion and function, including contraction, following treatment. LVP, left ventricular pressure; SL, segment length. (*Source:* Linke et al., 2005. Reproduced with permission from National Academy of Sciences, USA.)

porous ceramic cylinder. The cells used were chosen due to their osteoblastic potential, with a high capacity to form bone and easy isolated from bone marrow. Bone marrow aspirates were taken from 15 canines 16 days before surgery, with MSCs enriched and isolated from these aspirates in tissue culture utilizing density gradient centrifugation and selection of the nucleated fraction. MSCs at a volume of 30 million cells per implant were seeded into ceramic cylinders consisting of hydroxyapatite (65%) and β-tricalcium phosphate (35%). Limbs to be operated on were chosen randomly in the canines and cylinders were secured with two reabsorbable sutures. Considerable healing, bone growth, and union between host bone and implant was observed in MSC-seeded, implant-treated dogs as evident by the formation of a osseous **callus**, which is bony tissue forming around the ends of a broken bone (Figure 8.34). While some bone formation and rejoining was evident in animals with ceramic implants containing no MSCs, it was not as prevalent as in animals which received MSC-loaded ceramic implants and no callus was observed. No significant

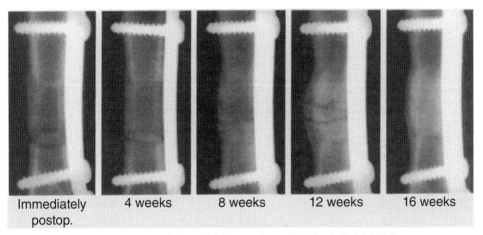

Figure 8.34 The application of a biodegradable ceramic scaffold loaded with MSCs to treat a canine bone fracture. Radiographs taken every week for 4 weeks post surgery illustrate a substantial callus formation around the implant and union between the host bone and implant. (Radiographs courtesy Scott P. Bruder and the *Journal of Bone and Surgery* (Bruder et al., 1998); reprinted with permission.)

healing was demonstrated in canines not receiving an implant (data not shown). These studies suggest that the addition of autologous MSCs to biodegradable ceramic implant scaffolds improves the healing of bone fractures. Bone formed throughout the entire network of implant pores that had been loaded with MSCs, with a thick callus of bone forming around most of the implants. It is speculated by the researchers that not only do the MSCs differentiate into osteoblasts to produce a pure bone matrix, but also secrete inductive factors that may recruit other MSCs toward an osteogenic fate (Bruder et al., 1998).

Spinal Cord Injury As discussed above, the regeneration of CNS tissue is extremely limited following injury. In most cases, SCI leads to debilitation and often death as a result of limited healing capacity in both humans and animals. Yet some stem cell platforms have shown promise in the treatment of SCI in canines. MSCs, for example, may not only be useful for the treatment of bone fractures but also SCIs in canines. MSCs derived from adipose tissue have been shown to differentiate into neural lineages with high efficiency and thus may be a valuable platform for the treatment of canine SCI. A group of Korean researchers in the Department of Veterinary Surgery and the Laboratory of Stem Cell and Tumor Biology at Seoul National University sought to exploit the utility and potential of adipose-derived MSCs for treating SCI. In this study, 11 canines with induced SCI were divided into either control groups or allogeneictransplant recipients. Adipose-derived stem cells were isolated from adipose tissue collected from a 2-year-old dog and dissociated enzymatically. Individual cells were isolated in tissue culture and expanded through several passages. These cells were confirmed for the presence of a MSC population and neurogenic differentiation potential in tissue culture by immunofluorescence staining for the presence of microtubule associated protein 2 (MAP2), which is a marker for neuronal differentiation. The cells were also assessed for the presence of the pluripotentmarker Oct4. Roughly 1 million cells were transplanted directly into the injured spinal cord region 1 week after experimentally induced SCI. Dogs were examined for SCI repair as evidence by both behavior and histology. For behavior, the researchers assessed the **Olby score** for each canine, which is a scoring system for dogs based on pelvic limb gait following lumbar disc

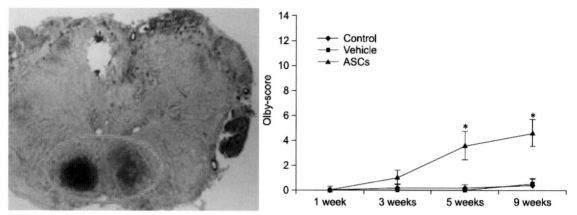

Figure 8.35 Evidence of new myelin and nerve growth and improved gait in adipose-derived stem cell treatment of canine spinal cord injury. (Left) Transverse histological section of a repaired canine spinal cord showing new myelin and nerve growth by Luxol staining. (Right) Olby scores for three groups of canine models of SCI. (*Source:* Ryu et al., 2009. Reproduced with permission from The Journal of Veterinary Science, Korean Society of Veterinary Science.)

herniations resulting from SCI. Histological analysis included an assessment of new myelin and neuronal growth using the staining agent **Luxol**, which is a fast blue stain that incorporates specifically into myelin. It stains myelin blue and nerve cells purple. New myelin and nerve growth was observed in ASC-transplanted animals 9 weeks post surgery that was not evident in control groups (data not shown). In addition, Olby scores significantly improved in ASC-treated canines (Figure 8.35) (Ryu et al., 2009). These studies suggest that allogeneic adipose-derived stem cells may be an effective therapeutic platform for treating SCI in canines.

STEM CELLS AS AN EMERGING INDUSTRY

The advancement of knowledge related to stem cell biology has yielded numerous opportunities to harness the potency of these cells for therapeutic benefit. As mentioned throughout this text, pluri-, multi- and even unipotent stem cells have the potential to impact many areas of medicine and clinical treatment of countless disorders. As with all potential valuable therapeutic platforms there is a financial aspect to the use of stem cells in the clinic. Over the past 20 years, an industry has emerged focused on the manipulation and application of a variety of stem cell-based technologies for medical benefit. Since W.T. Green's pivotal study in 1977 on articular cartilage repair using allografted chondrocytes, an industry has emerged, dominated by the study and application of stem cell potency for the treatment of countless disorders, examples of which are cited in the previous clinical trials section of this text. This final section outlines various areas of focus with respect to the use of stem cells in translational medicine and illustrates the growth of a new industry based on stem cell technologies. In some cases—for example, skin replacement—non-stem cell strategies have been employed.

Given advancements in the study of embryonic stem cell and adult stem cell types in the early 1990s, the regenerative medicineindustry matured quickly during this decade, capping out at $610 million in private investment in 2000 and over 3000 full-time employees in the industry. Table 8.2 illustrates the emergence and growth of the industry from the early 1990s through 2007.

TABLE 8.2 The emergence of the regenerative medicine industry.

	1994	1997	2000	2003	2007
Number of regenerative medicine companies	40	40	73	89	171
Number of full-time employees	1500	2380	3080	2610	6100
Total private investment capital (millions USD)	$246	$453	$610	$487	$2400

(*Source*: Adapted from (Nerem, 2010.))

During this time a great deal of industrial focus was on the application of tissue engineering and regeneration for the replacement of diseased or damaged skin. The primary platform was the manipulation of human fibroblasts. Advanced Tissue Sciences, Inc. developed both Transcyte[R] and Dermagraft[R] from human fibroblasts. Transcyte is a temporary substitute for skin derived from the culture of human fibroblasts and a polymer membrane. Dermagraft is a biodegradable polymer physically coated with human foreskin-derived fibroblasts and also contains extracellular matrix (ECM) components (Figure 8.36).

These are allogeneic grafts but, interestingly, immunorejection has not been an issue for either graft type. Epicel[R] is an **autograft**, which is a graft of tissue from one point to another point on a person's body, of epidermal fibroblasts. The fibroblasts are co-cultured with murine fibroblasts, thus it is considered a xenotransplantation product. Epicel thus carries the risk of viral transmission from mice to humans and therefore is only used under the most extreme of circumstances as a **humanitarian use device (HUD)** per FDA restrictions. In parallel with Epicel, Genzyme developed the isolation, expansion, and direct injection autologouschondrocyte-based cartilage repair platform Carticel[R] to address articular cartilageinjuries of the knee that have failed to properly respond to other modes of treatment such as arthroscopic surgery (Figure 8.37).

The Carticel technology was originally developed by researchers in the Department of Orthopedic Surgery, University of Goteborg, Sahlgrenska University Hospital, Sweden. It involves the isolation, culture and expansion of autologous chondrocytes in the presence of the patient's own serum. This allows for production of sufficient numbers of chondrocytes for cell transplantation utilizing the nutrient-rich base of the patient's serum and eliminating any chance of immunorejection. In a study assessing Carticel performance, chondrocytes

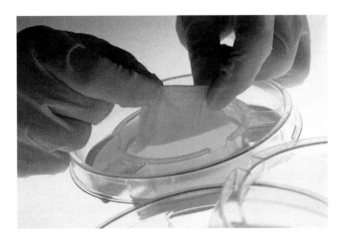

Figure 8.36 Dermagraft: a tissue engineered for the replacement of human skin. (Figure courtesy Advanced Tissue Sciences, Inc.; reprinted with permission.)

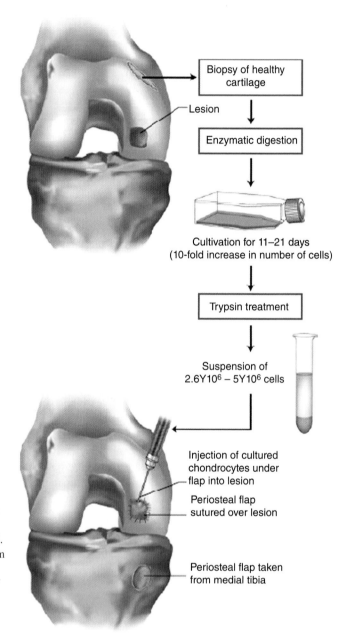

Figure 8.37 Diagrammatic illustration of the step-wise procedure for treating refractory articular cartilage damage with autologous chondrocytes, a.k.a. Carticel. See text for details. (Diagram courtesy Mats Brittberg and the *New England Journal of Medicine* (Brittberg et al., 1994); reprinted with permission.)

were isolated from 23 patients and expanded in tissue culture in the presence of the patient's own serum for 2–3 weeks. The cells were then concentrated and injected directly into the injured articular cartilage and assessments of therapeutic benefit made periodically for 66 months. The researchers demonstrated considerable healing of articular cartilage defects when autologous chondrocytes were injected and where previous attempts at surgical treatment had failed (Figure 8.38). It should be noted that Carticel was the only autologous chondrocyte cell transplantation technology approved by the FDA as of 1997. Carticel is

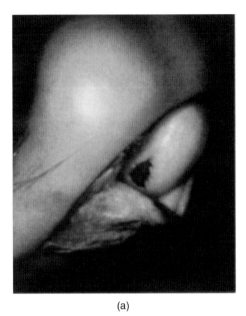
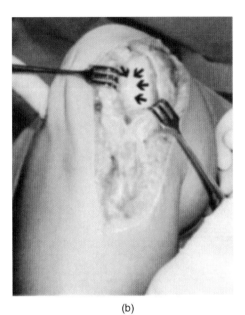

(a)   (b)

Figure 8.38 Direct chondrocyte injection into damaged articular cartilage results in repair.
(a) Photograph of an articular cartilage defect in a 22-year-old female patient before treatment.
(b) The same cartilage after treatment by direct injection of autologous chondrocytes. The arrows
denote the borders of chondrocyte transplant. (Figures courtesy Mats Brittberg and the *New
England Journal of Medicine* (Brittberg et al., 1994); reprinted with permission.)

considered to be one of the seminal cell-transplant technologies that achieved translational
value in the late 1990s and set the stage for the development of numerous other autologous
cell transplant strategies to treat a multitude of disorders.

Seminal Discoveries Driving the Growth of a New Industry

Toward the end of the 1990s, it was clear that cell-based tissue replacement strategies were
viable and reliable for the treatment of diseased and damaged tissue. It was at this time that
the intricate knowledge of stem cell biology and function merged with the concept of cell-
based tissue replacement to launch the dawn of a new industry. The following discoveries
are now considered as pivotal for driving the emergence of the regenerative medicine
industry:

- Isolation of human embryonic stem (hES) cells
- Discovery of neural stem cells in the adult brain
- Identification of dermal stem cells in adult skin tissue
- Discovery of cancer stem cells
- Identification of the four key pluripotency drivers
- Derivation of adult, fully mature dopaminergic neurons from hES cells
- Induction of pluripotency in adult somatic cells (iPS technology)

These and other scientific advancements in the field of stem cell research made it
conceptually possible to treat a variety of cell- and tissue-based disorders. As the basic

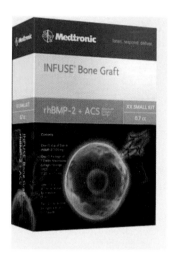

Figure 8.39 Medtronic's INFUSE bone graft therapy. (Image courtesy Wikimedia Commons; reprinted with permission.)

understanding and manipulation of stem cells matured, many newly formed as well as seasoned biotechnology and pharmaceutical companies sought to exploit this technology for therapeutic and financial gain. Thus the first decade of the new millennium saw an explosion in the number of companies with business models centered on stem cell-based regenerative medicine as a clinically viable way to reap financial gain. It was in 2010 that Geron Corporation initiated the world's first clinical trial (Phase 1) involving the application of hES cells for the treatment of SCI. Although Geron's hES cell was halted due to funding issues and a refocus on other technologies, this trial, along with numerous others as described earlier in this chapter, demonstrated the scientific, medical, and industrial communities' commitment to the field of regenerative medicine.

The regenerative medicine industry is not limited merely to the direct use of stem cells and their derivates for transplantation-based therapy. Other indirect business models have also emerged from advancements in stem cell research. These include the sale of **acellular** (devoid of any cells) products related to stem cell growth and differentiation, which topped out at $1.3 billion in revenue in 2007. For example, the Medtronic product **INFUSE**[R], which is a recombinant version of human bone morphogenetic protein 2 (BMP) used to treat lumbar spinal fusions, accounted for over half of this revenue (Figure 8.39).

Regenerative biomaterials, which are defined as synthetic or natural material designed to promote cell and tissue regeneration *in vivo*, accounted for $240 million of the $1.3 billion in product sales in 2007. Many of these biomaterials were based on the use of ECM formulations such as Depuy's Restore[R]. **Restore** is based upon an derived from porcine **small intestine submucosa (SIS)**. SIS is a naturally derived biomaterial rich in growth factors such as FGF2 and glycoproteins including fibronectin. Over 600 publications have illustrated SIS' utility as a regenerative biomaterial graft. Restore is a version of SIS developed as an orthobiologic implant intended to reinforce soft tissue. It is promoted as a natural alternative to more aggressive therapeutics strategies and acts as a catalyst to recruit both cells and secreted nutrients from adjacent tissues. Following recruitment of new cell populations and nutrients, the biomaterial is absorbed and damaged tissue is rapidly repaired (Figure 8.40).

In addition to stem cell transplants, growth factors, and regenerative biomaterials as revenue drivers for the regenerative medicine industry, the banking of patient-specific cells also garnered significant market demand. For example, the banking of cord blood stem

1   2   3   4

Figure 8.40 The healing process using Depuy's Restore regenerative biomaterial. (a) The 3D biomatrix. (b) Recruitment of cells to the implant site. (c) Further expansion and colonization of the implant. (d) Resorbed biomaterial and regenerated, bonded natural tissue. (Images courtesy Depuy; reprinted with permission.)

cells generated over $270 million in sales during 2007. The service has continued to see a considerable market demand and the business model has been expanded upon by a variety of companies to include the banking of other stem cell types such as dental pulp stem cells.

Thus, regenerative medicinehas now officially emerged as a legitimate industry, one with great potential to benefit humans via the development and clinical application of stem cell-based transplants, growth, and differentiation inducing factors and regenerative biomaterials. Advancements in these areas are not only driven by sound science and translational research but also by market demand. In fact, sales of stem cell-related therapeutics are expected to top $8.4 billion by 2016 (Figure 8.41).

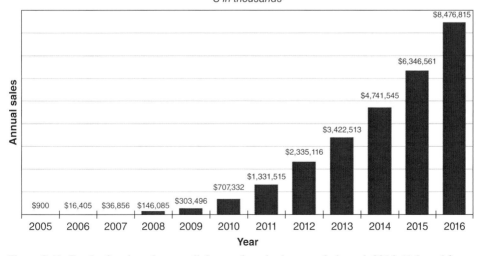

Figure 8.41 Graph of projected stem cell therapy-based sales growth through 2016. (Adapted from data presented at the 2nd Annual Stem Cell Summit, San Diego, California).

Regulation and Reimbursement of Stem Cell Commercialization

While clearly the regenerative medicine industry is here to stay, as with any fledgling commerce reimbursement and compensation strategies as well as government regulation must be developed and implemented. In the United States as in other countries, the advancement of science behind the use of stem cells and related products for therapeutic endeavors is currently outpacing the Food and Drug Administration's efforts to draft and implement a comprehensive regulatory plan. In addition, insurance companies often are hesitant to approve third-party reimbursement of stem cell-related therapies, in part due to a hesitance by physicians to accept this new mode of treatment as an alternative to other more traditional clinical strategies. Third-party payers must understand the cost effectiveness of the treatment regimen, which is often not the case for newly developed technologies. Health plan medical directors and coverage officials must have a firm grasp of both the clinical benefit and the value of the treatment in comparison to existing treatments. Finally, coverage decisions are based upon the disclosure of comprehensive scientific knowledge of the mechanisms and the long-term effects of the therapeutic regimen. As numerous stem cell-based treatments provide medical benefits, many of these are still not well understood at the mechanistic level. This must be addressed to increase the percentage of stem cell therapies that are covered by third-party payers. Often the most optimal opportunity to collect this information is during late-stage preclinical studies and during all phases of the clinical trial process.

A Word about Induced Pluripotency and Commercialization

The discovery of induced pluripotency has had a profound impact on the field of stem cell research. The ability to induce pluripotent characteristics in patient-derived adult somatic cells has enormous implications for the design of new stem cell-based treatment strategies. From an industrial and commercialization perspective iPS technology accomplishes three key goals:

- Elimination of immunorejection issues (autologous origin)
- Elimination of ethical issues associated with embryonic stem cells
- Provision of an "unlimited" source of cells for therapy and research

In addition, the very nature of induced pluripotency suggests that virtually any cell type can be generated and used in the clinic to treat countless cell-based disorders. Thus, it is anticipated that a new industry will emerge focused solely on the induction of pluripotency in patient-specific cells for purposes of autologouscell- and tissue-replacement treatments. Indeed, several companies have already taken the plunge into this new and exciting area of stem cell technology to develop new therapies. Real-world applications of iPS technology have now culminated with the initiation of the first clinical trial using autologous iPS cells derived from adult skin. To be conducted and approved by the Japanese Health Ministry, iPS cells derived from six patients suffering from age-related macular degeneration will be reprogrammed and differentiated into retinal pigment epithelial (RPE) cells. These iPS-derived cells will be transplanted directly into affected retinas and safety as well as vision repair will be monitored over 3 years (Figure 8.42). The trial will be conducted at the Riken Center for Developmental Biology and the

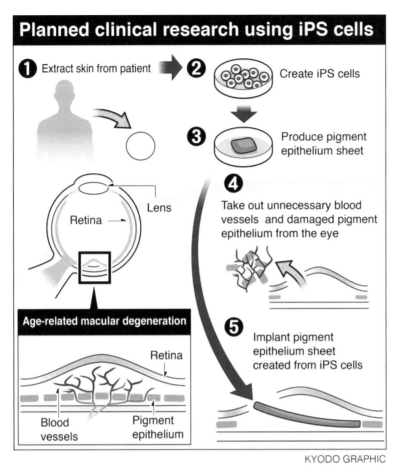

Figure 8.42 Diagrammatic illustration of the strategy for conducting the world's first clinical trial using iPS cells. (Diagram courtesy *The Japan Times*; reprinted with permission.)

Institute of Biomedical Research and Innovation Hospital in Kobe. Thus a new era has dawned in stem cell translational medicine and the growth of this exciting industry is surely to follow.

CHAPTER SUMMARY

History of Stem Cells and Therapeutics

1. The idea and concept of replacing diseased or dead tissue with healthy living material has been contemplated for greater than two thousand years and can be traced back to direct references in the Book of Genesis.
2. W.T. Green developed the first 3D system for cell propagation.
3. Early concepts in tissue engineering have now been combined with the utility of various stem cells platforms for the engineering of a variety of tissues and the regeneration of both tissues and specific cells within the body.

Disease-Specific Treatment and Patient Trials

1. The discovery of different types of stem cells exhibiting different levels of potency combined with advancements in the knowledge of key molecular and biochemical cascades defining potency has allowed for the use of stem cells in the clinic to become a reality.
2. Clinical trials involving stem cells are currently occurring worldwide.
3. HSC transplantation has been the most widely studied of all stem cell types at the clinical level.
4. Numerous different stem cell types have been shown to have positive therapeutic benefits in preclinical animal models of CV.
5. The CNS, specifically the spinal cord, does not allow for repair once damaged.
6. Stem cells show promise in the possible replacement of dead or damaged neurons and perhaps even the repair of damaged axons and encouraging results of SCI repair using stem cells have been demonstrated in preclinical animal models.
7. Mesenchymal stem cells (MSCs) have been used in clinical trials to treat numerous disorders including bone/cartilage disease, cancer, heart disease, diabetes, GI disease, autoimmunity disorders, and neurodegenerative disease.
8. Stem cells derived from a vast array of sources, from embryonic to adult, have been considered as candidates for stroke treatment.
9. Possible modes of action for stem cell treatment of stroke-induced brain damage include the generation of new neurons, decreased apoptosis,and promotion of angiogenesis.
10. The application of MSCs and HSCs to address autoimmune diseases may show great promise.
11. Defects of the cornea may be treated by direct injection of LSCs.
12. Hematopoietic disorders are perhaps some of the most amenable anomalies for stem cell-based therapy given ease of access to the hematopoietic system.
13. Sickle cell disease may be treated via the replacement of diseased red blood cells with stem cell-derived, healthy red blood cells.
14. Wiskott-Aldrich syndrome was treated clinically via the transplantation of genetically modified HSCs expressing the WAS gene.
15. Wiskott-Aldrich syndrome transplantation increases the recurrence-free survival rate of breast cancer patients undergoing chemotherapy.
16. Muscular Dystrophy may be treated through the injection of allogeneic myoblast satellite cells but immunorejection often occurs.
17. Bone marrow-derived stem cells (BMSCs) may be a good cell transplantation therapeutic platform for increasing liver mass in patients as they have been shown to contribute to liver cell proliferation.

Veterinary Applications

1. Stem cells have been actively used to treat various disorders in horses, dogs, and cats.
2. Autologous MSCs are by far the most widely used cell type with respect to veterinary applications.

3. The application of MSCs to treat equine ligament and tendon damage has shown much promise, both in the repair of the damaged tissue and in the minimization of scarring.

4. Advantages of using adipose-derived stem cells over bone marrow derived stem cells for veterinary applications include wider availability, higher numbers available for collection, and wider range of potency.

5. Osteoarthritis (OA) has been successfully treated in canines using autologous AD-MSCs.

6. Allogeneic cardiac stem cells have been used to treat myocardial infarction in canines.

7. Proper healing and repair of bone requires a stem cell source, growth and differentiation factors, scaffolding and active angiogenesis.

8. Implantation of autologous bone marrow-derived MSCs loaded onto a porous ceramic scaffold were effective in promoting bone remodeling and repair following fracture.

9. MSCs derived from adipose tissue have been shown to differentiate into neural lineages with high efficiency and thus may be a valuable platform for the treatment of canine SCI.

10. AD-MSCs transplanted directly into the injured spinal cords of canines yielded new nerve and myelin growth and improved pelvic gait.

Stem Cells as an Emerging Industry

1. Within the past 30 years or so the knowledge base and advancements in the study of stem cells have matured to the point where their use in tissue replacement is becoming a reality.

2. The regenerative medicine industry experience considerable growth during the 1990s due to advancements in the understanding and manipulation of embryonic and adult stem cells.

3. Skin replacement has been a major focus of the regenerative medicine industry with numerous products now on the market to treat diseased and damaged skin.

4. Given its widespread occurrence and treatability, the repair of damaged cartilage using stem cells is current being pursued by numerous regenerative medicine companies.

5. Some seminal discoveries driving the emergence of the regenerative medicine industry include the discovery of neural stem cells in the adult brain, the identification of the four key pluripotency factors and the discovery of induced pluripotency technology.

6. In 2010, Geron Corporation initiated the world's first clinical trial (Phase 1) involving the application of hES cells for the treatment of spinal cord injury.

7. Other indirect business models have also emerged from advancements in stem cell research including the sale of acellular products such as regenerative biomaterials and recombinant factors related to stem cell growth and differentiation.

8. Government regulation and reimbursement strategies have not kept up with the pace of scientific advancements in regenerative medicine.

9. Induced pluripotency technologies are expected to have a major impact on the regenerative medicine industry.

10. Real-world applications of iPS technology have now culminated with the initiation of the first clinical trial using autologous iPS cells derived from adult skin.

KEY TERMS

(Key terms are listed by order of appearance in the text.)

- **Regenerative medicine**—the process of replacing or regenerating human cells, tissues, or organs to restore or establish normal function.
- **Cell seeding**—the plating and culture of cells into an artificial matrix.
- **"Smart" scaffolds**—cell-seeded 3D-based systems for culture or cell/tissue replacement.
- **Tissue engineering**,—the combination of cells, materials, biochemical, or biophysical factors to generate living tissue.
- **Poland's Syndrome**—a congenital malformation of the rib cage and absence of the sternum.
- **Vacanti mouse, a.k.a. auriculosaurus**—a laboratory mouse with ear-shaped cartilage engineered onto its back by seeding cartilage cells into a biodegradable ear-shaped mold followed by subcutaneous transplantation.
- **Cardiovascular disease (CV)**—one or more disorders of the heart and blood vessels resulting in decreased blood flow.
- **Ischemic cardiomyopathy**—poor oxygen supply to the heart muscle and weakness of the cardiac muscle.
- **Cardiac stem cells (cardioblasts)**—a population of cells specified to the cardiac myocyte cell lineage prior to their (terminal) differentiation into a fully differentiated cardiac myocytes.
- **c-kit (stem cell factor receptor)**—a cell surface receptor tyrosine kinase that marks the surface of some multipotent stem cell lineages.
- **c-kit positive, lineage-negative cardiac stem cells**—stem cells derived from the vasculature and located specifically within vessel walls which commit to vascular endothelial or smooth muscle lineages upon differentiation *in vivo*.
- **Angioplasty**—a procedure implemented to widen blood vessels.
- **Spinal cord injury (SCI)**—injury to the spinal cord often resulting in the loss of nervous tissue and thus corresponding loss of both motor and sensory function.
- **Stroke**—the sudden death of brain cells due to inadequate blood flow.
- **Modified Rankin score**—a scale used for measuring the degrees of disability or dependence of individuals in daily activities that have suffered stroke or other forms of neurological damage.
- **National Institutes of Health Stroke Scale (NIHSS)** —a measurement of impairment caused by stroke.
- **Angiogenesis**—the production and growth of new blood vessels.
- **Parkinson's disease**—a progressive nervous system disorder resulting from destruction of the brain cells producing dopamine and characterized by muscular tremor, slowing of locomotion, partial facial paralysis, and general weakness.
- **Unified Parkinson's Disease Rating Scale (UPDRS)**—a scale that measures nervous system motor functionality, scaling such symptoms as tremors.
- **Multiple sclerosis (MS)**—a chronic demyelinating autoimmune disease of the central nervous system.

- **Amyotrophic lateral sclerosis (ALS)**—a neurodegenerative disease specifically affecting brain motor neurons and leading to respiratory and limb weakness.
- **Intrathecal injection**—defined as direct injection under the arachnoid membrane of the brain or spinal cord.
- **Kurtzke Expanded Disability Status Scale**—a method based on a functional system score to quantify disability in multiple sclerosis patients.
- **Juvenile idiopathic arthritis (JIA)**—a group of illnesses present in children exhibiting the common symptom of chronic joint inflammation.
- **Limbal stem cells (LSCs), a.k.a. corneal epithelial stem cells (CESCs)**—stem cells normally present in the basal layer of the corneal epithelium that act to replenish corneal epithelial cells which are continuously lost from the surface of the cornea.
- **Limbal stem cell deficiency (LSCD)**—a disorder caused by trauma in which the limbal stem cells become deficient and thus the corneal epithelium cannot be renewed.
- **Sickle cell disease**—an inherited blood disorder involving a single point mutation in the β-globin gene in which red blood cells exhibit an abnormal rigid, sickle shape.
- **Chimerism**—a mixture of both donor and recipient cells in the blood.
- **Graft vs. host disease (GVHD)**—a complication that can occur following a stem cell transplant in which the newly transplanted cells attack the recipient's body.
- **Anergic**—the inability to mount a complete immune response to a foreign antigen.
- **Granulocyte colony stimulating factor (G-CSF)**—a glycoprotein that stimulates the bone marrow to produce both granulocytes and stem cells and release them into the bloodstream.
- **Alemtuzumab**—a humanized monoclonal antibody that functions to deplete T cells and B cells from the patient's immune system, thus allowing the recipient to tolerate the introduction of foreign cells into the blood stream.
- **Wiskott-Aldrich syndrome (WAS)**—an X-linked inherited disorder involving deficiency in the immune system characterized by autoimmunity, eczema, and recurrent infections.
- **WAS protein (WASP)**—a protein that regulates the polymerization of actin in hematopoietic cells, abnormalities in which cause Wiskott-Aldrich syndrome.
- **Leukapheresis**—a laboratory procedure implemented to isolate white blood cells from whole blood.
- **Muscular dystrophy (MD)**—a group of genetic diseases resulting in damage to muscle fibers and related weakness.
- **Duchenne muscular dystrophy (DMD)**—a severe form of muscular dystrophy caused by a mutation in the X-linked gene dystrophin, which occurs in boys and is often fatal.
- **Dystrophin**—a protein involved in the organization and functioning of muscle tissue.
- **Satellite cells**—precursor cellscapable of myogenesis.
- **Myogenesis**—the generation of muscle cells, fibers and tissue.
- **MyoD (myogenic differentiation factor 1)**—a basic helix-loop-helix transcription factor and the first myogenic factor discovered with the capacity to convert fibroblasts into myoblasts in tissue culture.
- **Cyclosporine (CSA)**—an immunosuppressant drug widely used in organ transplant surgeries to prevent organ rejection by the patient.

- **Maximum voluntary isometric contraction (MVC)**—a measure of strength typically reported in units of force such as Newtons, and tetanic force.
- **Tetanic force (TF)**—a measurement of the force required to result in a maximal stimulation of a motor unit by its corresponding motor neuron.
- **Future liver remnant volume (FLRV)**—the volume of liver tissue that generates new liver growth.
- **Orthotopic liver transplantation (OLT)**—a graft of liver tissue occurring at its natural place.
- **Hepatectomy**—surgical removal of a portion of the liver.
- **Collagenase**—a naturally occurring enzyme that breaks down collagen.
- **Adipose-derived nucleated fractions (ADNC)**—collections of nucleated adipose cells that can quickly be used to inject into sites of injury.
- **Collagen oligomeric matrix protein (COMP)**—a key protein involved in tendon structure and function.
- **Osteoarthritis (OA)**—inflammation due to degeneration of joint cartilage and bone.
- **Articular cartilage**—the cartilage covering the surfaces of bone in synovial joints.
- **Olby score**—a scoring system for dogs based on pelvic limb gait following lumbar disc herniations resulting from spinal cord injury.
- **Luxol**—a fast blue stain that incorporates specifically into myelin.
- **Autograft**—a graft of tissue from one point to another point on a person's body.
- **Humanitarian use device (HUD)**—a medical device intended to benefit patients in the treatment or diagnosis of a disease or condition that affects or is manifested in fewer than 4000 individuals in the United States per year.
- **Acellular**—devoid of any cells.
- **INFUSE**[R]—a recombinant version of human bone morphogenetic protein 2 (BMP) used to treat lumbar spinal fusions.
- **Regenerative biomaterials**—synthetic or natural material designed to promotion cell and tissue regeneration *in vivo*.
- **Restore**[R]—a regenerative biomaterial based upon an extracellular matrix derived from porcine small intestine submucosa (SIS) used as an orthobiologic implant to reinforce soft tissue.
- **Small intestine submucosa (SIS)**—a naturally derived biomaterial rich in growth factors such as FGF2 and glycoproteins including fibronectin.

REVIEW QUESTIONS

(Answers to select review questions can be found at www.stemcelltextbook.com)

1. What is the first written account of the contemplation of tissue replacement?
2. What was W.T Green's contribution to tissue engineering?
3. How did researchers at Tokyo Women's Medical University repair a damaged pulmonary artery?
4. Name 10 potential clinical applications of stem cells.

5. What types of stem cells are most often used in clinical trials?

6. What are the criteria for defining the ideal cell type to be used in the treatment of cardiovascular disease?

7. Name two types of stem cells that have been used in clinical trials to treat cardiovascular disease.

8. Why is spinal cord injury so difficult to repair?

9. How did South Korean researchers improve sensory perception and mobility in a spinal cord-injured patient?

10. Name five disorders treatable using mesenchymal stem cells.

11. List three sources of stem cells that may be used for the treatment of stroke.

12. What are the proposed mechanisms of action for the repair of brain damage by stem cell transplantation?

13. How did Michael Rezak's team at the UCLA School of Medicine and Brain Research Institute treat a Parkinson's diseasepatient?

14. What autoimmune disorders might be treatable via stem cell transplantation and what types of cells are good candidates for this treatment?

15. How could defects of the cornea be treated using stem cells?

16. How could graft vs. host disease be addressed in sickle cell disease patients following allogeneic hematopoietic stem cell transplant?

17. What is one method to increase recurrence-free survival rates in breast cancer patients undergoing chemotherapy?

18. What type of stem cell might be used to treat Duchenne muscular dystrophy?

19. Why would transplantation of bone marrow-derived stem cells be a good strategy to increase liver mass?

20. What types of veterinary disorders are currently treatable by stem cell transplantation?

21. Why are adipose-derived stem cells so popular for the treatment of equine disorders in comparison to bone marrow-derived stem cells?

22. What are four canine diseases treatable using stem cells?

23. Name two clinically approved stem cell-based treatments for skin disorders.

24. Describe the basis behind the Carticel technology for healing cartilage defects.

25. List five seminal discoveries impacting the emergence of the regenerative medicine industry.

26. Why are insurance companies hesitant to approve reimbursement of stem cell-related therapies?

27. What three goals does iPS technology accomplish that makes it ideal for clinical applications?

THOUGHT QUESTION

Come up with the name of your own regenerative medicine company and devise a Phase II clinical trial for the stem cell-based treatment of a disease or disorder that you would like to see cured.

SUGGESTED READINGS

am Esch, J. S., 2nd, W. T. Knoefel, et al. (2005). "Portal application of autologous CD133+ bone marrow cells to the liver: a novel concept to support hepatic regeneration." *Stem Cells* **23**(4): 463–470.

Banerjee, S., D. A. Williamson, et al. (2012). "The potential benefit of stem cell therapy after stroke: an update." *Vasc Health Risk Manag* **8**: 569–580.

Black, L. L., J. Gaynor, et al. (2007). "Effect of adipose-derived mesenchymal stem and regenerative cells on lameness in dogs with chronic osteoarthritis of the coxofemoral joints: a randomized, double-blinded, multicenter, controlled trial." *Vet Ther* **8**(4): 272–284.

Bolli, R., A. R. Chugh, et al. (2011). "Cardiac stem cells in patients with ischaemic cardiomyopathy (SCIPIO): initial results of a randomised phase 1 trial." *Lancet* **378**(9806): 1847–1857.

Boztug, K., M. Schmidt, et al. (2010). "Stem-cell gene therapy for the Wiskott-Aldrich syndrome." *N Engl J Med* **363**(20): 1918–1927.

Brinkman, D. M., I. M. de Kleer, et al. (2007). "Autologous stem cell transplantation in children with severe progressive systemic or polyarticular juvenile idiopathic arthritis: long-term follow-up of a prospective clinical trial." *Arthritis Rheum* **56**(7): 2410–2421.

Brittberg, M., A. Lindahl, et al. (1994). "Treatment of deep cartilage defects in the knee with autologous chondrocyte transplantation." *N Engl J Med* **331**(14): 889–895.

Bruder, S. P., K. H. Kraus, et al. (1998). "The effect of implants loaded with autologous mesenchymal stem cells on the healing of canine segmental bone defects." *J Bone Joint Surg Am* **80**(7): 985–996.

Cao, Y., J. P. Vacanti, et al. (1997). "Transplantation of chondrocytes utilizing a polymer-cell construct to produce tissue-engineered cartilage in the shape of a human ear." *Plast Reconstr Surg* **100**(2): 297–302; discussion 303–294.

Daley, G. Q. (2012). "The promise and perils of stem cell therapeutics." *Cell Stem Cell* **10**(6): 740–749.

Honmou, O., K. Houkin, et al. (2011). "Intravenous administration of auto serum-expanded autologous mesenchymal stem cells in stroke." *Brain* **134**(Pt 6): 1790–1807.

Hsieh, M. M., E. M. Kang, et al. (2009). "Allogeneic hematopoietic stem-cell transplantation for sickle cell disease." *N Engl J Med* **361**(24): 2309–2317.

Kang, K. S., S. W. Kim, et al. (2005). "A 37-year-old spinal cord-injured female patient, transplanted of multipotent stem cells from human UC blood, with improved sensory perception and mobility, both functionally and morphologically: a case study." *Cytotherapy* **7**(4): 368–373.

Karussis, D., C. Karageorgiou, et al. (2010). "Safety and immunological effects of mesenchymal stem cell transplantation in patients with multiple sclerosis and amyotrophic lateral sclerosis." *Arch Neurol* **67**(10): 1187–1194.

Kolli, S., S. Ahmad, et al. (2010). "Successful clinical implementation of corneal epithelial stem cell therapy for treatment of unilateral limbal stem cell deficiency." *Stem Cells* **28**(3): 597–610.

Levesque, M. F., T. Neuman, et al. (2009). "Therapeutic microinjection of autologous adult human neural stem cells and differentiated neurons for Parkinson's Disease: five-year post-operative outcome." *The Open Stem Cell Journal*, **1**, 20–29.

Linke, A., P. Muller, et al. (2005). "Stem cells in the dog heart are self-renewing, clonogenic, and multipotent and regenerate infarcted myocardium, improving cardiac function." *Proc Natl Acad Sci U S A* **102**(25): 8966–8971.

Miller, R. G., K. R. Sharma, et al. (1997). "Myoblast implantation in Duchenne muscular dystrophy: the San Francisco study." *Muscle Nerve* **20**(4): 469–478.

Nerem, R. M. (2010). "Regenerative medicine: the emergence of an industry." *J R Soc Interface* **7**(Suppl 6): S771–S775.

Nixon, A. J., L. A. Dahlgren, et al. (2008). "Effect of adipose-derived nucleated cell fractions on tendon repair in horses with collagenase-induced tendinitis." *Am J Vet Res* **69**(7): 928–937.

Ryu, H. H., J. H. Lim, et al. (2009). "Functional recovery and neural differentiation after transplantation of allogenic adipose-derived stem cells in a canine model of acute spinal cord injury." *J Vet Sci* **10**(4): 273–284.

Shin'oka T, Imai Y, Ikada Y. "Transplantation of a tissue-engineered pulmonary artery." N Engl J Med. 2001 Feb 15;**344**(7):532-3.

Tallman, M. S., R. Gray, et al. (2003). "Conventional adjuvant chemotherapy with or without high-dose chemotherapy and autologous stem-cell transplantation in high-risk breast cancer." *N Engl J Med* **349**(1): 17–26.

Tedesco, F. S., A. Dellavalle, et al. (2010). "Repairing skeletal muscle: regenerative potential of skeletal muscle stem cells." *J Clin Invest* **120**(1): 11–19.

Thomson, M., S. J. Liu, et al. (2011). "Pluripotency factors in embryonic stem cells regulate differentiation into germ layers." *Cell* **145**(6): 875–889.

Trounson, A., R. G. Thakar, et al. (2011). "Clinical trials for stem cell therapies." *BMSC Med* **9**: 52.

Woodbury, D., E. J. Schwarz, et al. (2000). "Adult rat and human bone marrow stromal cells differentiate into neurons." *J Neurosci Res* **61**(4): 364–370.

Yousef, M., C. M. Schannwell, et al. (2009). "The BALANCE Study: clinical benefit and long-term outcome after intracoronary autologous bone marrow cell transplantation in patients with acute myocardial infarction." *J Am Coll Cardiol* **53**(24): 2262–2269.

ABOUT THE AUTHOR

Rob Burgess is a scientist, entrepreneur, businessman, and author. He has held numerous academic and industry-related positions including Research Fellow at the University of California, San Diego; Founding Scientist at Lexicon Genetics; Founder and President, Genome Biosciences; Vice President, Research and Development at Zyvex Corporation; and Vice President, Business Development at Stem Cell Sciences. He has published key research findings in top-tier peer-reviewed scientific journals such as *Nature and Science*. In graduate school, at the age of 23, he discovered and characterized a master regulatory protein, termed *paraxis*, involved in the control of cell fate and anterior--posterior axis formation during embryonic development. He has founded or co-founded two successful stem cell-based biotechnology companies and recently published the world's first college textbook on nanomedicine (*Understanding Nanomedicine: An Introductory Textbook*; Pan Stanford Publishing, 2012). He is currently Vice President of Global Business Development at RayBiotech Incorporated, and an adjunct professor in the Department of Molecular and Cell Biology at the University of Texas, Dallas. As a die-hard Texas Longhorns fan, he holds a Bachelor of Arts in biochemistry from the University of Texas, Austin, and a PhD in molecular biology from the University of Texas M.D. Anderson Cancer Center in Houston. He grew up in Silsbee, Texas, and currently resides in the Austin, Texas area, with his wife, Jane, daughter, Zoie, and son, Bobby. He can be reached at rob@robburgess.com.

www.stemcelltextbook.com

Stem Cells: A Short Course, First Edition. Rob Burgess.
© 2016 John Wiley & Sons, Inc. Published 2016 by John Wiley & Sons, Inc.

INDEX

f=figure and t=table